AF813779

Entrepreneurship in Indian
AGRICULTURE

MJP
PUBLISHERS

Entrepreneurship in Indian
AGRICULTURE

C Sekhar

Professor and Head
Department of Social Sciences
Horticultural College and Research Institute
Tamil Nadu Agricultural University
Periyakulam – 625 604
Theni District

MJP PUBLISHERS

Chennai New Delhi Tirunelveli

Honour Copyright
&
Exclude Piracy

This book is protected by copyright. Reproduction of any part in any form including photocopying shall not be done except with authorization from the publisher.

MJP PUBLISHERS

ISBN 978-81-8094-298-3

MJP Publishers
No. 44 Nallathambi Street,
Triplicane,
Chennai 600 005

All rights reserved
Printed and bound in India

MJP 251
© Publishers, 2016

Publisher : J.C. Pillai

Preface

In the present day scenario, the government cannot come forward to provide employment to everyone in the country unless otherwise the control in population growth is stabilized and the policy measures in support of providing employment is thought off for implementation effectively. This called for stiff competition for employment. The uneducated or less educated have the attitude of risk taking naturally because of their survival. But the well educated or the graduates of universities are lacking the self-confidence and the initiatives and drives in their mind because of lack of practicalities even though they are knowledgeable. Besides, they are not in possession of risk taking skills.

A grocery shop vendor quickly judges to open a branch in a particular street amidst so many competitors available. He has the confidence. He carves his own market segments, does business and earns profits through his own strategies of attracting few consumers as his segments. But the graduates are depending on monthly paid jobs and mostly they are underemployed. This is applicable even to the professional stream of graduates. The engineering graduates have the skill of entering into the agribusiness. But the agriculture and horticulture graduates are not entering very boldly into the stream of establishing an agribusiness. This called for infusing some ideas on methods of establishing an agribusiness which is the focal point of writing this book entitled "**Entrepreneurship in Indian Agriculture – Texts and Cases**".

This book is designed into three sections. Section one deals with general aspects of establishing, registration formalities of business besides discussing on the market potential, human resource aspects, recruitment and selection and the cost details associated with the business. Section two discusses on the aspects of successful agripreneurs who were zero yesterday but have become heroes today by establishing and running agriculture and horticulture as their principal business activity. In the section two of the book, methods of running the business with success are highlighted to motivate the readers and users to enter into the business venture for their prosperity following the case study approach. The success of the business is visualized in person and enough details are documented for confidence building besides giving the details of cost of production. This book is poised to give some ideas on establishing the business and associated constraints. However, some recommendations are also given at the end of each case and the farmers are following the write-ups on their business as the guide for further development.

Section three of the book is focusing on the marketing channels for the principal produce, trade scenario and export potentials. I have a strong and positive idea that the contents of the book will definitely motivate the young minds to capture an innovative business for adoption and execution not in the environment of tomorrow but today itself.

I also strongly recommend each university to infuse syllabi on methods of establishing the business, besides focusing on similar cases following the guidelines of the book and the methods of preparation of bankable projects for availing the support of financial institutions for establishing the business. I request the colleagues and aspiring agripreneurs who are reading the book to motivate other graduates too to start reading the book so as to establish the business on their own or in partnership with someone as working partners and financial partners.

I wish everyone to become successful entrepreneurs.

(Dr. C. SEKHAR)

About the Author

Dr. C. Sekhar, M.Sc. (Ag), M.B.A (PM), MBA (HR), Ph.D.(F.E), F.AB.M (USA), is the Professor and Head in Department of Social Sciences, Horticultural College and ResearchInstitute, Tamil Nadu Agricultural University, Periyakulam, Theni. He has 27 years of rich experience in the field of Forestry, Fisheries, Agriculture, Horticulture and in the Management discipline. He has served as Agricultural Economist in the State Planning Commission, Chennai for Ninth Five Year Plan Preparation related works and also served as Agricultural Economist and Member in the World Bank Project entitled "Tamil Nadu Irrigated Agriculture Modernization and Water Bodies Restoration and Management Project" and played key role in appraisal and implementation of the sub-basin proposals. He has visited New York during 2006 for a short program on Food and Agribusiness Management organized by Cornell University, USA. He is the recipient of "Best Extension Worker" award from Tamil Nadu Agricultural University, Coimbatore during 2007.

Contents

Section – I

KEY AREAS IN AGRIBUSINESS

Section – II

PRODUCTION FOR AGRIBUSINESS

Section – III

MARKETING AND TRADE FOR AGRIBUSINESS

Section–I

KEY AREAS IN AGRIBUSINESS

Chapter 1

Introduction to Agribusiness Management

INTRODUCTION

The "agribusiness" concept was developed in the 1950s by Ray A. Goldberg, the founding father of the field and George M. Moffett, Professor of Agriculture and Business, Emeritus, at Harvard Business School. The idea for an agribusiness seminar for senior executives soon grew out of the school's elective MBA course in Agribusiness Management, which was based on a conceptual framework that Goldberg co-developed with John H. Davis, the former U.S. Assistant Secretary of Agriculture.

The Harvard Business School (HBS) seminar has helped to create a growing profession. While "agribusiness" once was not even a word, more than 100 agribusiness programs currently are offered around the globe. Today, the International Food and Agribusiness Management Association (IFAMA), created in 1990, has more than 950 members in more than 50 countries.

With much to learn in this advancing field, new materials are created for each annual offering, which prompts many executives to participate in the program year after year. The agribusiness seminar will continue to build on its impressive legacy to educate global leaders for years to come.

Through the study of agribusiness at Harvard Business School, the notion of the value-added food chain was born. Both the study and practice of agribusiness have evolved significantly over the decades. During that time, agribusiness has come to be seen not just as economically important, but as a critical part of society. Agribusiness touches on health, nutrition, safety, science, and government.

One particular step in the evolutionary process is a transformation that occurred in the past decade. Some industry players have gone from being low-margin commodity producers to higher-margin providers of "solutions."

The future for this massive industry will be both exciting and complex. To feed the world, production must double by 2050, while dealing with political issues, water constraints, price volatility, and the need for more sustainable practices. But, due to more efficient operating practices, new technologies, and increased levels of partnership and collaboration across the supply chain, the future for the industry is very attractive. The Honorable Chief Minister of Tamil Nadu, Ms. J. Jayalalitha has also advised the State Agriculture Machinery with the phrase "Doubling the Agricultural Production and Trebling the Income of the Farmers". Accordingly the Tamil Nadu State Agricultural Machinery is poised to achieve the food production by ensuring the income earning possibilities of the farmer by announcing various developmental schemes.

THE CONCEPT OF AGRIBUSINESS

The term 'Agribusiness' was first used publicly in 1956 by John H. Davis in a paper presented at the Boston Conference on Distribution as "The sum total of all operations involved in the production and distribution of Food and Fiber" (Fusione, 1986; 2 Fusione, 1995). A few months later the concept of agribusiness was further refined by Davis and Goldberg (1957) as follows.

"Agribusiness means the sum total of all operations involved in the manufacture and distribution of farm supplies; production operations on the farm; and the storage, processing, and distribution of farm commodities and items made from them" and thus agribusiness essentially encompasses today the functions which the term agriculture denoted 150 years ago.

Davis and Goldberg (1957), in their book 'A Concept of Agribusiness', developed a framework to analyze and understand the multiple relationships between the many different sectors and businesses engaged in production and distribution of food and fiber products. The aim of that study, assigned to them by the Food Foundation, was "to better understand the mutually supporting relationships between agriculture and business and as possible, to improve fundamental economic relationships among all engaged in the production, distribution and use of such products."

THE CHARACTERISTIC FEATURES OF AGRIBUSINESS

- Primary forces of production
- Farming is a small size of business and thus provides little scope for division of labor
- Frequency of decisions are very often and speedy in nature
- Seasonality in production
- Inelastic supply in nature
- Yield variability is very wide
- Standardization of products is not possible
- Agricultural produce are highly perishable in nature
- Imparts innovation, new ideas and processes
- Agribusiness is rural based mostly
- Partnership mode in agribusiness is common
- Lack of public and private initiatives to support agribusiness

TYPES OF AGRIBUSINESS

- ***Production*** Crop Production, Livestock, Dairying, Poultry, Forestry, Fishery, Floriculture, etc.
- ***Wholesaling and Distribution*** Packaging, Labeling, Transporting, Refrigerating, Freezing Processing, Storage, Performing any service, etc.
- ***Retailing*** Florists, Input Stores, Grocery Store
- ***Professional Services*** Insurance Agents, Investment Brokers

- ***Personal Services*** Hotels, Restaurants, Agroservice Centers
- ***Financial Services*** Commercial Banks, Insurance Companies and Thrift Societies
- ***Franchising*** KFCs, KDFCs, Quick Service Food Drive-Inn, etc. (Elanchezhian, 2007)

In the last four decades of the twentieth century, the agriculture sector has turned from traditional to commercial, production-oriented to market-oriented, which got spurt with the advent of Green Revolution, White Revolution, Blue Revolution, Brown Revolution and Yellow Revolution etc.

Agriculture has now blossomed into a mature industry which included the following:

- Input supply system
- Agricultural production system and
- Output marketing system

Each system is dependent on other two for its development and growth of agriculture. The recently drafted National Agricultural Policy has accorded priority for promoting agribusiness activities at different levels for catapulting growth in agriculture (Talathi, 2011).

THE AGRIBUSINESS COMMODITY SYSTEM

Goldberg (1974) expanded the concept of agribusiness introducing the notion of Agribusiness Commodity System to refer to all firms and institutions involved in the manufacture and distribution of farm supplies; production operations on the farm; and the storage, processing, and distribution of a given commodity and the items made from it.

In Goldberg's view, the Agribusiness Commodity System approach, while drawing from disciplines of macro and micro economics, agriculture economics, agriculture science, and management, aims to

- analyze the agribusiness environment;
- set forth the structure and operations of a particular commodity system in the larger environment; and
- relate the specific operations of a firm or institution to the total vertical commodity system and to the ultimate purpose of that

system, namely, to provide food in an efficient, nutritionally acceptable, and socially desirable manner.

As suggested by Goldberg (1974, 1979), the overall mission of agribusiness as a profession is to "inform private and public policy and strategic decision making with the ultimate purpose of providing food in an efficient, nutritionally acceptable, and socially desirable manner."

Goldberg applied this systemic view of agribusiness to analyze the interrelated functions of all agents horizontally and vertically across and along the value chain. This approach helped him study firms and institutions in its commodity system and with reference to other commodity systems as well as to the national and global environment. Under this approach he recognized that each agent, be it private or public, has two functions; one which refers to its specific operations, and the other to the manner in which each agent coordinates its function to the total commodity system of which it is part. (Shelman M.L., 1998)

THE COORDINATION PROBLEM

Recognition of the coordination problem in agribusiness started when, due to technological advances and the industrial revolution during the 19th century, activities previously carried out on the farm were transferred to businesses outside the farm. Agriculture entered a long process of specialization and vertical disintegration. On-farm production functions and off-farm functions such as upstream agriculture inputs supply and downstream manufacturing and distribution were separated. Agribusiness is considered to be a product of the dispersion of functions historically performed on-farm. The separation of functions, the extreme fragmentation of agriculture, the specific and unique biological and seasonal nature of agriculture, its dependence on climate, ever changing consumer demands, regulations and technological innovations, all worked together to exacerbate the "coordination problem". Davis and Goldberg, conscious of the changing nature of agriculture and of its increasing interdependence with what they at that time referred to as "business" argued about the need to develop a new discipline to address "the need for improved coordination between agriculture and business to achieve a more balanced economy conducive to progress". In their view, there was a need for "the art of getting diverse groups to work together" and of research aimed at "reconciling differences between groups or to workout policies drawing on the ideas of all interested groups as source material".

THE INTERNATIONAL AGRIBUSINESS MANAGEMENT ASSOCIATION

In 1990, the International Agribusiness Management Association (IAMA) was founded by a group of academic, government and industry leaders. Its original mission was to stimulate strategic thinking across the full spectrum of the agribusiness system. It was aimed to serve as an effective worldwide networking organization, acting as a functional bridge between the agribusiness industry, researchers, educators, government, consumer groups and non-governmental organizations. IAMA was to play a key role as a worldwide leadership forum bringing together top executives, academics, policy makers, students and stakeholders to network and stimulate strategic thinking across the global Food, Fiber, Fuel, Floral, Fodder, Floss and Forestry systems (7Fs). Its overall goal was to make the global food and agribusiness sector more efficient, effective, exclusive, responsive, reasonable and sustainable. IAMA was to be dedicated to an efficient system that is sensitive to the needs of consumers, safe, environmentally responsive and has a high degree of business integrity.

Study of the food sector showed it to be characterized by interdependencies among parties throughout what came to be seen as a value-added "food chain." (Both the concept of a value-added food chain and the "food chain" terminology came to be adopted by numerous other sectors.)

Despite the interdependencies, Professor Goldberg was struck by the high level of distrust that permeated the sector. A similar lack of trust had hindered the study of agribusiness across academic disciplines, including Harvard's Medical School, Kennedy School, economics department, biochemistry department, and more. Given how intertwined food is with health/nutrition/safety, culture, science, government, and economic development, Professor Goldberg was able to convince the deans of other Harvard schools and departments to collaborate on an innovative interdisciplinary approach.

Agribusiness itself was propelled forward about ten years ago by a newly emergent partnership mentality. One food industry company discovered it could create far greater value by becoming a "solutions provider" up and down the value chain. Taking a partnership approach toward customers and suppliers, helping them better serve their stakeholders, was suddenly seen as a win/win for all. The result was a transformation for this company and the industry. Instead of making "cents on the bushel," players as providers of value-added solutions were able to earn "dollars per bushel."

Agribusiness is seen today as more than a commodity industry; it is seen as a driver of societal and economic health. Food system players are starting to address not only the needs of their partners and stakeholders but the needs of society—the ultimate stakeholder.

INDIAN AGRIBUSINESS SCENARIO

India has 52 per cent of total cultivable land as against 11 per cent in the world. All the 15 major climates of the world from snow-bound Himalayas to hot humid southern peninsula and the Thar desert to heavy rain areas exist in India. There are 20 agro-climatic regions and nearly 46 out of 60 soil types available in the country. Sunshine hours and day length are ideally suited for round the year cultivation of crops. India is the centre for biodiversity of plants, animals, insects, micro-organism and accounts for 17 per cent animal, 12 per cent plants and 10 per cent fish genetic resources of the globe.

Using the above comparative advantages, India has been the second highest fruit and vegetable producer in the world, second highest producer of milk with a cold storage capacity of 70000 tonnes, fifth largest producer of eggs, and the sixth largest producer of fish with harvesting volumes of 5.20 million tonnes. However, the exports from India are poor. Compared to other countries, the land productivity in India is low. Even within the country, there are tremendous variations among the states in terms of farm productivity. There is tremendous scope for increasing the farm outputs using state of the art pre-harvest technologies and harvesting techniques. Value added agriculture has not yet caught on within the Indian scene. Post-harvest technologies are at a nascent stage in India. Pre-cooling, cold chain, packaging and transport using small containers are all absent, resulting in substandard delivery of groceries and enormous amounts of wastage. It is reported that the value addition in processing has been only about seven per cent of grains, 30 per cent of fruits and vegetables and 10 per cent of seed spices the remaining of which are lost due to poor post-harvest technologies. Similarly over Rs 500 billion worth of output is lost annually due to poor quality in the marketing, storage and handling facilities. Despite these heavy losses, the value of agro-processed products is reckoned at Rs 4200 billion. Considering the value of primary agricultural products, farm inputs and processed products, the total value of agribusiness can be as high as Rs 10,000 billion (Maria Saleth, 2012).

Indian is one of the leading producers of agricultural and livestock commodities—grains, fruits and vegetables, milk, etc. However the share of these products going for processing is very small. Also, India's share in world trade of processed food is 1.60 per cent, and value addition is 20 per cent (Ministry of Food Processing, Annual Report 2006).

THE ASPECTS OF ENABLING AGRIBUSINESS ENVIRONMENT IN INDIA

1. Changes In The Socio Economic Environment

The changes in the socio-economic environment is leading to higher demand for processed foods. This change is characterized by fast urbanization, increase in income level, less time devoted for cooking, and exposure to global markets; the globalization increased the access to market and variety of product flow and also increasing number of nuclear families called for higher demand for value added foods. Besides these, increase in the number of super markets and organized retailing of food products also commanded change in the buying behavior on food items and higher preference of consumers towards convenience foods creates opportunities for door delivery of value added outputs to the consumers.

2. Changes In The Policy Environment

Following liberalization in 1991, government has followed a liberalized overall policy regime, with specific incentives for high priority food processing sector to provide a conducive environment for investments and exports in the sector. The government has accorded a high priority to the food processing sector to encourage commercialization and value addition to agricultural and horticultural produce; for minimizing pre and post harvest wastage, generating employment and export growth. Recognizing that the robust and dynamic food processing sector is to play a vital role in the new emerging global economy, all policies and plans for the food processing industries is coordinated by the Ministry of Food Processing Industries. Because of establishment of food processing sector, thrust is given to produce value addition of low value crops to high value products. Agribusiness firms in Tamil Nadu have started converting minor millets into value added products such as breakfast cereals, which are well received among different consuming segments

3. Changes in the Availability of Credit and Finance

Though several steps have been taken to enhance credit availability in the past few years, the Union Budget for financial year 2006-07 was hailed as the first ever Union Budget that has focused attention on Food Processing and Food Technology. Major recommendations include use of warehouse receipts, amendments in priority sector lending, restructuring government schemes, modifying evaluation parameters by banks / financial institutions (MOFPI 2005). Implementation at field level in respect of farm credit and finance is still facing certain problems among the small and marginal farm segments. In the Union Budget of 2013, the Government has given emphasis on establishing Women Development Bank for the first time in India to give focus on women development and to address their specific issues through financial empowerment. Similarly, to address the problems of small and marginal farmers, separate policy measures need to be adopted to establish a bank for small and marginal farmers.

4. Emphasis for Infrastructure Development

Upgrading infrastructure for food processing, is a key priority for the Ministry of Food Processing. Several initiatives for upgrading the infrastructure are presented in the annual report of the Ministry of Food Processing Industries, 2005-06 to enable food processing which in turn enable the agribusiness development.

5. Supporting Institutions

Several supporting institutions and mechanisms have been put in place to meet the changing needs of the agro processing sector to enhance farmer information about markets and modern farming methods, developing linkages with private sector, providing venture capital, providing support for exports, etc.

6. Empowerment of Commodity Interest Groups

Tamil Nadu Agricultural University has established a department entitled "Department of Market Extension" which is first of its kind among State Agricultural Universities in India directed to provide farm linkages with markets during May 2008. This department has established so far around 10 interest groups in respect of flowers, moringa, amla in various districts and it is functioning well. The precision farming farmers in Dharmapuri and Krishnagiri also have established one such group which is also planning to establish its own retail outlet in Coimbatore city to fulfill the needs of customers in fruits and vegetables.

Another example is an Information Technology scientist working for a software firm has resigned his job and established a unit called **E-Farm** which is planned to establish a direct link with producers and the consumers through the E-Farm personnel and they are conducting periodic training on agribusiness initiatives and emulating the success stories to the farmer friends and the E-Farm type of selling the farm fresh is also in progress. Let us give a brief introduction of E-Farm to those who are not exposed to this enterprise.

E-Farm E-Farm is a young social enterprise firm based in Chennai. The proprietor of the E-Farm is an Information Scientist serving for a multinational concern but got motivated and established this E-Farm in consultation with many departments of related field including the Department of Agricultural and Rural Management of Tamil Nadu Agricultural University, Coimbatore. E-Farm is India's first end-to-end agricultural supply chain platform, providing a combination of technology solutions and on-ground distribution mechanism to enable farmers reach end markets in an effective manner. E-Farm has tie-ups with farmers, intermediaries like wholesalers, distributors, wholesaler-cum-retailers and the retailers, logistics providers, small retailers and cart vendors all integrated into a single chain to deliver fresh, clean, and low priced farm produce.

One of the key problems for the farmers is the marketing of their produce to end consumers. It often goes through several middlemen, in a very inefficient manner, which contributes to over 30 to 40 per cent wastage. It depends on the type of the produce. Being the farmers with poor literacy, they lack proper planning and procurement and application of untimely market inputs in the cultivation process resulted in poor yield and higher cost. Often the farmers decision on what to grow, when to grow, when to harvest are all based on the ad-hoc decisions rather than well planned approach. This information should flow from the technical departments of Tamil Nadu Agricultural University and the Department of Agriculture of Government of Tamil Nadu.

The ultimate consumers are often hit by huge variations in prices and quality and availability which is one of the biggest problems facing our nation today. Though several modern retailers have been operational for over a decade, owing to a combination of factors such as high operational costs, localized presence in metros, and low margins have forced even many large chains to close down their operations. There is hence a critical need for an efficient, low cost agricultural supply chain mechanism to connect farmers with the end markets.

In a simplc word, E-Farm is a hi-tech concern which is connecting the farmers, business consumers and the intermediaries such as transport operators, storage providers, etc., using a more organized supply chain mechanism. The efficiency of functioning of E-Farm is found to be good and hence more number of farmers were operating in their fold. However, the institutions like Tamil Nadu Agricultural University through the Directorate of Centre for Agricultural and Rural Development Studies should conduct research on the efficiency of functioning of E-Farm, its merits and demerits, the beneficiary fold and the share of farmers in the supply chain need to be documented. If the E-Farm system is found effective, then the concept of E-Farm type of functioning may be taken to all the regions for effective functioning of the supply chain.

CHALLENGES IN AGRIBUSINESS

- Due to the increasing population, the requirement of food will be doubled. Agricultural and allied sciences sector personnel need to concentrate keenly on the ways and means of meeting the doubling and trebling of food production through technological development and strategic time bound actions.

- Biofuels, particularly sugar-based ethanol, make sense economically and calorifically and are here to stay; they will become a greater part of the energy supply. However, Mr. Weisser thinks that recent market incentives for ethanol's production were too aggressive and helped precipitate the food price crisis.
- Many attempts are being made in many locations for regional self-sufficiency which are inefficient and altogether impossible. Demand is growing in areas where food simply cannot be grown. In contrast, globalization increases competition and efficiency.
- Sustainability is becoming more important now-a-days, which will affect all the areas of agribusiness.
- There will always be cycles of supply and demand that move prices up and down; market speculation played less of a role in the current crisis than many believe.

STRATEGIC ACTIONS NEEDED

- Radical reorientation of agricultural strategy from its narrow focus on conventional crop production to a broader focus on modern agribusiness with all its manifestations. For such a reorientation, the focus should shift from mere production expansion to quality improvement and value addition through processing, packaging and marketing.
- Production efficiency needs to be achieved through use of supply chain technology and regional specialization.
- The excessive reliance on public investment should give space for the promotion of private investment especially from the corporate sector.
- Reforms in the institutional structure within which agriculture operates are also critical. In this respect, the institutions responsible for supplying the inputs and extension services and the marketing of agricultural products assume greater importance.
- Trade development should be enhanced further in agribusiness with the limited interference of the intermediaries and the role of trade promotion organization is vital in this regard.
- Transportation and logistics efficiency need to be improved and imparted by prioritized investment in agribusiness infrastructure like roads, link roads, port, airways, etc.

- Exploring new markets for agribusiness produce, trade promotion opportunities, value addition, standardization, quality control and export are the other areas equally looked into with utmost care to promote agribusiness.
- Countries like India and China should make more reforms in agricultural and allied sectors to make them agriculturally sustainable and should have vision and mission based actions to feed the millions away from India and China.

OTHER IMPORTANT POINTS

Self Protection

When food crisis occurs, the developing countries rushed to follow the strategies to protect themselves and were poised to produce food grains on their own and in meeting the self-sufficiency. In this situation, developing countries should have the following for its immediate execution by utilizing the scarce resources even without wasting the wastages for feeding the people.

- Land distribution is skewed. Large size of holding are with few and few size of holding with the large section of people resulting in mechanization at one end and another end facing the problem of mechanization and resource use warranted a corporate farming and group farming approaches so as to impart mechanization and efficient resource use in the farm so as to feed the millions.
- Another important resource is water which is seen as the key limiting factor. Water is the scarcest resource. So water development is a key strategy. Flood is flowing at one end. Desert is at another end. Drought is spread across most of the states which warranted the action of linking the regional and national waters to feed the desert and the drought affected areas. Water development organizations should be brought under one umbrella to manage the national water to feed the millions. Water Resources Organization should be more responsive, responsible, accountable and admirable in actions. Regional strategies in this regard will pay dividend in water development. In Tamil Nadu, the Chief Minister's order to erect Water Harvesting Structures at Micro Level and Macro level had created an impact in raising ground

water level. Similarly, surface water development, storage structure and its maintenance, erection of new storage structure like dams and reservoirs, chain link system of tanks and canals are the immediate requirements for water development to promote agribusiness.

References

- Maria Saleth R. Agribusiness Potentials in Tamil Nadu (Madras Institute of Development Studies and Indian Society of Agricultural Marketing: Nagpur) preface, 2012.
- Elanchezhian, T., Samuel, A.D.S and P. Paramasivam (2007). "Agribusiness Management" (Gowri Printers and Publishers: Madurai)
- Talathi, J.M., V.G. Naik and V.N. Jalgaonkar (2011). "Introduction to Agricultural Economics and Agribusiness Management" (Ane Books Private Limited: New Delhi)

Web References

- http://www.exed.hbs.edu/programs/agb/Pages/seminar.aspx
- http://www.hbs.edu/centennial/businesssummit/business-society/agribusiness.html

Chapter 2

AGRICULTURE AND AGRIBUSINESS

Agriculture has always been the mainstay of Indian economy because of its high share in employment and livelihood creation notwithstanding its reduced contribution to the nation's GDP. The share of agriculture in the gross domestic product has registered a steady decline from 36.4 per cent in 1982-83 to 18.5 per cent in 2006-07. Yet this sector continues to support more than half a billion people in India providing employment to 52 per cent of the workforce. It is also an important source of raw material for many industrial products, particularly fertilizers, pesticides, agricultural implements and a variety of consumer goods. The importance of agriculture to the national economy is crucial in terms of meeting the growing demands for food and raw materials of the non-farm and urban population, expanding the secondary and tertiary sectors of the economy, widening the extent of markets for industrial produce, earning the much-needed foreign exchange, maintaining price stability and mobilization of resources for capital formation. Agricultural development thus facilitates a cumulative growth process. The significance of agriculture in the national economy is best explained by considering the role of agriculture under different heads namely product contribution, factor contribution and market contribution.

INDIAN AGRICULTURE POLICY

"The policy of excessive protection of Indian industry hurt Indian agriculture in several ways. It raised the prices of industrial products relative to agriculture products that hurt the rural sector of the economy as consumers of industrial products. Increased profitability of industrial production compared to agriculture production drew away investable resources from agriculture. This shift took place in various ways. Low returns in agriculture activity reduced the ability to pay economic prices for many inputs such as power, water and credit. Instead of continuing with the subsidies for these inputs it is better to reduce the protectionist bias against agriculture by lowering tariffs favoring industry and altering relative prices in favor of agriculture. This would create potentially a more profitable agriculture which would be able to bear the economic costs of technological modernization and expansion". - Dr Manmohan Singh.

Several policy initiatives have been undertaken to improve the socio-economic condition of the agricultural community; provide access to services like banking, health and education; strengthen local institutions and provide infrastructural support in terms of markets, transport, irrigation and communication. Policy insights have led to changes in cropping pattern resulting in economic benefits to farmers. The era of globalisation has now shifted the focus from internal problems like dependence on monsoons, marginal land holdings, high cost of inputs like fertilizers and power, erratic input usage, inadequate processing and marketing infrastructure, dispersed processing facilities leading to diseconomies of scale and a lack of professionalism in the sector to external threats as a result of opening up of the markets.

PROGRAMMES AND SCHEMES

Over several decades, efforts were made by the Government of India to overcome weaknesses in the agricultural sector like an over-dependence on monsoons; subsistence orientation; predominance of small and marginal farmers; uneconomic cultivation practices and extremely low levels of productivity, through the implementation of several rural and agricultural development programmes and schemes. Policy initiatives were also aimed at improved infrastructural facilities like irrigation, providing a network of agricultural financial institutions; focus on agricultural education, marketing, storage and transport facilities; supply of good quality inputs; flood control, land reforms, price incentives, changes in cropping patterns, commercial cultivation, soil and water conservation. This led to increased production, but improved productivity still remains unsatisfactory on account of inefficient usage of limited resources.

Agribusiness Management

Agribusiness encompasses agricultural and allied sectors like dairy, poultry, fisheries, forestry etc; portion of industrial sector which produces farm inputs or utilizes farm supplies and the processors of farm products. As this comprises a major component of the Indian national economy, agribusiness management has a major role to play in India's development. Being a developing country which is traditionally agricultural in nature, India needs to infuse professionalism into the agricultural sector in order to cash in on its inherent competitive advantage.

Production Management

With the emergence of agriculture as an industry, farmers no longer function in isolation and most of the production decisions are governed by marketing firms and consumer preferences. Agribusiness management would be successful only when farmers produce what the market wants rather than what they are good at producing. Technology has been a major factor governing production management as it is yield enhancing infusing vast improvements in varietal differentiation, fertilization and pest resistance. However very often increased production in the absence of effective marketing results in falling prices and increased government support measures. Farmers, agriculture suppliers and marketing firms have to work in cohesion so that enterprises are mutually beneficial.

Input Management

Farmers combine land, water, commercial inputs, labor, and their management skills into practices and systems to produce food and fibre. On account of rising population levels, it is difficult to sustain the needed increase in food production based on expansion of area under cultivation or low-cost expansion of irrigation facilities. Even in places where such opportunities exist, unrestrained expansion would probably involve a high risk of long-term damage to the natural resource base and the environment. It is therefore imperative that agricultural input markets should work efficiently and effectively. The agricultural input supply sector comprises of firms that manufacture and distribute the myriad of inputs that fuel the production in agriculture sector. Seeds, fertilizers, machinery and equipment, fertilizers, crop protection, credit, etc., are a few of the major agricultural inputs. Input management should focus on agricultural productivity. Agricultural productivity measures the difference between output growth and input growth rates. If productivity growth is positive, then the same output can be produced with fewer inputs and lower costs. Increased efficiencies in agricultural productivity have been an important factor in maintaining the balance between food supply and world demand. Productive agriculture should increase its output of food and raw materials and at the same time improve its efficiency of resource use. No single, simple solution can regulate agricultural productivity but rather, there is a need for the employment of site specific best management practices (BMPs) within improved systems of agriculture. The input supply sector can be briefly categorized into the manufacturing sector, the distribution sector and the services sector.

Agricultural Marketing Management

Agricultural marketing management includes all business activities involved in the flow of agricultural products and services from the point of initial agricultural production to the final consumer. It involves a large number of activities from monitoring and influencing demand patterns, decision making related to production parameters, transportation, grading, refrigeration, packaging, labeling, advertising and processing so as to add value to farm products by creating time (refrigeration and storage), form (processing), place (transportation, retailing) and possession utility (advertisements/ supermarkets). Agrimarketing is defined as all the activities associated with agricultural production and with food, feed and fibre assembly; processing and distribution to final consumers, including analysis of consumer needs, motivations and purchasing and consumption behaviour.

Supply Chain Management

"In view of globalization, the main players of the future would no longer be conventional land owners, but agribusinesses linked directly to multinational food corporations" - *Bunyard, 2002*

The agricultural supply chain relates to productive relationships between producers, processors, manufacturers and retailers. Competitive advantage cannot be achieved by firms in isolation but through the engineering and management of effective supply chains in collaboration with suppliers and customers. This led to cooperation in purchasing, finance, strategic operations, logistics, manufacturing and marketing management areas. The concept of Efficient Consumer Response (ECR) was introduced in the US grocery industry which recognized the fact that many business practices within the food industry were counterproductive with firms trying to maximize their own profitability by passing costs and problems up and down the supply chain. ECR aims at applying a systems approach to encourage firms within supply chains to work together as a whole. Retailers through their brands will be able to deliver quality only if quality is maintained throughout the supply chain right from the producer, manufacturer, wholesaler and retailer.

Information Management

International trade environment in terms of globalization and market orientation has posed severe challenges to the Indian agricultural sector with regard to quality, costs and distribution efficiency . With the increasing population, agriculture in general and food security in particular are now dependent for sustainability on the efficiency and efficacy of networking amongst various nations. Networking facilitates easy access to information and effective use of limited resources. Gathering, storing, analyzing and communicating relevant information helps to increase efficiency, enhance competitiveness and access business opportunities. In view of globalization, the main players of the future would no longer be conventional land owners, but agribusinesses linked directly to multinational food corporations.

Information and Communication Technology (ICT) in agriculture has therefore become a pre-requisite for professionalization of the agricultural sector. Introduction of ICT applications in the agricultural sector is a must, especially in case of developing nations. ICT applications will facilitate the easy availability and usage of requisite information to all end users. The development of ICT infrastructure focuses on village connectivity linking agricultural universities and colleges, scientific and research organizations as also local, state and central government departments with the end users through ICTs and the establishment of community access points. Similar linkages are available in Tamil Nadu Agricultural University to link the farmer's quest of answering the problems of agriculture through the scheme E-Velanmai (E-Agriculture). Through this, the pest and disease damage is photographically sent to the experts, the experts are answering the problems on a day to day basis and the problems are set right at field by the farmers for enhancing productivity. Enhanced productivity adds to agribusiness initiatives too.

Risk Management

Risk management involves choosing among alternatives that have uncertain outcomes and varying levels of expected returns. Risk management is dependant on the certainty or uncertainty related with various activities and events. Exposure to risk in agriculture has increased due to climatic change, agricultural trade liberalization and stringent quality control measures. Trade liberalization aims at increasing price transmission from world markets to domestic markets and therefore bringing domestic prices more under the influence of fluctuating world market prices. Decision-making is difficult as it depends very often on imperfect information. Risk management continues to be one of the most neglected areas in Indian agriculture on account of various reasons such as low risk perception by the farmers, over reliance on government safety net programmes, especially the price support mechanisms that have been playing a role in reducing price risks for key products, limited coverage of agricultural insurance systems in terms of products and participation rates and a lack of know-how about risk-management tools like futures and options markets, as also high start-up transaction costs.

INSTITUTIONAL LINKAGES TO AGRICULTURE IN INDIA

India has one of the most developed institutional frameworks for both backward as well as forward integration in the agricultural sector. Some of the major institutions providing services to the agricultural sector are the Ministry of Agriculture, GOI; financial institutions like the Reserve Bank of India, public sector banks, regional rural banks, private banks, foreign banks, insurance agencies, etc.; and other organisations such as National Bank for Agriculture and Rural Development (NABARD), National Cooperative Development Corporation (NCDC), National Dairy Development Board (NDDB), National Cooperative Union of India (NCUI), Agricultural and Processed Food Export Development Authority (APEDA), Indian

Council of Agricultural Research (ICAR), National Agricultural Cooperative Marketing Federation of India (NAFED), Food Corporation of India (FCI), Central Warehousing Corporation (CWC), National Horticulture Board (NHB), Central Food Technological Research Institute (CFTRI) and Export Promotion Organizations.

References

- Economic Survey 2007-2008 , website: http://indiabudget.nic.in, accessed on 4 March, 2009.
- Address by Mr. Manmohan Singh, Finance Minister, Government of India to the Indian Association of Agricultural Economists, 2004.
- J.H. Davies and Goldberg R. A Concept of Agribusiness, Harvard University, Boston. 1957.
- Ibid E. Paul Roy, Exploring Agribusiness, Interstate pinters and publishers, 1980.
- www.ers.usda.gov/Briefing/AgChemicals/accessed on January 2, 2008.
- S.P Erickson et al, Agribusiness Management, McGraw-Hill Publications, 2002.
- S.P Erickson, J.T.Akridge, F.L.Barnard and W.D.Downey. 2002. Agribusiness Management. McGraw-Hill Publications.
- Brandenburger, A.M. & Nalebuff, B.J, Competiton. Doubleday, New York .1996.
- Mitchell, A. Efficient Consumer Response: A new paradigm for the European FMCG Sector. Financial Times Retail and Consumer Publishing, London. 1997.
- Weigel, G., and D. Waldburger (Eds.), *ICT4D – Connecting People for a Better World.* Berne, Switzerland: Swiss Agency for Development and Cooperation and Global Knowledge Partnership, 2004.
- Peter Bunyard , 'When the Cheapest is not cheerful' The Ecologist, April 2002, pp. 30-32 . 2002.
- What Is Risk?, Agricultural Marketing, PENN State College of Agricultural Sciences ,http://agmarketing.extension.psu.edu/Business/WhatIsRisk.html accessed on 4 April, 2009.

- European Commission, Risk Management Tools for EU Agriculture with a special focus on insurance, http://ec.europa.eu/agriculture/publi/insurance/text_en.pdf., accessed on 5 April, 2009

Chapter 3

AGRIBUSINESS POTENTIALS IN INDIA

AGRICULTURAL GROWTH

Agricultural growth has been exhibiting a reversal from prolonged deceleration of growth. Agricultural growth rate is expected to accelerate to 3.30 per cent per annum during the eleventh plan period as compared to 2.20 per cent in the tenth plan period. Higher growth is likely to be sustained as the plan outlay on agriculture has been increased substantially in the tenth plan. However, the question is how far high growth would strengthen the livelihood base of the farming community and improve the food security of the poor. This is mainly due to the gap between the planning and implementation of plan proposals. How the gap should be addressed needs clear documentation. For that, the monitoring mechanism need to be strengthened and the number of grass root level officials need to be enhanced for effective implementation of government programs.

According to the study conducted by Professor G.S. Bhalla and Gurmail Singh, the agricultural growth in all the districts of the country during 1962 and 2008 brings out that the pace of growth of crop production achieved in the green revolution period could not be sustained in the post reform period. Except Gujarat, Maharashtra and to some extent in Jammu and Kashmir, the crop production growth rates decelerated in all the states during the post reform period. Agricultural diversification towards allied sectors such as dairy, fisheries, meat and eggs and horticultural products have also contributed to overall agricultural growth (Radhakrishna, 2012).

PROBLEMS CONFRONTING FARMER CUM AGRIPRENEURS

- The middlemen are more advantageous in respect of gain and fall of international price of farm produce.
- Instruments available for stabilization of price are blunt due to complex market structure embedded in a variety of supply routes including local fragmented markets.
- The demanding produce, quantity demanded at both domestic and international markets, and form of demand are not available to the farmer groups or farming community.
- Cost of production is comparatively higher in respect of certain farm produce adding poor access to the international markets and thereby competition need to be overcome.
- Government intervention is often outdated as well as grossly inadequate.

There is a growing disconnect between agricultural growth and the farmers income. The value of agricultural produce at retail prices are exorbitant when compared to the price received by the farmers at their door steps. The margin accrued to middlemen is the highest. The increasing capital and credit intensity of agricultural marketing technology restricts small and marginal farmers' access to it. In this context, building institutions such as farmers' self help groups, farmer federations and producer cooperatives and imparting skills to them by integrating the small and marginal farmers with agribusiness knowledge are equally important. The concept of Agribusiness Centre (ABC) to produce and trade (seed to seed) the farm produce and qualifying them to export and redistribute the farm produce is essentially important.

CHANGING DEMAND

It is essential to recognize that the consumption patterns are changing due to higher growth as well as globalization exposure to the Indian consumers. As a consequence, the food basket is getting diversified and demand for quality food is on the increase. Consumers are knowledgeable and depend on the traders for quality produce and hence the production and distribution of qualitative output is the need of the hour. Changing lifestyles, market integration and trade liberalization at the global level have led to the increasing demand for processed food products particularly the meat and meat products, milk and milk products, and fish and fishery products. In future, both diversification of farming and quality produce will definitely influence agricultural growth and thereby agribusiness will gain further momentum.

AGRO-PROCESSING SECTOR IN INDIA

The agro-processing in India is still in the nascent stage. Processing of:

- fruits and vegetables is around only 2 per cent
- milk and milk products is around 35 per cent
- meat and meat products is around 21 per cent
- poultry products is around 6 per cent (Jayanti Kajale, 2012)

The demand for agro-based non-food processing sector like textiles, wood, paper, and leather based products is also increasing due to urbanization, increasing incomes and spread of education. India has paid little attention to overseas supply even though we import substantial quantities of various agricultural commodities such as timber, wood products, paper pulp, oil seeds, raw cashew nuts, pulses, etc. Producers in the exporting countries are well organized for exporting farm produce to India. To cite an example, it is reported that Australia earmarks thousands of acres of land to produce pulses for Indian consumers (Radhakrishna, 2012). In India, such efforts are totally lacking and hence the efforts on the agricultural and horticultural department should be intensified with vision- and mission-based actions so as to deliver the goods.

EXPORT POTENTIAL FOR MEAT

India has a huge livestock population, which includes buffaloes, cattle, sheep, goat, horses, pig, poultry and other species. Efficient utilization of these promising resources is very important to earn increased returns and sustenance of rural livelihood activities. Buffalo is the animal of triple purpose usage for milk, meat and draught power which has a share of about 60 per cent in total milk production and 30 per cent in total meat production of India and it has the immense potential for poverty alleviation and rural entrepreneurship. They contribute an estimated 1.59 billion tonnes of meat, amounting to 25 per cent of total meat produced in the country (FAO, 2008). Even though goat has higher annual growth rate than buffalo, buffalo is the only potential animal that can boost meat industry in India as the quantum of meat produced per buffalo is several times higher than that of goat or sheep and hence it commands higher market potential.

India is one of the major exporters of buffalo meat. The top five major customers of Indian buffalo meat at present are Vietnam, Malaysia, Egypt, Saudi Arabia and Phillippines. It is observed that Vietnam is importing about 102000 tonnes of buffalo meat worth Rs 12647 million from India which constitutes about 15 per cent of our total exports. (Jagathnath Krishna, 2012).

POTENTIALS IN POULTRY SECTOR

Poultry is one of the fastest growing segments of agriculture sector in India with an average growth rate of 6 per cent in egg production and 12 per cent per annum in broiler production. Integrated production, market transition from live birds to chilled and frozen products and policies that ensure supplies of competitively priced domestic or imported corn and soybeans are keys to future poultry industry growth in India. Similarly, some of the poultry meat products viz., chicken patties, chicken pizza, chicken nuggets, chicken tikka and chicken sauces, chicken loaf and chicken chunkoloma, also have great export potential (Ramanathan, 2012). The total egg production in Tamil Nadu during the year 2009-10 was 10747.58 millions, which was mainly contributed by improved eggs which accounted for 96.56 per cent.

PROSPECTS OF DAIRY BUSINESS

India is the largest milk producing country in the world. Annual milk production in India has grown more than six times since independence. The average annual growth rate in the production of milk in recent years has been close to 4 per cent. The domestic demand for milk is growing at about six million tonnes per year whereas annual incremental production over the last ten years has been about 3.50 million tonnes per year. As per the Planning Commission report, the domestic demand for milk by 2021-2022 is expected to be 172.20 million tonnes. As projected under the National Dairy Plan, the production of milk in the country is required to increase to 180 million tonnes by 2021-2022 to meet the demand. With higher growth of the economy, increase in population, and increased health consciousness among the population, it is only natural that the demand for milk and milk products will increase leading to the proportion of income spent on milk and milk products to increase. Keeping the above facts in mind, there are enormous scope for milk production and processing in the nation as well as the state. Thus starting and maintaining a dairy farm and plant is a choice of business to gain ample profit.

PROSPECTS IN GOAT FARMING

Every year 4.55 million goats are being slaughtered to meet out the demand of chevon in India. In Tamil Nadu, about 22.16 lakh goats were slaughtered and 124 lakh kgs of chevon was produced during the year 2009-10. Mutton and chevon contributed to more than 60 per cent of the total meat produced in Tamil Nadu. It implied that chevon is the choicest meat and has huge domestic demand. Thus the price of chevon is continuously increasing from Rs 100 to Rs 400 during the last decade and the price of chevon currently rose to Rs 500 per kg. With more than 124 million population, goats account for more than 25 per cent of the total livestock in the country and contribute Rs 106335 millions annually to the Indian economy. Goats provide food and nutritional security to millions of landless, marginal and small farmers. Responding to the market signals, the goat production system in India has been slowly moving from extensive to intensive system of management for commercial production (K.N.Selvakumar, 2012).

PROSPECTS IN SWINE INDUSTRY

Pork production in India represents 7 per cent of the country's animal protein sources. Production is concentrated mainly in the north eastern corner of the country and consists primarily of backyard and informal sector producers. Pigs convert inedible feeds, forages, certain grain, byproducts obtained from mills, meat byproudcts, damaged feeds and garbage into valuable nutritious meat. Most of these feeds are either not edible or not very palatable to human beings. With small investment on building and equipment, proper feeding and sound disease control program the farmer can profitably utilize his time and labor in this commercial venture. Pig grows fast and is a prolific breeder, capable of yielding 10 to 12 piglets at a time. It is capable of producing two litters per year under optimal management conditions. Creation of awareness among the farmers, about scientific farming and technological empowerment through training in pig management will ultimately provide gainful income and employment as well as nutritional security to the farmers and progressive business entrepreneurs (K.N.Selvakumar, 2012).

PROSPECTS OF SEED INDUSTRY

In India, more than 80 per cent of the farmers rely on farm saved seed leading to a low seed replacement rate. Indian seed program includes the participation of central and state governments, Indian Council of Agricultural Research, state agricultural universities and the cooperatives and the private sectors. It is estimated that about 46 per cent of the seed commercially sold in the country is by private seed companies. The domestic hybrid seed market is placed at INR 4.9 billion and is annually growing at 10 per cent, against 5 per cent global growth rate. The major players are Monsanto India and Syngenta India who dominate the hybrid seed market. In Tamil Nadu, the vegetable seed market is led by few firms among which Griffin Crop Sciences Private Limited and Rasi Seeds Private Limited are the key players. With respect to paddy seeds, Tharani seeds, Agri Seeds, Rockfort Seeds, TUCAS, and Annalakshmi seeds are the agencies involved in a big way in the state of Tamil Nadu. Though there are many players, few of them face success and few face difficulties in continuing hybrid seed as an agribusiness venture. If a company wants to increase the sales volume, it should give special attention to farmers' expectations and their preferences towards that particular hybrid seeds.

PROSPECTS IN HORTICULTURE AND MEDICINAL PLANTS

Export of horticultural produce is on the increase in a decade of time. It is mainly due to the quality produce realized out of hi-tech practices practiced in horticulture. Mango, vegetables, processed fruits and vegetables, spices including curd chilli, flowers particularly jasmine and cut flowers are traded abroad and it commands good price in the international markets also. However, the farmers' share in the consumers' price is still low and hence it commands role of farmers' organization in producing and trading the produce abroad so as to get higher share in the consumers' price. Potential exists for fresh and processed horticultural produce trade abroad and hence one can enter into this venture.

The World Health Organization (WHO) has estimated that the present demand for medicinal plants is approximately US $ 14 billion per annum. The demand for medicinal plant based raw material is growing at the rate of 15 to 25 per cent annually and according to an estimate of WHO, the demand for medicinal plants is likely to increase more than US $ 5 trillion in 2050. In India, the medicinal plant related trade is estimated to be approximately US $ 1 billion per year. An estimated 4000 to 10000 species of medicinal plants face potential local, national, regional or global extinction, with subsequent serious consequences for livelihoods, economies and health care systems. Hence raising these potential medicinal plants in the farm conditions, and preserving the same suitably for extraction of medicinal compounds is gaining importance (Sacratees, 2012).

PROSPECTS OF COMMERCIAL NURSERY

Commercial nurseries are the best agribusiness ventures in Tamil Nadu. Many of the small agripreneurs have entered into this business venture and minting money from this venture by raising different ornamental and commercial nursery seedlings which are of heavy demand in the urban and semi urban environs. The Commercial Case Nursery based at Dindigul District of Tamil Nadu had produced 22.70 lakh seedlings in which 85 per cent of the seedlings belonged to forestry importance and the 15 per cent of the seedlings produced were of horticultural importance. The only one dominant species produced and sold in the nursery is sandal seedlings accounting for 52.86 per cent to the total seedlings produced in the nursery outlining that potential exists for the sandal seedlings which is mostly of demand in the Forest Departments of different states. Andhra Pradesh is found to be a good market for the sandal seedlings followed by Karnataka and Kerala. Tamil Nadu is found to be a poor taker of sandal seedlings from private nurseries as they have their own interest in establishing sandal nursery in select pockets like Mettupalayam of Coimbatore District and Sathyamangalam of Erode District. In toto, sandal seedlings are produced and distributed to the tune of 12 lakhs and other forest seedlings to the tune of 10.70 lakhs from the case nursery of Tamil Nadu. The Forest Department of Bangalore city, Karnataka is found to be the major supplier of seeds of both sandal and Melia to the commercial nursery. With the investment of Rs. 71.43 lakhs per annum (US $ 115210), the owner manager is producing 22.70 lakh seedlings (2.27 million) and the cost of production of one seedling is arrived at Rs. 3.15 and the output-input ratio from the enterprising commercial nursery is arrived at 3.29 revealing that for a rupee of investment in the commercial nursery, the owner manager could reap the gross return of Rs. 3.29 and a net return of Rs. 2.29 revealing that the commercial nursery is highly profitable and hence the young agripreneur can enter into similar investment on a small scale.

PROSPECTS IN AQUACULTURE

Fisheries sector contribute considerably to the Indian economy. Through mariculture efforts, the coastal fishery development and exploitation promoted the export earnings to the level of 19500 crores during the year 2012-13. The per day earnings through export of marine products is arrived at 52.43 crores. Another important area is export of ornamental fishes which is gaining momentum in India. Through ornamental fish rearing and export, the earnings have touched to the level of 6.50 crores during the year 2012-13. The per day earning of foreign exchange through ornamental fishes alone is arrived at 0.017 crores. Besides, the fisheries development in India is made through the aquaculture initiatives wherein, the inland resources like tanks, ponds, reservoirs and backwaters have been effectively utilized to meet the domestic demand for food fishes. Hence aquaculture sectors have ample scope to develop the derelict water bodies for effective use.

SUMMARY

Peeping into the above mentioned agribusiness ventures, we can arrive at a conclusion that they are capable of offering higher incentives on its establishment and effective run. Hence, the agripreneurs are encouraged to take up any one of the venture in consultation with the technocrats of the concerned sector and the same may be initiated for their prosperity.

References

- Radhakrishna, R. (2012). "Opening Remarks on Agribusiness Potentials in Tamil Nadu" In: Agribusiness Potentials of Tamil Nadu (Madras Institute of Development Studies and Indian Society of Agricultural Makreting: Nagpur). Pp 1-4.

Chapter 4

Procedure and Constraints in Establishing Agribusiness

The entrepreneur gets information regarding various numerable investment opportunities from the magazine, internet, financial institution, government, commercial organization, friends, and relatives and so on. To choose the best business opportunity from the information collected requires ingenuity, skill and foresight on the part of the entrepreneur. As entrepreneurs, they have to select feasible and rewarding opportunity to choose such good business opportunities. For this purpose they have to evaluate the following ideas and understand the gap between demand and supply.

1. Study government rules and regulations regarding different business opportunities.
2. Extensive study of promising investment opportunity.
3. SWOT analysis of business potential (strength, weakness, opportunities and threats).
4. Market survey.

The following are the procedures to be followed in setting up agro-based industries:

1. Assessment of agricultural resource potential in the desired areas.
2. Qualification of agriculture output and inputs.
3. Present utilization of resources in existing units.

4. Surplus produce left out.
5. Requirements to preserve agricultural produce. Even for local consumption of food grains, processing is necessary.
6. Selection of certain items of considerable importance.
7. Follow certain approach for actual location of the units to avoid wastage of resources and maximize utilisation of existing infra-structure.
8. Preparation of industrially profitable and feasibility studies.
9. Identification of entrepreneurs.
10. Suggesting appropriate technology and imparting suitable training.
11. Finance and other problems.
12. Marketing assistance.

The following steps are necessary for establishment of agro-based in-dustries.

SOURCES OF BUSINESS OR PRODUCT IDEAS

1. *Market Characteristics* Unfulfilled demand of a product will open the door for new product. Supply and demand of various products and demand for new products should also be analysed.

   For example, the success of Kellogs as morning breakfast as a quick and nutrious food with less calories is an example of success of new idea. The acceptance of 'SINTEX' water tank is another example.
2. *Import and Export* The Government of India is encouraging exports and various EXIM policies encourage the entrepreneur to think about the new option.
3. *Emerging New Technology and Scientific Know-how* Commercial ex-ploitation of indigenous or imported technologies and know-how is another source of project idea.
4. *Social and Economic Trends* Social and Economic status of people are always dynamic in nature and offer wide opportunities. An en-trepreneur should observe such changes. For example there is now shift towards readymade garments; possessing consumer durables, western outfit, priority and preferences for cosmetics, etc.

5. *Product Profile* An analytical study of the end products and byproducts can throw light on new project ideas. For example byproduct of sugar industry gave rise to one more large-scale industry called paper industry.
6. *Changes in Consumption Pattern* Changes in consumption pattern of the people in the home country and foreign countries also require the entrepreneur's attention.
7. *Revival of Sick Units* A sick unit gives ample investment opportunities in the hands of a dynamic entrepreneur. He/she can revitalize and turn a sick unit into a profitable one.
8. *Trade fairs and Trade journals* Magazines, journals, industries or trade fairs offer wide scope for business opportunities.

IDENTIFYING A BUSINESS OPPORTUNITY

An entrepreneur is said to be an opportunity seeker. For the potential entrepreneurs, their first task is to identify, explore and then select an attractive business opportunity.

Opportunity can be defined as an attractive and excellent project idea which entrepreneurs search for and accept, as such ideas serve as a basis for their investment decision. A good business opportunity must be capable of being converted into feasible projects.

Two major characteristics of a business opportunity should be highlighted.

1. Good and wide market scope, i.e., gap between present or likely demand and supply.
2. An attractive, acceptable and reliable return on investment.

Business opportunity need to be analyzed from the view points of production, commercial, managerial, potential and prospective demand for the product, technical viability, etc.

Steps in Identification of Business Opportunity

Identification of opportunity involves following steps.

I. Preliminary evaluation

II. Selection of product or service

III. Conduct a market survey

IV. Contact programmes to collect sufficient information about proposed venture

V. Succeeding in the market.

I Preliminary Evaluation

As soon as the entrepreneurs realize the business opportunity, they have to evaluate the investment opportunities against a set of specific criteria to select those project ideas which are commercially feasible. The criterias are:

1. Is the opportunity compatible with the promoter.
2. Is the opportunity compatible with government regulations and priorities.
3. Whether raw materials are easily available.
4. What is the size of the potential market.
5. Whether cost justifies the project.
6. What is the risk inherent in the project.

Compatible with the Promoter The entrepreneur must conform that the project undertaken should be compatible to men, material, and money market available at their disposal. Project beyond the capacity of the entrepreneur are bound to fail.

Compatible with Government Regulations and Rules Entrepreneur should not violate government regulations and priorities. He must carefully take into account all related rules and regulations of the government regarding investment, license, reservation of certain categories of items, etc.

Easy Availability of Raw Materials Cost and availability factors of raw materials should be considered carefully. Scarcity of raw materials will cause delay in production process.

Potential Market Potential market, nature of competition, competitors, availability of substitutes, barriers and the possibility of entry of substitutes and technological developments taking place in the industry should be assessed and evaluated by the entrepreneur.

Cost of the Project The cost of the project should be reasonable in the sense that a desired profit margin can be realized from a competitive price. A study of the cost structure under raw material cost, labour cost, direct expenses, factory overheads, administrative expenses, selling and distributive overheads, and after sales services costs will give a good idea regarding different types of cost.

Inherent Risk in the Project should be Analyzed Risk related to the project such as changes in demand, technological developments, entry of substitutes, competition and seasonal variations should be assessed before working on a project.

II Selection of Product or Service

Entrepreneur should identify the product, which they wish to manufacture. While deciding about the product following points should be considered:

1. Potential demand for the product or service.
2. Estimated volume of demand for the product.
3. Assess potential of existing competitor and estimate about probable competitors.
4. Study the scope for future demand.
5. Infrastructural facilites—power, transport, etc.
6. Current status of technology and scientific development in the field.
7. Availability of raw material and required labour.
8. Government policies, legislation, controls.
9. Environmental factor.
10. Degree of profitability for the product.
11. Information regarding particular line of product.
12 Locational advantage of the product.

13. If product belongs to an ancillary unit and serves as major component for the parent industry. It provides a ready demand hence selection of this type of product entails easy marketability.
14. Selection of a product would also be assessed in favour of availability of skilled and unskilled labour.
15. Study the various characteristic of the proposed product to be produced.

III. Conduct a Market Survey or Purpose of Market Survey

Market survey with reference to the availability of raw material, equipments, marketing and distribution and consumer behaviour should be conducted.

1. *Raw Material Availability*

a. Search for leading suppliers of raw materials required for the concerned product.
b. Study the price policy of various suppliers and analyse impact of price fluctuations on production.
c. Fix time for order execution.
d. Study local and outside source of raw materials—the advantages and disadvantages.
e. Thorough analysis of credit facilities, advance payments, terms and conditions for suppliers.

2. *Equipment Availability*

a. Identify major manufacturer here and abroad.
b. Comparative features of various manufacturers.
c. Price structure of different brands.
d. Repair maintenance and after sales service facilities.
e. Guarantees and warranties by suppliers.
f. Technical and skilled staff requirement.
g. Machinery and delivery schedules.

3. *Marketing and Distribution*

a. Selection of best channel of distribution.

b. Advertising and publicity programme for the product.

c. Product positioning.

d. Outstanding features of product or service.

e. Market features and practices—credit facility, minimum order, incentives.

f. Business terms, commission, stocks, warehouse facilities.

4. *Consumer Behaviour*

a. Motivate buyers to buy new product.

b. Analyse the buyer's purchasing power.

c. Analysis of consumption pattern to capture the major market share.

d. Understand the preference for durability, service, and economy.

e. Understand consumer characteristics of different region and devise appropriate sales message, accordingly.

IV Contact Programmes to Collect Sufficient Information about Proposed Venture

Entrepreneur often need information and guidance, particularly in the initial stages, on product potential, raw materials, policies, facilities, procedures, finance formalities, incentives, etc. It can be collected through state government agencies.

Contact with central and state level agencies set can be helpful in collecting sufficient information about proposed venture. Industrial Finance Corporation of India (IFCI) in collaboration with Industrial Development Bank of India (IDBI), Industrial Credit and Investment Corporation of India (ICICI), state organizations and banks, have set up a network of state level technical consultancy agencies. They offer a package of professional and consultancy services to stimulate industrial growth.

Entrepreneurship development programmes of 6 weeks duration are designed to impart following information:

1. How to develop entrepreneurial capabilities.
2. Identify viable projects for potential entrepreneurs.
3. Impart managerial skills.
4. Help and secure necessary financial and infrastructural related assistance. These technical consultancy agencies also monitor the progress of the unit.

V Succeeding In the Market

There is no way your business will earn money and profit unless customers buy product or service. The secret formulae for an entrepreneur's success is to produce what customers will buy.

The following are the important characteristics, which help the entrepreneur to succeed in the market:

1. Study people and their needs before starting any project.
2. Identify unsatisfied needs.
3. Design product in such a way that it should satisfy the customer better than the competitor's product.
4. Ensure that what customer feel about the product which entrepreneur is offering.
5. Always look for newer and more effective ways of reaching a customer.
6. Entrepreneurs must have clear vision, goals and objectives, well-defined mission, and employee's participation about the proposed project.
7. Constant feedback of results as well as setting and adherence of high standards gives an organization a cutting edge over others. Planning, foresight and analysis are also important qualities.
8. The process of systematic market research is used to develop products or process and to provide value for money to the customers. This helps to gain the market share.
9. Another important characteristic for successful business is a commitment to innovation which is vital in keeping ahead of the competition and perhaps the most difficult one to achieve yet most of the organizations just do not give due importance for the same.

CONSTRAINTS IN ESTABLISHING AGRO BASED INDUSTRIES

1. Proper guidance is not available to entrepreneurs.
2. It involves some element of risk taking.
3. Change in crops/cropping pattern.
4. Change in variety of crop due to technological improvement.
5. Failure of monsoon may hit the raw material supply.
6. Proper guidance, training for modern and sophisticated agro-industries are not available.
7. As modern small industries are capital intensive, supply of finance will be a considerable problem.
8. Promotional activities such as conducting intensive campaigns, identifying candidate industries and explaining to entrepreneurs about prospects are inadequate.
9. Uncertainty about future market demands.
10. Absence of information about quantity and quality of market.

Chapter 5

Human Resource Planning and Management

MEANING

Human Resource Planning (HRP) is the process of forecasting an organization's future demand for, and supply of, the right type of people in the right number. It involves analyzing and identifying the need for and availability of human resources so that the organization can meet its objectives.

HUMAN RESOURCE PLANNING – THE PLANNING PROCESS

HRP essentially involves forecasting personnel needs, assessing personnel supply and matching demand–supply factors through personnel-related programmes. The planning process is influenced by overall organizational objectives and the environment of business. The process is illustrated in Figure 1.

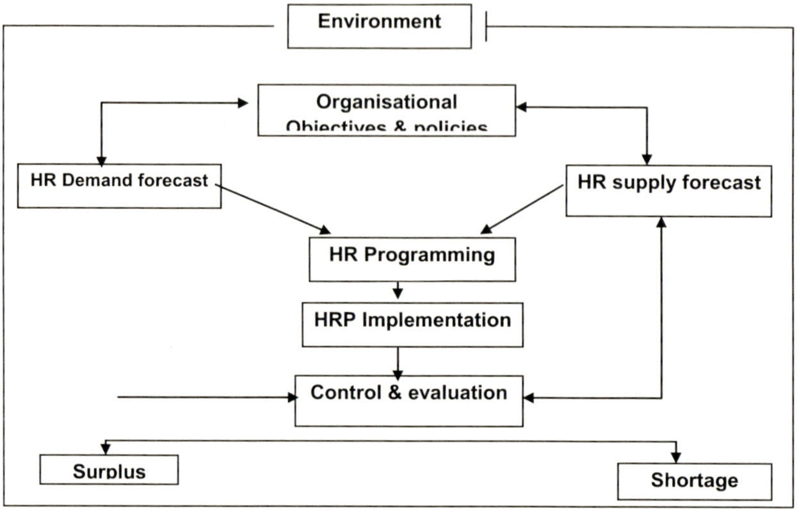

Environmental Scanning

It refers to the systematic monitoring of the external forces influencing the organization. Managers monitor several forces but the following are pertinent for HRP:

- Economic factors, including general and regional conditions.
- Technical changes, including robotics and automation.
- Demographic changes, including age, compensation and literacy.
- Political and legislative issues, including laws and administrative rulings.
- Social concerns, including child care and educational facilities and priorities.

By scanning the environment for changes that will affect an organization, managers can anticipate their impact and make adjustments early.

Organizational Objectives and Policies

HR plans must be derived from organizational objectives. Specific requirements in terms of number and characteristics of employees should be derived from the organizational objectives.

Organizational objectives are defined by the top management and the role of HRP is to subserve the overall objectives by ensuring availability and utilization of human resources. Once the organizational objectives are specified, communicated and understood by all concerned, the HR department must specify its objectives with regard to HR utilization in the organization. In developing these, specific policies need to be formulated to address the following:

- Vacancies to be filled by promotions or hiring?
- How to enrich employee's job?
- How to ensure continuous availability of adaptive and flexible workforce?
- Extent of production and operations automation

HR Demand Forecast

Demand forecasting is the process of estimating the future quantity and quality of people required. The basis of the forecast must be annual budget and long-term corporate plan, translated into activity levels for each function and department. Demand forecasting should consider both internal and external factors. Among the external factors are competition, economic climate, laws and regulatory bodies, change in technology, and social factors. Internal factors include budget constraints, production levels, new products and services, organizational structures and employee separations.

Forecasting Techniques Forecasting techniques vary from simple to sophisticated ones. Some of the techniques used are

1. *Managerial judgement* Managers sit together, discuss and arrive at a figure which would be the future demand for labour. The technique may involve a 'bottom-up' or a 'top-down' approach. In the first, line managers submit their departmental proposals to top managers who arrive at the company forecasts. In the 'top-down' approach, top managers prepare company and departmental forecasts.
2. *Ratio-trend analysis* This is the quickest forecasting technique. This involves studying past ratios, say between the number of workers and sales in an organization and forecasting future ratios, making some allowances for changes in the organization or its methods.

3. *Regression analysis* Similar to the ratio-trend analysis in that forecast is based on the relationship between sales volume and employee size. A diagram depicting the relationship between sales and workforce size is drawn. It then calculates regression line—a line that cuts right through the center of the points on the diagram. By observing the regression line, one can find out the number of employees required at each volume of sales.

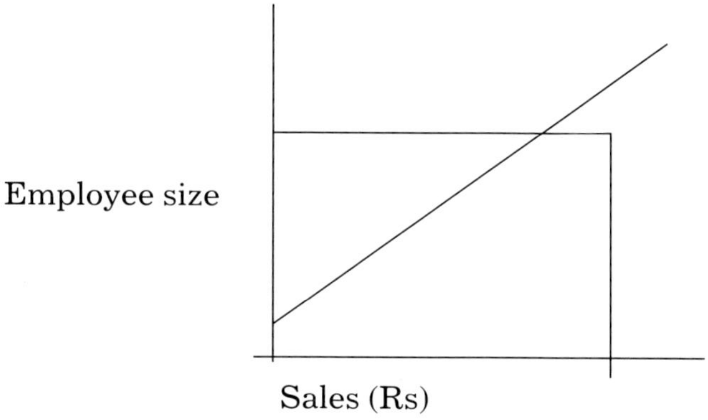

In the above diagram, the regression line shows the relationship between sales and employee size.

4. Work study techniques Can be used when it is possible to apply work measurement to calculate the length of operations and the amount of labour required.
5. Delphi technique It solicits estimates of personnel needs from a group of experts, usually managers. The HRP experts act as intermediaries, summarize the various responses and report the findings back to the experts. The experts are surveyed again after they receive this feedback. The distinguishing feature of the Delphi technique is the absence of interaction among experts.

HR Supply Forecast

Measures the number of people likely to be available from within and outside an organization, after making allowance for absenteeism, internal movement and promotions, wastage and changes in hours, and conditions of work.

Supply analysis covers

- Existing human resources
- Internal sources of supply
- External sources of supply

HR Programming

Supply forecast and demands forecast must be reconciled or balanced in order that the right employees can fill vacancies at the right time.

HR Plan Implementation

Implementation requires converting HR plan into action. A series of action programmes are initiated as a part of HR plan implementation namely, recruitment, selection and placement, training and development, retention plan, succession, planning, etc.

Control and Evaluation

HR plan includes budgets, targets and standards. It should also clarify responsibilities for implementation and control, and establish reporting procedures which will enable achievements to be monitored against the plan. These may simply report on the numbers employed against establishment and on the numbers recruited against the recruitment targets.

ROLE OF HR MANAGER IN ORGANIZATION

It refers to the tasks performed in an organization to provide for and coordinate human resources. Human resource functions are concerned with a variety of activities that significantly influence almost all areas of an organization and aim at:

1. Ensuring that the organization fulfills all of its equal employment opportunities and other government obligations.

2. Conducting out job analysis to establish the specific requirements for individual jobs within an organization.
3. Forecasting the human resource requirements necessary for the organization to achieve its objectives—both in terms of number of employees and skills.
4. Developing and implementing a plan to meet these requirements.
5. Recruiting and selecting personnel to fill specific jobs within an organization.
6. Orienting and training new employees.
7. Designing and implementing management and organizational development programmes.
8. Designing systems for appraising the performance of individuals.
9. Assisting employees in developing career plans.
10. Designing and implementing compensation system for all employees.
11. Appraising performance and communicating.
12. Managing wages and salaries.
13. Providing incentives and benefits.
14. Employee health and safety.
15. Grievances and labor relations.

ASSIGNMENT

1. Prepare a human resource plan for your own firm.

Chapter 6

Recruitment and Selection of Staffs in Agribusiness Firm

AIM

- To introduce the methods of selection of right employees for the agribusiness firms
- To highlight the importance of job design, job analysis, job specification and job description

INTRODUCTION

The managerial function of staffing is defined as filling and keeping filled positions in the organization structure. This includes identification of workforce requirements, inventorying the people available, and recruiting, selecting, placing, promoting, appraising, planning the careers of, compensation, and training or otherwise developing both candidates and current job holders to accomplish their tasks effectively and efficiently. It is clear that staffing must be closely linked to organizing, that is, to the setting up of intentional structure of roles and positions.

As far as searching the personnel for different positions in the organization is concerned, a manager has certain tools that are important part of staffing job and is essential to ensure effective human resource management. These tools are

- Job Analysis
- Job Description and
- Job Specification

Job Analysis

Job analysis is the process of collecting, analyzing and recording information about the particular job or jobs. The primary purpose of job analysis is to determine the duties and responsibilities of the job and what qualifications an individual should possess to perform the job effectively.

Job Description

The job description is a written statement of the duties and responsibilities of a job. It explains what the job is, what are the duties, responsibilities, general working conditions, to whom the person should report, and what is expected of them in the performance of their duties. The statement is called job description at the operational level, whereas it is called position description at the managerial level.

Job Specification

The job specification concentrates on the characteristics needed by an employee to perform the job. It describes the educational qualifications, training, and experience, physical and mental capabilities that the employee must possess to perform the job effectively.

RECRUITMENT

Recruitment involves seeking and attracting a pool of people from which qualified candidates for job vacancies can be chosen. It is the process of searching for prospective employees and stimulating and encouraging them to apply for jobs in an organization. Recruitment, thus, involves the creation of pool of labor from which the organization can draw when it needs additional workforce. The relationship between human resources planning, job analysis, recruitment and selection process is furnished below.

Human Resource Planning → Number of Specific Jobs to be Filled → Nature and Requirements of Specific Job → Recruitment (What? How? Who?) → Pool of Qualified Applicants → Screening Test / Interview → Selection → Placement Training → Appointment to the Position.

FACTORS AFFECTING THE NUMBER AND KINDS OF MANAGERS REQUIRED

The number of managers needed in an enterprise depends not only upon the size but also depends upon the complexity of the organization structure, the plans for expansion, and the rate of turnover of managerial personnel. The ratio between the number of managers and the number of employees does not follow any law. It is possible, by enlarging or contracting the delegation of authority to modify a structure so that the number of managers in a given instance will increase or decrease regardless of the size of an operation. Although the need for determining the number of managers required has been stressed here, it is clear that numbers are only part of the picture. Specifically, the qualifications for individual positions must be identified so that the best suited managers can be chosen.

Sources of Recruitment

The sources of recruitment can be conveniently classified into internal and external sources.

Internal Sources of Recruitment

Internal recruitment seeks applicants for positions from those who are currently employed. Internal sources include present employees, employee referrals, former employees and former applicants.

External Sources of Recruitment

When an organization has exhausted internal sources or when it finds internal sources to be unsuitable, the following external sources are used.

- Professional or trade associations
- Advertisements floated in dailies
- Employment exchange both in government and private
- Campus recruitment
- Walk-in interviews
- Consultants of the organization
- Contractors associated with the organization
- Radio and television ads
- Mergers and acquisitions
- Competitor organizations
- Miscellaneous sources

Selection

Selection is a process of weeding out unsuitable candidates and finally arriving at the most suitable candidates. Thus, while recruitment is considered as a positive process of attracting job applicants, selection is a negative process. Selection is a crucial process in management. In big enterprises, selection is a time consuming process. Selection process includes the following sequence of cardinal steps.

- Recruitment determines the total number of applicants applying for each position in the organization.
- Preliminary screening tests / Interview
- Blank Application Form—how a person completes the form also might give some clues about the person's organizational ability and intelligence.
- Psychological tests to measure mental alertness, special aptitude and physical ability
- Evaluation interview—purpose is to have a good overview of the candidate's strength and weaknesses for the position.

- Background investigation with the referees either by phone or by written communication.
- The physical examination of the individual
- Placement—it is the final step in the selection process
- Providing Employment Letter—it should incorporate the information on place of work, gross emoluments, hours of work, working conditions if any. Besides this, one can inform the promotional avenues, the rate of increment over a particular period, etc.

MODEL ADVERTISEMENT FOR RECRUITING THE STAFF

Tamil Nadu Horticultural University, Periyakulam, 625 604

EMPLOYMENT NOTICE

Advertisement Number 01/TANHU/HR/Recruitment/2014-15/ Date: 03.06.2014

The University invites online applications from the Indian Citizens for the following **G**roup 'A' Positions

Sl. No	Position	Number of Positions
01	The Registrar	01
02	Controller of Examinations	01
03	Deans	05
04	Directors	03
05	Librarian	01
06	Professors	10
07	Associate Professors	15
08	Assistant Professors	27
09	Farm Managers	41

For details of various positions, eligibility criteria and procedure for sending the applications, please visit http://tanhu.ac.in

The University has every right to fill the position or left unfilled.

Place: Periyakulam **Sd. XXX (J.S. PAARRY)**

03.06.2014 **Registrar - in - Charge**

Exercise

Using the above information, recruit and select the suitable candidates for the above positions and provide the offer letter to the selected candidates on or before 31st July 2014 and give a narration report for the selection of different candidates

MODEL INTERVIEW LETTER FOR THE POSITION OF REGISTRAR

TAMIL NADU HORTICULTURAL UNIVERSITY

Date: 03.06.2014

Dear Sir,

Sub: TANHU- Filling up the post of Registrar – Selection Committee Meeting fixed on

18th June 2014 at **9.30 a.m.** – Reg.

Ref:

1) Our University Advertisement dated 05.03.2014
2) Your application for the post of ***Registrar*** of the University, Dt. 30th March 2014

****

With reference to your application cited (2), in response to our University advertisement cited (1) under reference, I am directed to inform you that the Selection Committee Meeting for the position of ***Registrar*** of the university will be held at **9.30 a.m**, on **Friday**, the **18th July 2014** in the **Chamber of the Vice Chancellor, Tamil Nadu Horticultural University, Kamatchipuram, Periyakulam East, Theni – 625 604.** You are requested to appear before the Selection Committee Meeting on the said date, time and at the said venue.

You are requested to arrive at the venue on **18th July 2014** by **8.30 a.m** along with all the original certificates including **"No Objection Certificate"** from your employer if you are employed recently.

You are also requested to bring all your certificates from SSLC/Matric onwards and testimonials in support of your age, educational qualifications, administrative experience and also teaching/research/industrial experience, if any. Needless to say, you must come **well prepared** for the in-depth interview, which will be conducted in **English** only.

Please note the following carefully

i. *Please bring Seven Copies of your updated curriculum vitae at the time of interview.*

ii. *Before the conduct of personal interview, a* ***Written Test*** *in English, relating to Government of Tamil Nadu Rules and Regulations, University / Higher Education in Tamil Nadu, etc., will be conducted.*

Travelling Allowance is admissible to the candidates called for the interview as per the rules by the shortest route on submission of tickets of the journey. Please bring the tickets along with you for reimbursement.

You may inform to the undersigned either by mail (http://tanhu.ac.in) or fax (04546 - 456789) or by a speed post letter immediately indicating **whether you plan to attend the interview or not.** Please do not make telephone calls in this connection.

You may also note that canvassing in any form will disqualify you from being considered for the said position.

Ms. J.S. Thenndrall, Deputy Registrar

VC's Secretariat

To,

The Individuals by Name as per the List.

TAMIL NADU HORTICULTURAL UNIVERSITY,
PERIYAKULAM -625 604

F/No. TANHU / R1 / 001 / 2014

Ph: 04546-456789

Fax: 04546-345678

To

The Registrar
Tamil Nadu Agricultural University
Coimbatore – 641 003

Sir,

The University has advertized in 3rd June 2014, among others, the Post of ***Registrar, Tamil Nadu Horticultural University.*** The appointment for this post would be made in the pay band of Rs. 37400 – 67000 (AGP Rs 10000/-) for a tenure of Three years on deputation terms with a provision of renewal of appointment for a further tenure of Three years, if needed.

The Selection Committee constituted for this purpose recommended that ***Dr. C. Sekhar***, Professor in Agricultural Economics, Department of Social Sciences, Horticultural College and Research Institute, Periyakulam may be appointed to the post of Registrar and the Board of Management of the University at its First Meeting held on 20.07.2014 approved the recommendation made by the said committee.

The period of deputation will be initially for three years from the date of his joining to the Post of Registrar, which could be considered for extension after mutual consultation with your institution. The University will pay the Leave Salary and Pension Contributions as per the University norms.

The Tamil Nadu Horticultural University shall be grateful if ***Dr. C. Sekhar*** could please be relieved to take up this assignment of The Registrar of this University at an early date. It is also requested that the amount of Leave Salary and Pension Contributions for the period of deputation of ***Dr. C. Sekhar*** to the University may be indicated.

Yours Faithfully

Ms. S. Dhanalakshmi
Director, Human Relations

Copy Communicated to

Dr. C. Sekhar
Professor in Agricultural Economics
Department of Social Sciences
Horticultural College and Research Institute
Periyakulam – 625 604

Chapter 7

THEORY OF COSTS AND THE PRODUCTION DECISIONS

Knowledge on various costs and their relationships that exist between cost and the output is necessary to comprehend the concepts of equilibrium conditions of different firms under different market situations. The students should know how the different costs are related to different output levels of the firms under different levels of technology used by the business firms. The different types of the costs are delineated below.

- Fixed Costs (FC) or Total Fixed Cost (TFC)
- Variable Costs (VC) or Total Variable Costs (TVC)
- Total Costs (TC)
- Average Fixed Costs (AFC)
- Average Variable Costs (AVC)
- Average Total Costs (ATC) and
- Marginal Cost (MC)

One has to assume here that every business firm has a certain state of Technology or know how. That is, how the firm is producing various levels of outputs with the given prices of inputs and outputs. With data on those variables, one can understand the relationships between costs and output levels. Basically we require data on output, fixed costs, variable costs and the prices of inputs and outputs. From this data one can derive all the costs delineated above.

Costs are broadly classified into short run costs and long run costs. Short run is a period during which some factors are variable while others are fixed. Long run is a period over which all the factors become variable. Consideration of cost curves is essential and forms the basis for entry and exit of the firms in the industry. Profit maximizing rule is determined with the help of cost curves, cost functions and production functions. There are seven costs, which explain the behavior of the firms in the production of farm produce or the produce of industrial importance.

SHORT RUN COST CURVES

Fixed Costs or Total Fixed Costs (TFC)

Total Fixed Cost denotes those costs, which do not vary with the level of production and are incurred even when production is not undertaken. In other words fixed costs are independent of output. That is fixed cost do not change as the quantity of output changes. Fixed costs include the cost items like taxes, insurance, cess, depreciation on machinery, implements, tools, buildings, salaries of personnel working in the firm, etc. These are also known as indirect costs, sunk costs and overhead costs. The summation of all these costs is called total fixed costs. TFC is a horizontal straight line parallel to the 'X' axis.

Variable Costs (VC) or Total Variable Costs (TVC)

Variable costs are those costs that vary with the level of output and are incurred only during the time of production of any output. In other words, variable costs are dependent on output. These include costs of raw materials, labor, power, repairs, maintenance charges of machinery in respect of manufacturing sector; seed, fertilizer, plant protection chemicals, labor, manure and bio fertilizers in respect of farm related production activity. These are also known as working costs, operating costs, direct costs, prime costs, circulating costs and running costs. These are second phase costs. The summation of these costs refers to total variable costs.

Algebraically it is denoted as $TVC = Px_1.X_1 + Px_2.X_2 + ... + Px_{n.}X_n$

Where,

Px is the price of input X,

X is the quantity of input used in the production process, and

1,2,.......,n are the number of inputs used in the production activity.

Total Costs (TC)

Total costs included total fixed costs (TFC) as well as total variable costs (TVC). Its shape is similar to that of total variable costs.

$$TC = TFC + TVC$$

Average Fixed Cost (AFC)

It is worked out by dividing the total fixed cost by the amount of output produced in the farm/firm. It is fixed cost per unit of output. AFC will vary for each level of output. As output increases, AFC decreases. When output is zero, AFC = TFC, AFC always slopes downwards regardless of production function. Its general location on the graph depends upon the magnitude of total fixed costs.

Average Variable Cost (AVC)

AVC is worked out by dividing total variable costs by the amount of output. AVC varies with the level of production. The shape of AVC curve depends upon the shape of the production function. The height of the AVC depends upon the unit cost of the variable input. Like AFC, AVC cannot be computed when the output is zero. AVC is inversely related to Average Physical Product (APP). That is when APP is increasing, AVC is decreasing. When APP is maximum, AVC attains minimum. When APP is decreasing, AVC is increasing. AVC first falls due to the economies of large scale production and then rises due to diseconomies of scale in production of commodities. Thus, as on a production function, APP measures the efficiency of the variable input, for cost curves AVC provides the same measure.

- When AVC is decreasing, the efficiency of the variable input is increasing

- When APP is at maximum, AVC is at a minimum
- When APP is decreasing, the AVC is increasing

Average Total Cost (ATC)

Average Total Cost can be computed in two ways; by dividing total costs by output or by adding AFC and AVC. The shape of the ATC curve depends upon the shape of the production function. ATC decreases as output increases, attains minimum and increases thereafter. ATC is often referred to as the unit cost of production. That is the cost of producing one unit of output.

The initial decrease in ATC is caused by the spreading of fixed costs among an increasing number of units of output and the increasing efficiency with which the variable input is used. As output increases further, ATC attains a minimum and begins to increase, as increase in AVC can no longer be offset by decrease in AFC.

Characteristics of Average Cost Curves

- AFC curve will continue to decline and never show upward movement because after maximum output is achieved, input addition beyond this becomes irrational.
- As production expands, the AVC tends to decline. It reaches the lowest point and then bends upwards. At this point APP is the highest.
- ATC curve has the same shape as AVC. Difference is that the lowest point in case of AVC reaches earlier as compared to ATC.

Marginal Cost (MC)

Marginal cost is the change in total cost in response to a unit increase in output at the margin. It is the change in cost associated with an increase or decrease of an additional unit of output and is computed by dividing the change in the total cost by the corresponding changes in the output.

Characteristics of Marginal Cost Curve As output increases, MC first falls due to more efficient use of the variable factors of production and then it slopes upwards due to less efficient use of the variable factor. Marginal cost is related to marginal product in the same manner that AVC is related to AP.

RELATIONSHIP BETWEEN AVERAGE COST AND MARGINAL COST

Marginal cost has definite relationship with ATC and AVC. The relationships are furnished as follows.

- AVC and ATC curves will slope downwards and keep falling as long as MC curve is below them. Conversely AVC and ATC curves will move up when MC curve is above them.
- MC curve will intersect AVC curve at the lowest point from below. As MC curve moves on, it intersects the ATC curve from below at the lowest point.

DETERMINATION OF NET REVENUE AS A FUNCTION OF OUTPUT

The principle of profit maximization can be discussed in two ways.

- As a relationship between marginal costs and marginal revenue, and
- As a marginal product ratio in relation to factor-product price ratio (MC = MR)

Assignment

Each and every student should draw neat and labeled diagrams on graph papers to show the following concepts used in cost function analysis.

Relationship between TFC, TVC and TC

Units of Mango	TVC incurred in Mango Orchard in Lakhs	TFC incurred in the Mango Orchard in Lakhs	TC in Lakhs
00	00	10	10
02	02	10	12
05	04	10	14

Units of Mango	TVC incurred in Mango Orchard in Lakhs	TFC incurred in the Mango Orchard in Lakhs	TC in Lakhs
09	06	10	16
14	08	10	18
19	10	10	20
23	12	10	22
26	14	10	24
28	16	10	26
29	18	10	28

Relationship between Average and Marginal Cost Curves

Units of Mango	AVC	AFC	ATC	Change in the TC	Change in the Output	MC
00	0.00	0.00	0.00	0		
02	1.00	5.00	6.00	2	2	1.00
05	0.80	2.00	2.80	2	3	0.67
09	0.67	1.11	1.78	2	4	0.50
14	0.57	0.71	1.38	2	5	0.40
19	0.53	0.53	1.06	2	5	0.40
23	0.52	0.43	0.95	2	4	0.50
26	0.54	0.38	0.92	2	3	0.67
28	0.58	0.36	0.94	2	2	1.00
29	0.62	0.34	0.96	2	1	2.00

Net Revenue as a Function of Output

Detail three regions of cost function for profit maximization. Use the data on AC and MC as given in the table above. Use enlarged scale. Assume MR = AR = 0.75 and draw the inferences.

Chapter 8

Cost of Cultivation and Cost of Production of Farm Produce

COST OF CULTIVATION

Cost of cultivation of principal crops is the permanent scheme approved by Directorate of Economics and Statistics, Ministry of Agriculture, Government of India and the same is in operation in each and every state. In Tamil Nadu, the scheme is functioning in the Department of Agricultural Economics, Directorate of Centre for Agricultural and Rural Development Studies (CARDS), Tamil Nadu Agricultural University, Coimbatore. Similarly, the scheme is functioning in different state agricultural universities of India. Every three years, the Government of India, will notify the principal crops for which the cost of cultivation details should be taken up in a fixed format designed and tested by the Directorate of Economics and Statistics and the analyzed data on cost of cultivation will be used for fixing the minimum support price (MSP) for the principal crops cultivated by the farmers. The cost concepts prescribed by the Government of India is delineated as follows.

Cost A1

Cost A1 included the following costs. They are furnished as follows.

- Hired human labor
- Value of hired and owned bullock labor
- Value of hired and owned machine labor
- Value of owned and purchased seed
- Value of insecticides and pesticides
- Value of owned and purchased manures
- Value of fertilizers
- Depreciation cost on machinery and buildings
- Cost of irrigation
- Land revenue
- Cess and other taxes
- Interest on working capital and
- Miscellaneous expenses

Cost A2

Cost A2 = Cost A1 plus rent paid for leased in land

Cost B1

Cost B1 = Cost A2 plus imputed interest on the value of owned fixed capital assets (excluding land)

Cost B2

Cost B2 = Cost B1 plus rental value of owned land

Cost C1

Cost C1 = Cost B1 plus imputed value of family labor

Cost C2

Cost C2 = Cost B2 plus imputed value of family labor

Cost C2 is the total cost incurred in establishing principal crops like onion, paddy and maize which are included in the list of principal crops under the cost of cultivation scheme. These costs are referred to as the costs involved in cultivation of crops like onion and vegetables.

With respect to trees and fruit trees, the cost of cultivation referred to the following costs.

- Establishment Cost
- Annual Maintenance Cost and
- Harvesting Cost

Establishment Cost

Establishment of mango plantations included land preparation, cost of seedlings, pitting, planting, manuring, basin formation, watering charges. It is referred to often as fixed cost.

Annual Maintenance Cost

It included the costs on gap planting, irrigation charges, manuring, plant protection chemicals, cost of fertilizer and its application, growth regulators, pruning, basin formation charges.

Harvesting Cost

It included the costs incurred on harvesting of mango, grading, packing and forwarding charges. Total cost included establishment cost plus annual maintenance cost and the harvesting costs.

Cost of Production

Cost of production is referred to as the total cost of cultivation divided by the yield obtained from the production activity. For example, a brief illustration is given for better understanding in respect of production of onion.

Details of Cost of Cultivation and Cost of Production of Onion

Sl.No	Particulars of Cost	Cost Incurred by Small Farm	Cost Incurred by Large Farm	Percentage to Total Cost
		Cost of Cultivation		
01	Cost A1	8925	9125	42.84
02	Cost A2	8990	10100	47.41
03	Cost B1	9415	10730	50.38
04	Cost B2	16970	18350	86.15
05	Cost C1	17200	19150	89.91
06	Cost C2 (TC)	19400	21300	100.00
07	Yield of Onion in Quintals (Y)	48.00	49.50	
		Cost of Production		
	Cost of Production per Quintal = TC /Y	404.17	430.30	

Here, the cost of production of onion per quintal is arrived at Rs 404 in respect of small farms and Rs. 430 in respect of large farms. Cost of production per kg of onion is arrived at Rs 4.30. The market price is normally in the range of Rs 20 paid to the producer while the consumer price is at Rs. 40 per kg.

Example II

The pattern of investment made in the mango pulp industries of Tamil Nadu is detailed below.

Table 1 Total Investment Cost Made in the Mango Pulp Processing Industry

Sl. No	Particulars of Investment	Amount in Lakhs	Percentage of Investment
01	Land	07.00	01.85
02	Buildings	40.00	10.55
03	Furniture Items	02.00	00.53
04	Machineries	300.00	79.15
05	Deposits / Margin Money	30.00	07.92
	Total	379.00	100.00

Table 1 reveals the status of investment in asset mobilization for the pulp industry. An industry on an average requires around 380 lakhs for purchase of assets for establishing the industry. By sending suitable bankable project proposals with commercial bank, one can get the investment money for establishing the pulp industry.

Table 2 Variable Cost Incurred by the Sample Mango Pulp Industry

Sl. No	Particulars of Investment	Amount in Lakhs (Rs)	Percentage of Investment to the Total
01	Mango Fruits	140.00	32.94
02	Packing Materials	75.00	17.65
03	Chemicals and Consumables	80.00	18.82
04	Wages Paid to the Labor Force	40.00	09.41
05	Electricity / Fuel / Water / Fuel Wood	55.00	12.94
06	Other Expenses	35.00	08.24
	Total	425.00	100.00

The variable expenses need to be incurred are on raw materials, packaging materials, chemicals and consumables, wages paid to the casual laborers, electricity expenses, fuel, water and fuel wood expenses and the miscellaneous expenses incurred towards various unforeseen items.

These are arrived at Rs 425 lakhs per pulp industry. Among these variable costs, the raw material costs alone accounted for around 33 per cent to the total variable cost followed by packing materials cost.

Table 3 Fixed Cost Incurred by Mango Pulp Industry in Tamil Nadu

Sl. No	Particulars of Investment	Amount in Lakhs (Rs)	Percentage of Investment to the Total
01	Salary Paid for Permanent Staff	15.00	14.15
02	Repairs and Renewals	12.00	11.32
03	Rental Value of Land	03.00	02.83
04	Duties and Taxes	03.50	03.30
05	Depreciation Charges	47.50	44.81
06	Miscellaneous Expenses	25.00	23.59
	Total	106.00	100.00

The details of fixed costs incurred by the pulp industry is delineated in Table 3. It reveals that the depreciation charges on buildings and machineries accounted for 44.81 percent to the total fixed cost followed by salary paid to the permanent staff attached with the pulp industry accounting for 14.15 per cent to the total. The total fixed cost arrived per pulp industry is Rs. 106 lakhs.

Table 4 Costs and Returns from Mango Processing Industry

Sl. No	Particulars	Amount in Lakhs (Rs)	Percentage of Investment to the Total
01	Total Establishment Cost	379.00	41.65
02	Total Fixed Cost	106.00	11.65
03	Total Variable Cost	425.00	46.70
04	Total Cost	910.00	100.00
05	Gross Income by Sale of Pulp	1980.00	
06	Net Returns	1070.00	

Table 4 reveals that the total establishment cost incurred by the pulp industry arrived at 41.65 per cent to the total cost. The total fixed cost arrived at 11.65 per cent while the total variable cost arrived at 46.70 per cent indicating that the variable expenses per industry consumed higher costs. The total cost incurred by the pulp industry was arrived at Rs 910 lakhs leaving the net revenue generation to the level of 1070 lakhs per annum revealing the truth of why mushrooming of pulp industries in Tamil Nadu is possible.

INFERENCE

Through this exercise, a detailed aspect of various types of costs with suitable examples has been discussed. The case of mango pulp industry revealed the details of costs incurred under various items and created confidence in the minds of young agripreneur in establishing such industry for generating higher profit margin.

Chapter 9

FINANCIAL SOUNDNESS OF AGRIBUSINESS

WHAT IS FINANCE?

Imagine that you are going to set up an agribusiness unit in your locale or in the locale which has the potential prospects for the same. One of the main instrument or element is finance to establish the agribusiness unit. You have to answer the important questions of the following.

1. What long term investments should you consider? To be specific, what type of buildings, machinery and equipment you need immediately to establish the venture?
2. Where will you get the long term financing for meeting the requirements of agribusiness unit aimed at?
3. How will you manage your every day financial routine such as collection of money form retailers or customers and paying money to the suppliers of inputs for the agribusiness firm?

These are the most important questions one should ask. Corporate finance broadly speaking is the study of ways to answer these three questions.

The owners or stockholders of large business establishments are usually not directly involved in making business decisions particularly on a day-to-day basis. Instead, the corporation or business establishments will engage the managers to represent the owner's interests and make the decisions on their behalf. In large corporations, the financial manager would be the person in-charge of answering the above three questions raised in the preceding section. He may be the Vice President (Finance) or the Chief Financial Officer of the agribusiness concern.

FINANCIAL MANAGEMENT DECISIONS

The Vice President (Finance) or the Chief Financial Officer of the agribusiness firm should be concerned with the three basic types of questions. The first question is to find the ways and means of meeting the long term investments. The process of planning and managing the firm's long term investments are called capital budgeting.

Capital Budgeting

The financial manager of the agribusiness firm tries to identify the investment opportunities that are worth more to the firm than they cost to acquire. It means, the value of the cash flow generated by an asset exceeds the costs of the asset.

The types of investment opportunities that would typically be considered depend on the nature of the firm's business. For a large retailer like Wal-Mart, deciding on whether to open another outlet in a developing country like India or not would be an important capital budgeting decision. Here, the financial manager must not only be concerned with how much cash they expect to receive, but also with when they expect to receive it and how likely they are to receive it. Evaluating the size, timing, and risk of future cash flows is the essence of capital budgeting.

Capital Structure

The second question raised in the preceding section is, the way in which the financial manager obtains and manages the long term financing needs to support the long term investments of the agribusiness concern. The firm's capital structure is the specific mixture of long term debt and equity the firm uses to finance its operations. The financial manager has two concerns in this area. First, how much money should the agribusiness firm borrow? That is what mixture of debt and equity is best? The mixture chosen will affect both the risk and the value of the firm. Second, what are the least expensive sources of funds for the agribusiness firm?

In a firm, the capital structure means, how the earnings of a rupee or an investment of a rupee is sliced into different requirements of the business. That is what percentage of the firm's cash flow goes to creditors and what percentage goes to shareholders. In addition to deciding the financing mix, the financial manager has to decide exactly how to raise the money and from where to raise the money. The expenses associated with long term financing can be considerable, so different possibilities should be evaluated. Choosing the money lenders and among the loan types for generating long term investment need is another task of financial manager.

Working Capital Management

The third question concerns working capital management. The term working capital refers to a firm's short term assets such as inventory, and its short term liabilities such as money to be paid for the input suppliers to the agribusiness firm. Managing the firm's working capital is a day-to-day activity that ensures that the firm has sufficient resources to continue its operations and avoid costly interruptions. This involves a number of activities related to the receipt and disbursement of cash of the agribusiness firm.

CASH FLOW STATEMENT

An analysis of cash flows is useful for short run planning in an agribusiness firm. A firm needs sufficient cash to pay the debts maturing in the near future, to pay interest and other expenses and to pay dividends to the shareholders. The firm can make projections of cash inflows and cash outflows for the near future to determine the availability of cash. This cash balance can be matched with the firm's need for cash during the period, and accordingly, arrangements can be made to meet the deficit or invest the surplus cash temporarily. A historical analysis of cash flows provides an insight to prepare the reliable cash flow projections for the immediate future.

Cash flow statement summarizes the causes of changes in cash position between dates of the two balance sheets. It indicates the sources and uses of cash. The cash flow statement is similar to the fund flow statement except that it focuses attention on cash instead of working capital or funds. Thus, this statement analyses changes in non-current accounts as well as current accounts to determine the flow of cash.

Sources of Cash

- Cash can be mobilized through the profitable operations of the agribusiness firm,
- Keeping only the need based and useful assets in the firm. Other assets can be disposed for revenue generation,
- Increase in liabilities (including debentures or bonds), and
- The floating of ordinary or preference shares.

Uses of Cash

- The loss generated from business can be rejuvenated using the cash,
- Increasing the quantity and quality of assets using the cash,
- Decrease in liabilities (including redemption of debentures or bonds),
- Redemption of redeemable preference shares, and
- Cash dividends.

Significance of Cash Flow Statement

It is an essential tool of financial analysis for short term planning. The advantages of cash flow statement are furnished as follows.

- It is very much useful in the evaluation of cash position of an agribusiness firm.
- It is useful to know the future cash position of the firm.
- A series of intra-firm, inter-firm cash flow statement reveals whether the firm's liquidity position is improving or deteriorating over a period of time and in comparison to other firms in a given period of time.
- It helps in planning the repayment of loans, replacement of fixed assets and other similar planning of cash. It is also significant for capital budgeting decision.

- Cash flow statement provides information of all activities classified under operating, investing and financing activities.

Limitations of Cash Flow Statement

- It ignores the basic accounting concept of accrual basis.
- Cash flow statement is not suitable for judging profitability of a firm as non-cash charges are ignored while calculating cash flow from operating activities.

Procedure for Preparing a Cash Flow Statement

The basic information required for the preparation of cash flow statement is obtained from the following three sources:

1. Comparative balance sheets of two points of time (beginning and at the end of the accounting period).
2. Income statement of the current accounting period or the profit and loss account.
3. Some selected additional data to extract the hidden transactions.

Steps in Preparing the Cash Flow Statement

1. Compute the net increase or decrease in cash equivalents by making comparison of these accounts given in the comparative balance sheets.
2. Calculate the net cash flow used in operating activities of the firm by analyzing the profit and loss account, balance sheet and additional information. There are two methods of converting net income into net cash flows from operating activities; the direct and the indirect methods.
3. Calculate the net cash flow from investing activities.
4. Calculate the net cash flow from financing activities.
5. Prepare a format cash flow statement highlighting the net cash flow from operating, investing and financing activities separately.
6. Make an aggregate of net cash flows from the three activities and ensure that the total net cash flow is equal to the net increase or decrease in cash and cash equivalents as calculated in step 1.

7. Report significant non-cash transaction that did not involve cash or cash equivalents in a separate schedule to the cash flow statement. Example: Purchase of machinery against issue of share capital or redemption of debentures in exchange for share capital. A model cash flow statement is depicted as under for easy understanding.

Table 1 Cash Flow Statement of Agribusiness Firm during 2013 and 2014

Sl.No	Particulars of Cash Flow	2013 (in lakhs)	2014 (in lakhs)
1.0	Cash Balance	4500000	8000000
2.0	Cash Inflow		
2.1	Sale of Agribusiness Output	12000000	16000000
2.2	Capital Sales	1500000	4000000
2.3	Miscellaneous Cash Income	2000000	6000000
2.4	Total Cash Inflow	20000000	34000000
3.0	Cash Outflow		
3.1	Operating Expenses	3000000	5000000
3.2	Capital Purchases	6000000	9000000
3.3	Miscellaneous Expenses	1500000	2000000
3.4	Total Cash Outflow	10500000	16000000
4.0	Cash Balance (2.4) –(3.4)	9500000	18000000
5.0	Borrowed Funds Needed	3000000	000
6.0	Loan Repayment	000	2500000
7.0	Closing Cash Balance (4)+(5)-(6)	12500000	15500000
8.0	Debt Outstanding	3000000	000

Please see the model cash flow statement of an agribusiness firm from Tamil Nadu. In the year 2013, there is Rs 200 lakhs inflow of cash and the same arrived at Rs 340 lakhs during the year 2014 leaving a projected cash balance of 95 lakhs and 180 lakhs respectively. During the year 2013, around Rs 30 lakhs is needed for running the agribusiness firm and the same was not needed in respect of 2014 due to considerable cash balance available during the year 2014 which had permitted the firm to repay the loans outstanding with the agribusiness firm.

FINANCIAL RATIO ANALYSIS

A ratio is defined as an arithmetic relationship between two figures. The study of ratios between various items or groups of items in the financial statements is known as financial ratio analysis. It is the principal tool for analysis of financial statements.

Financial Ratios have been classified in several ways. For our purpose we will divide these into four types.

- Liquidity Ratios
- Turn Over Ratios
- Leverage Ratios and
- Profitability Ratios.

These ratios are used to analyze the performance of agribusiness firms.

FINANCIAL PERFORMANCE OF AGRIBUSINESS FIRMS

To assess the financial position and performance of industries, the financial ratios were computed. For the purpose, the data available with the industries over years were collected and the mean values used for calculating financial ratios. The ratios arrived at are outlined in Table 2.

1. Liquidity Ratios

Liquidity refers to the ability of the agribusiness firm to meet its obligations in the short run. Liquidity ratios are generally based on the relationship between current assets and current liabilities. The important liquidity ratios are the current ratio, acid-test ratio, and the bank finance to working capital gap ratio.

1.1 *Current Ratio* Current assets include cash, marketable securities, debtors, inventories, loans and advances, and repaid expenses. Current liabilities consist of loans and advances received, trade creditors, accrued expenses and provisions. The current ratio is a very popular financial ratio, measures the ability of the firm to meet its current liabilities.

Implications

- Higher the current ratio, greater the solvency.
- A ratio less than one is unacceptable to a lending banker because it means there are no sufficient current assets to meet the current liabilities of the agribusiness firm.
- A ratio of 1.33 and above is greater according to the Tandon Committee.
- Ideal ratio is stated to be 2:1.

Significance of Current Ratio

- Too much emphasis need not be placed on this ratio for a growing agribusiness concern.
- It is not in touch with reality.

The average current ratios were 2.94, 1.77 and 0.66 in respect of newsprint industry, pulp industry and paper industry. The higher ratios associated with the newsprint and pulp industries are indicative of their capacity to remain solvent in the event of any adversity in the near future. The current ratio for paper industry which was less than unity reveals that the company is plagued by more current liabilities than current assets. The company has to reduce its current liabilities by adopting cost effective technologies in paper manufacturing or by increasing the current asset position by staggered disposal of paper produced to ensure that the current ratio is maintained above one throughout the accounting year.

1.2 *Acid Test Ratio* The acid test ratio also called as quick ratio, is a fairly stringent measure of liquidity. It is based on those current assets which are highly liquid. Inventories are excluded from the numerator of this ratio because they are deemed to be the least liquid component of current assets.

From Table 2, it could be seen that the acid test ratios were 0.37, 1.55 and 2.07 for paper, pulp and newsprint industries respectively. An acid test ratio of more than one in case of pulp and newsprint industry reflects the positive aspects of their short term solvency and liquidity position. As per the standard, an acid test ratio of 1:1 is regarded as satisfactory. But higher ratios of 1.55 and 2.07 underscore the ability of the industries to meet the liabilities without relying on the sale of inventory in the event of immediate liquidation. A low ratio of 0.37 in case of paper industry suggests that it may not be in a position to meet any liabilities. Sometimes, the industry might have to sell out the inventories to meet the liabilities. Hence, it is recommended that the paper industry should increase its asset position by staggered disposal of paper produced to push the ratio above one.

1.3 *Inventory Utilization* The inventory utilisation ratios were 8.27, 4.74 and 4.01 in respect of paper, pulp and newsprint industries respectively. This indicates that the pulp and newsprint industries are not having excessive stocks of inventory as in the case of paper industry. Inventory position is better in paper industry as evidenced by the ratio which is twice as much as that of the other industries.

1.4 *Fixed Assets Utilization* The fixed assets utilisation ratios were 4.31, 2.61 and 0.93 in respect of paper, pulp and newsprint industries respectively. This suggests that private firms like paper and pulp industries are more effectively utilising their fixed assets than the newsprint industry which is government owned. Among the three, paper industry is more efficient in fixed assets utilisation.

Table 2 Financial Performance of Pulpwood Consuming Industries in Tamil Nadu

Sl .No.	Ratios	Paper industry (1990-2004)	Pulp industry (1993-2004)	Newsprint industry (1985-2004)
1	Current ratio = Current assets / Current liabilities	0.66	1.77	2.94
2	$\text{Acid test ratio} = \frac{(\text{Current assets} - \text{Closing stock})}{(\text{Current liabilities})}$	0.37	1.55	2.07
3	Inventory utilisation = Sales / Inventory	8.27	4.74	4.01
4	Fixed assets utilisation = Sales / Net fixed assets	4.31	2.61	0.93

SI .No.	Ratios	Paper industry (1990-2004)	Pulp industry (1993-2004)	Newsprint industry (1985-2004)
5	Total assets utilisation = Sales / Total assets	2.97	2.42	0.85
6	$\text{Times interest earned} = \frac{\text{Earning before interest and taxes}}{\text{Interest charges Net}}$	1.99	1.78	0.91
7	$\text{Return on total assets} = \frac{\text{Profit after taxes}}{\text{Total assets}} \text{x } 100$	3.78	26.85	0.27
8	$\text{Gross ratio} = \frac{\text{Total expenses}}{\text{Cross income Net profit}}$	0.80	0.67	-
9	$\text{Return on investment} = \frac{\text{Net profit}}{\text{Total capital employed}}$	25.46	48.39	-

* Financial ratios are mean ratios for the year specified

1.5 *Total Assets Utilization* The total assets utilisation ratios were 2.87, 2.42 and 0.85 in respect of paper, pulp and newsprint industry respectively revealing that the paper and pulp industries were utilising their total assets effectively to produce the given quantity of output. The total assets utilisation in newsprint industry was poor as evidenced by a ratio of less than one. This implies that the newsprint industry is not generating sufficient volume of business commensurate with the size of its investment. The poor asset utilisation of the newsprint industry is the outcome of its large overheads.

1.6 *Times Interest Earned* The times interest earned ratios were 1.99, 1.78 and 0.91 in respect of paper, pulp and newsprint industries respectively. The ratio was comparatively better for paper and pulp industries than for newsprint industry. The interest earned towards the investment could meet the industries additional requirement of funds. But it was not so for newsprint industry. The interest earned may not be sufficient enough to acquire additional loan facilities from funding agencies.

1.7 *Return on Total Assets* The ratios in respect of return on total assets were 3.78, 26.85 and 0.27 in respect of paper, pulp and newsprint industries respectively. Among the three, pulp industry was doing well. Its ratio was more than 6 times as much as that of the paper industry which ranked next to it. The return on total assets was very poor in respect of newsprint industry which may be attributed to low profit margin on sales and increase in the total expenditure by way of large establishment.

1.8 *Gross Ratio* It is nothing but the ratio of amount of total expenses to gross income. The ratios were 0.80 and 0.67 in respect of paper and pulp industry respectively. The data on total expenses in respect of newsprint industry was not available.

1.9 *Return on Investment* Profitability is the primary measure of the overall success of a business enterprise. It is a necessary condition for the survival of business itself. It can be defined as the ability of a business unit to generate profit. The return on investment was 25.46 and 48.39 in case of paper and pulp industries respectively. The parameter was not worked out for the newsprint industry due to non-availability of sufficient data.

To sum up, among the three industries, pulp industry's performance was better than that of the other two. It was supported by higher percentage of return on total assets and return on investment. The paper industry was also equally better when compared to newsprint industry. Among the various ratios, only the current ratio and acid test ratio were in favour of newsprint industry. It might be due to non-profit motive of the organisation. Being a government organisation, its primary objective is to provide service to the people rather than generating profit. The other two industries come under private sector. Their first objective will be profit-making rather than service as evidenced by the financial ratios. Even though the paper and pulp industries are concentrating on profit, their return reveals that these could invest a little more on service to the people in the vicinity of their location.

Chapter 10

Management Education Scenario in India

GENESIS OF THE WORD 'MANAGEMENT'

Management is a ten letter word which encompasses the following words which are important conceptually in managing a firm or an organization and are considered to be the four pillars of 'Management'. They are:

- Man
- Age
- Men
- T stands for Time

The Word 'Man'

Man refers to a Manager or General Manager or Managing Director of the firm or the organization who plays a major role in making strategic decisions based on the goal of the organization.

The Word 'Age'

Age refers particularly to the age or experience of the person who leads or runs the firm or the organization with or without formal education.

The Word 'Men'

Men refers to the workers or laborers or supervisors working in the firm or the organization based on the directions of the Manager or the General Manager of the firm.

The Word 'Time'

Time (T) refers to accomplishing the tasks of the firm or the organization within the stipulated time by utilizing the given resources of the firm. If we pool all the words in a sequential manner, one could get a clear definition for management

Thus, **Management** is defined as the 'Man' either by virtue of his 'Age' or expertise leading or guiding the group of 'Men' to fulfill the tasks or goals of an organization by utilizing the available resources optimally within the stipulated 'Time' to realize the expected quality output to gain the customer satisfaction.

SIGNIFICANCE OF THE WORD MANAGEMENT

In general, management is defined as getting things done by other people.

- **How effectively** an individual is executing a task to fulfill the objectives or goal of an organization. Here the concept of Man is related.
- **How efficiently** the task is executed with the available resources? Here the concept of Age in terms of expertise and Men in terms of labor and supervisory forces are applied.
- **How fast** the assigned task is performed to the expectations of the institutions and the consumers? Here the concept of Time is applied.
- **What sequence, precision and standards are** followed in delivering the product to maintain quality?
- **How one has to fulfill** different stakeholders in the business to have satisfaction on consumption of the product?

The above questions are very much relevant in designing a management program. Here identification of need based courses, the composition of theory and practical components of management subjects, the art of infusing managerial skills to manage the organization as stated are important and assumes significance in designing a program.

HISTORY OF MANAGEMENT EDUCATION IN INDIA

Management education in India has a long history. The history of management education in India goes back to the pre-independence era. The detail of evolution of management education in India is outlined in Table 1. It reveals that the first college level business school in the country was founded in 1913 in Mumbai and was soon followed by another in Delhi in 1920. These business colleges imparted basic skills about the principles of trade and commerce to clerks and supervisors from fields as diverse as banking, transport, and accounting (Indian Management, Sept 2004). They were found to be the forerunners in imparting the management education in India. When we compare the experiences, it is the time to celebrate hundredth year of management education in India. In India, only one general management education was adopted and practiced during pre-independent era and now it has been branched out to nearly 60 areas of specialized management education including the area of Forest Business Management. Such an important discipline need to be analyzed with the objective of documenting its history, specialization, contents, quality and career development opportunities.

METHODOLOGY

For fulfilling the objective enunciated in the previous section, many of the papers published in this respect were reviewed and the main contents were drawn to develop the findings. Besides, secondary data from government sources like All India Council for Technical Education, New Delhi and other reports were relied upon to document the results. In addition to these, the curricula and syllabi prepared for establishing the MBA degree program in Forest Business Management at Forest College and Research Institute, Tamil Nadu Agricultural University, Mettupalayam formed the basis for drawing some results. The syllabi for Forest Business Management were prepared by establishing Syllabus Preparation Committee with the support of internal and external experts in the management education. The results were analyzed and discussed using the conventional percentage analysis.

RESULTS AND DISCUSSION

The results are analyzed and discussed under the following major headings. They are

- History of Management Education in India
- Number of Management Institutions Established in India over a Period of Time
- Diversification of Management Education in India
- Forest Business Management Education in Tamil Nadu
- Future Programmes in Management

THE HISTORY OF MANAGEMENT EDUCATION IN INDIA

Tracing the history of development of management education in India will provide basic understanding on how to proceed, and how to plan for the development of management education so as to groom the young minds to make them as eminent administrators in the business firms. The details of history of management education in India are analyzed and the same are presented in Table 1.

Table 1 History of Management Education in India

Sl. No	Particulars of Institution	Year of Establishment	Place
1.	First College Level Business School	1913	Mumbai
2.	Second Business School	1920	Delhi
3.	Indian Institute of Social Welfare and Business Management	1953	Kolkatta

SI. No	**Particulars of Institution**	**Year of Estab-lishment**	**Place**
4.	MBA at University Level	1960	All States
5.	Indian Institute of Management	1961	Kolkatta
6.	Indian Institute of Management	1961	Ahmedabad
7.	Indian Institute of Management	1971	Bangalore
8.	Indian Institute of Management	1974	Lucknow
9.	Indian Institute of Forest Management	1982	Bhopal
10.	All India Council for Technical Education	1987	New Delhi
11.	Indian Institute of Management	1990	Indore and Calicut
12.	Tamil Nadu Agricultural University (MBA in Agribusiness Management)	1990	Coimbatore
13.	Indian Institute of Management	2011	Trichy
14.	Forest College and Research Institute (MBA -FBM)	2012	Mettupalayam, Tamil Nadu

All India Council for Technical education constituted a board of management studies to co-ordinate and develop management education in 1954. The recommendations of the board led to the start of management programs in Bombay (Mumbai), Delhi, Calcutta (Kolkatta) and Madras (Chennai) during the year 1955 (Pylee, 1967). The first program in Business Management was offered in 1954 by the Indian Institute of Social Welfare and Business Management in Calcutta (Dayal, 2002). It was a part-time course for practicing executives. The All India Institute of Management and Labour Welfare and the Department of Management, Delhi University followed suit in 1968. A number of universities set up their MBA program in 1960s. The IIMs were established based on the recommendation of Dean Robbins of the University of California, who was invited by the planning commission. The first IIM was set up in 1961 at Kolkatta followed by IIMs at Ahmedabad in 1961, Bangalore in 1971, Lucknow in 1974 and in the late 1990s at Indore and Calicut. The flagship management education program, MBA, is widely popular as it offers quick gateway to the riches and to the top echelon of corporate world. The early 90's saw the boom of founding new management schools, most of them in private sector. In the last three years alone 400 business schools came into existence. Few business schools have also established collaboration with some western universities. Indian management institutions produce over 30,000 full-time MBAs and 10,000 part-time MBAs every year. Many business schools are also running MBA equivalent program such as Post Graduate Diploma in Management (PGDM) on distance mode, Masters' Program in International business (MIB, etc. Even some of the leading business houses are establishing their own business schools like Infosys, Birla, Xavier's etc. Efficiency and equality are two major issues in management education (Pillana, 2007). Moreover management education should train the students conceptually but the student should be trained to use the theory in day-today activities (Dayal, 2002).

Management education gives a holistic picture to the students about how to manage the four 'M's of any organization, i.e., Men, Money, Material and Machine. Whether the knowledge about management principles is obtained through formal study programs at universities or internally through on-the-job training or through external seminars or programs, it is of utmost importance for any person in any job in life, even for the self-employed and for the entrepreneurs. With the changing dynamics of the business environment, indigenous industry realized the need for employees having a formal education packaged with the experiences of real life case studies. Hence the shift started taking place from IQ (Intelligence Quotient) to EQ (Emotional Quotient). The new industries like software industries and the consultancy group started demanding "Knowledge Workers". Managing people became a buzzword. The CEO now had to be more of a generalist than a specialist. Harvard and Stanford Universities have reported that 85 per cent of the times, the reason a person gets a job and gets ahead in that job is due to attitude and only 15 per cent is because of technical or specific skills.

NUMBER OF MANAGEMENT INSTITUTIONS ESTABLISHED IN INDIA

Table 2 Number of Management Institutions Established in India over a Period of Time

Sl. No	Period of Establishment	Number of Years	Number of Institutions	Number of Institutions Established per Year	Percentage to Total
01	1950 – 1980	30	118	04	04.78
02	1981 – 1995	15	304	20	12.32
03	1995 – 1999	04	260	65	10.54
04	1999 – 2000	01	092*	92	03.73
05	2000 – 2005	05	358	72	14.52
06	2005 – 2010	05	808	162	32.75
07	2010 – 2013	03	527	176	21.36
08	1950 – 2013	63	2467**	39	100.00

(*Source: Dayal I, 2002; **AICTE, New Delhi – 2011-12)

The Challenges in Management Education in India

- Is 'Management Education' in India in crisis?
- Will the mushrooming of management institutions help the stream tide over the crisis?
- How sensitive are these institutions and authorities concerned to the problems and challenges that have already started bothering it?

The challenges before management education, or rather managing management education

vis-à-vis the demands posed by the industry and other socio-economic and cultural factors, need to be properly analyzed and understood from an altogether fresh approach, according to experts in management.

The Rani Channamma University (RCU), Belgaum, the second largest varsity in Karnataka, had recently organized a seminar on the issue 'Management Education - Road Ahead' with a view to revisit the road already traversed, assess the present situation and explore the ways and means to restructure the system to make it meaningful, effective and productive. C.M. Thyagaraja, Professor and Chairman, Department of Post-Graduate Studies in Business Administration at RCU's Vidyasangama campus in Belgaum, shared his views with *The Hindu Education Plus* on the sidelines of the seminar. Management education is passing through a critical phase. There is a crisis of Identity, Character and Quality. This is the apt time to discuss and deliberate the matter by taking stock of the situation.

Management education has grown *quantitatively but not qualitatively*. It had contributed too little to the labour-rich but skill-poor economy. Too many attractive, lucrative and competent jobs are chasing very few individuals, who are highly talented, skilled and dynamic. An average student finds it difficult to sail through the acid tests of the corporate sector. The element of professionalism, an important ingredient towards building managers, is not taught by institutions imparting management education.

Saleable Product

Like professional education, management education too has become a saleable product. From Kashmir to Kanyakumari and from Amritsar to Singur, more than 4,000 institutions have emerged in the management education landscape. Among these the approved institutions are to the tune of only 2467 in numbers (AICTE, 2012).

The origin of management education dates back to ancient times. It is a unique art developed by mankind along its journey of evolution. Greeks, Chinese and Indian thinkers contributed to this art of getting things done by others. Strategic decisions and art of administration were adopted during the times of Koutilya. Management education is an offshoot of the industrial revolution which created the factory system, thereby providing a ledge to the art of management. In the Indian subcontinent, management education has come over a period of past 100 years, whereas Europeans are teaching this education since the last 400 years.

Due to the slow rate of economic growth after independence till 1990 the opportunities created by industry were too few. Globalization gave a big boost to the economy while the service sector came to dominate other sectors. This has necessitated the demand for management education.

The IIMs (Indian Institutes of Management) top the list in imparting management education which pick the very best. Management institutes affiliated to universities, autonomous institutions approved by the All India Council for Technical Education, institutions without the approval of AICTE, and foreign universities are also offering degree and postgraduate degrees in management education in India. But its quality is questionable.

Role of All India Council for Technical Education (AICTE)

The Union Government, during its winter session of 1987, accorded permission to establish the AICTE, which is a body for professional education in this subcontinent. According to AICTE data, the number of institutions imparting management education mushroomed in just five years. The intriguing question confronting all the stakeholders is: "Why was permission given to so many institutions?"

In the last five years, the AICTE, by granting permission to more institutions, also allowed increase in the intake of students by more than 300 per cent, revealing that the AICTE adopted a quantitative expansion strategy but paid inadequate attention to quality of management education. Further, among 4,000 plus institutions sanctioned by AICTE last year, Andhra Pradesh has 945, Uttar Pradesh 459, Maharashtra 419, Tamil Nadu 389, Madhya Pradesh 215 and Karnataka 209. Experts feel that AlCTE's policy is liberal in according permission to start institutions and increase intake. Secondly, institutions enter the management education scenario with the intention of making quick profits as against contributing their bit to the field with some genuine concern. Thirdly, institutions imparting management education negated quality and concentrated on quantity.

Lapses in Management Education

The first and foremost lapse is designing the curriculum; curriculum upgradation at periodic intervals and banking on some core subjects and niche electives added to the problems. The element of 'skill quotient' is not appropriately addressed to add value to the management education.

Faculty members with industrial and field experience are less in numbers to share their expertise. Former Chairman of the University Grant Commission, Arun Nigavekar, observed that under the changed circumstances it became inevitable to shun the old curriculum in management education and prepare CEOs by directing them to set goals and develop necessary skills to pursue them. He also quoted from the survey to say that three per cent of the sample population who set goals had more net worth after 25 years than those who did not.

The post graduates in management are still facing lack of confidence, lack of communication skills, and presence of shallow knowledge on management education which are inter-related and has chain of actions on their placement and career development.

- The management graduates do not know how to start a business firm?
- What are the required formalities to establish a business?
- What are the organizations to be approached for registration of the business?
- How to identify and assess the potential business/agribusiness?

For that, the syllabi should have both theory and practical components with attachment to the business firms to establish the procedures and process involved in establishing subsidiary of the organization, merger and acquisition of an organization. Even if some of the management institutions do practice the concept of industrial attachment, the core areas of business and other aspects were not exposed to the students who are attached with the firm. But the firm is interested in using the services of students mainly in unrelated aspects of the business concern. To be specific, a manual prepared for the students of management education who were attached with certain business firms showed that the firms were not at all interested in placing them or exposing them to the core area of the business due to competitive threats from the students and the students were finding lot of difficulties in filling the manual of industrial work experience. What is needed here is the management institutions should establish a memorandum of understanding with proven industries or business firms to fulfill their basic knowledge in industry and their business, scope and potentials associated with the business and their stakeholders and consumers welfare.

Management scholars view good knowledge management practices as an important focus area for sustainable competitiveness. Today, the only way to have a competitive advantage is through innovation and upgrading (Porter, 1998). The major problems in knowledge distribution in Indian B-Schools are as follows:

- Functional silos within B-Schools (Pillana, 2007)
- Lack of collaboration in B-Schools (Bhattacharyya, 2004)
- Lack of B-Schools-Industry collaboration (Bhattacharyya, 2004)
- Lack of industry interests in management development programs (MDP) and consultancies (Pillana, 2007).

DIVERSIFICATION OF MANAGEMENT EDUCATION IN INDIA

When we look into diversification of management or specialization of management program, it is vast and is still growing. A postgraduate degree in management is considered a prestigious one and it opens up various career opportunities that are satisfying both in terms of remuneration as well as professional livelihood. Thc postgraduate degree also offers better insight into diversified aspects of business such as marketing, human resources, finance, systems and production. Besides these, specialized areas of management education were also developed and implemented which rose to around 60 now. The Masters in Business Administration (MBA) course offered by various universities provides students an overall approach to a business and its management, which most of the companies expect in an aspiring candidate. Today's industries require skilled professionals, so an MBA degree from a reputed university has become a norm. From the student's point of view, there are various fields of specialization available in the industry today (Table 3).

Table 3 Diversification in MBA Degree Programs

SI. No	Particulars of Institution	Institutions
1.	MBA in General Management	All State Universities
2.	MBA in Marketing Management	All State Universities
3.	MBA in Production / Operations Management	All State Universities
4.	MBA in Human Resource Management	All State Universities
5.	MBA in Financial Management	All State Universities

SI. No	Particulars of Institution	Institutions
6.	MBA in Hospital Management	Bharathidasan University, Trichy
7.	MBA in Investment Management	Alagappa University, Karaikudi
8.	MBA in International Business	University of Madras, Chennai
9.	MBA in Retail Management	Anna University, Chennai
10.	MBA in Information and Systems Management	Bharathiar University, Coimbatore
11.	MBA in Financial Services	Bharathiar University, Coimbatore
12.	MBA in Family Business and Entrepreneurship	Hindustan University, Chennai
13.	MBA in Services Management	Bharathiar University, Coimbatore
14.	MBA in Export - Import Management	University of Madras, Chennai
15.	MBA in Technology Management	Anna University, Chennai and Bharathiar University, Coimbatore, Alagappa University, Karaikkudi
16.	MBA in Project Management	Alagappa University, Karaikudi and Bharathiar University, Coimbatore
17.	MBA in Banking and Finance Management	Bharathiar University, Coimbatore
18.	MBA in Logistics and Supply Chain Management	Tamil Nadu Open University , Chennai
19.	MBA in Logistics and Shipping Management	Vels University, Chennai
20.	MBA in Telecom Management	Vinayak Missions University, Salem

SI. No	**Particulars of Institution**	**Institutions**
21.	MBA in E- Commerce	Vinayak Missions University, Salem
22.	MBA in Travel and Tourism Management	Bharathiyar University, Coimbatore
23.	MBA in Systems Management	Bharathidasan University Jrichy
24.	MBA in Educational Management	University of Madras, Chennai
25.	MBA in Apparel Management	University of Madras, Chennai
26.	MBA in Garment Industry Management	Karpagam University, Coimbatore
27.	MBA in Textile Management	Sardar Vallabai Patel Institute of Management , Coimbatore
28.	MBA in Aviation Management	Hindustan University, Chennai
29.	MBA in Airport Management	Bharathiar University, Coimbatore
30.	MBA in Shipping and Port Management	Vinayak Missions University, Salem
31.	MBA in Hospital and Health Care Management	Hindustan University, Chennai.
32.	MBA in Nutrition Management	Bharathiar University, Coimbatore
33.	MBA in Advertising and Sales Management	SASI School of Advertising, Coimbatore
34.	MBA in Media and Entertainment Management	Hindustan University, Chennai
35.	MBA in Pharmaceutical Management	Indian Institute of Pharmaceutical Marketing, Lucknow
36.	MBA in Agribusiness Management	Tamil Nadu Agricultural University, Coimbatore

SI. No	Particulars of Institution	Institutions
37.	PGDFM in Forest Management	Indian Institute of Forest Management, Bhopal
38.	MBA in Forest Business Management	Forest College & Research Institute, Mettupalayam
39.	PGDM in HR, Finance, Operations, Marketing Management	Indian Institutes of Management Ahmadabad, Bangalore, Calcutta
40.	PGDM in General Management	All India Management Association, New Delhi
41.	PGDHRM in Personal Management & Industrial Relations	XLRI, Jamshedpur
42.	PGDRM in Rural Management	Institute of Rural Management , Anand, Gujarat
43.	PGDDM in Development Management	Dhan Foundation, Madurai
44.	PGDM in National Economic Planning and Entrepreneurship	Indian Institute of Planning and Management, Chennai
45.	MBA in NGO Management	Amity University, Noida
46.	MBA in Hospitality Management	Hindustan University, Chennai
47.	MBA in Environmental Management	Bharathidasan University, Trichy
48.	MBA in Insurance Management	Tamil Nadu Open University, Chennai
49.	MBA in Fashion Design Management	Manonmaniam Sundaranar University, Tiruneveli
50.	MBA in Cosmetology and Fitness Management	Manonmaniam Sundaranar University, Tiruneveli

SI. No	**Particulars of Institution**	**Institutions**
51.	MBA in Power Management	Management Development Institute, Gurgoan
52.	MBA in Oil and Gas Management	Management Development Institute, Gurgoan
53.	MBA in Construction Management	Management Development Institute, Gurgoan
54.	MBA in Product Management	Management Development Institute, Gurgoan
55.	MBA in Forest and Environmental Management	Management Development Institute, Gurgoan
56.	MBA in Event Management	Gems Business School, Bangalore
57.	PGDM in Biotechnology Management	MITCON Institute of Management, Pune
58.	PGDM in Media Management	Galaxy Institute of Management, Chennai
59.	MBA in HR, Finance, Marketing, Operations Management and General Management	Indira Gandhi National Open University, New Delhi
60.	PGDM in Agribusiness Management	National Institute of Agricultural Extension Management, Hyderabad
61.	PGDM in Agricultural Management	National Academy of Agricultural Research Management, Hyderabad
62.	PGDM in Agribusiness and Plantation Management	Indian Institute of Plantation Management, Bangalore

Table 3 reveals that the diversification in management education increased many-fold from general management education to more than 60 specialized areas. Recent specialization in management education is summarized as follows.

- MBA in Agribusiness Management
- MBA in Event Management
- MBA in Eco Business Management
- MBA in Energy and Environment Management
- MBA in Educational Management
- MBA in Forest Business Management
- MBA in Media and Entertainment Management
- MBA in Services Management
- MBA in Technology Management
- MBA in Fashion Design Management
- MBA in Cosmetology and Fitness Management

Students choose their specialization based on their interest and, have the drive to become entrepreneurs after gaining some professional experience. Students select their specialization under the influence of market demand, job scenario and the influence of their parents. From Table 3 one could understand that the diversification in management education is alarmingly increasing to meet the requirements and expectations. To cite a few, the Technology Management is a subject almost conceived by Indira Gandhi National Open University, New Delhi and later on the Anna University, Chennai has brought the specialization area in Technology Management followed by Bharathiar University, Coimbatore by identifying need based core and supporting courses for the MBA in Technology Management. Besides, fashion design, cosmetology and fitness management, power management, event management, media and entertainment management are new disciplines which has motivated the growing institutions to search for specialist graduates in a different dimension to overcome the competition. But it is contradicting to the view of Harvard and Stanford Universities where the attitude alone speaks on efficiency rather than technicality. In India, number of specializations offered is documented program wise in Table 4.

Table 4 Details of Management Programs in India

Sl. No	Programs in Management	Number of Specializations	Percentage to Total
01	MBA Degree Programs	36	39.13
02	PG Diploma in Management	51	55.43
03	PG Certificate Program in Management	05	05.44
	Total	92	100.00

(Source: AICTE Approval Process Hand Book 2011-12)

Table 4 reveals that MBA degree programs are offered in 36 different specializations across India through different management institutions of national and regional repute. Post Graduate Diploma in Management programs are offered, the diploma awarded in 51 different specializations and the Post Graduate Certificate programs are offered, certificates awarded only in five institutions which respectively accounted for 39 per cent, 55 per cent and 5 per cent respectively. From this one could infer that the Indian institutions imparting management education are driven by specialization based employment and hence the spiraling effect of initiation of different specializations in management education.

Students select their specialization keeping in mind his/her degree in the undergraduate level. The first category decides their career mainly based on their attitude and the people belonging to this category are keen to start an industry of their own after gaining some experience.

The second category decides a career mainly under the influence of their parents, market demand, job profile and attractive pay packets. The people belonging to this category overweigh the other two categories; and the third category decides the career based on their under graduation.

The criteria that all B schools, management institutes and universities follow in formulating their electives stream list are based on the current trends in literature and emerging technologies in management. It also depends on the organization's requirements. For example, many organizations now-a-days follow some heuristic partitions. It is a type of algorithm followed in product development. Most importantly the electives are formulated to balance the students' needs and interest. Sometimes it is based on the undergraduate program the students studied. For example, commerce students can prove their expertise and skill in finance compared to forestry, agriculture and engineering students. But engineering students fare better in computer skills, modeling, experimenting and other technical aspects. Similarly, the graduates in agriculture, horticulture, forestry, fisheries, veterinary science fare better than any other graduates due to their specialized curriculum design. The curriculum designed for these sciences were based on Indian Council of Agricultural Research which infused almost all skills required for the graduate to outperform. This is true with Tamil Nadu Agricultural University. Its graduates were almost toppers in Civil Services Examinations, Indian Forest Service Examinations conducted by the Union Public Service Commission.

The share of TNAU students in getting selected into the Agricultural Research Service Examinations conducted by Agricultural Scientists Recruitment Board is also found to be on the higher side. Besides these, the graduates of TNAU excel in banking, communication, tourism, landscaping and architecture, environment and climate change. The novel disciplines in agriculture included agricultural economics, agribusiness management, marketing, agricultural information technology, food process engineering and energy and environmental engineering which has also added higher confidence in the young minds.

A student must select his specialization based on what he likes and what he intends to do in the future. His choice must not be based on others, as identity is more important. It is the attitude that takes the student to greater heights.

Table 5 Growth of Management Programs in India

Year	Growth of Management Programs	Growth of Approved Number of Seats	Growth of Seats / Year	Number of Seats Per Program
2005–06	1888	122663	000	065
2006–07	2031	144372	21709	071
2007–08	2062	185780	41408	090
2008–09	2734	227989	42209	083
2009–10	3482	273732	45743	079
2010–11	3858	378907	105175	133
Mean	643	63151	51195	080

Table 5 reveals that the growth in management programs in India is found to be 643 with the approved number of seats in various discipline of management programs arriving at 63151 with a permitted intake of 80 seats per MBA specialization. During the year 2010-11, the permitted intake of seats in management education increased to 133 which accounted for 66 per cent higher than the average number of seats per management program.

MBA IN FOREST BUSINESS MANAGEMENT

Indian Institute of Forest Management (IIFM) was established in 1982 in response to the growing need for managerial human power and development of managerial capacity in the forestry and allied sectors. Since its establishment, IIFM has emerged as a premier educational institute in the area of Forest Management in India and southeast Asia. IIFM has also been successful in getting considerable global recognition for its educational and research programmes. Forest College and Research Institute, Tamil Nadu Agricultural University, Mettupalayam is offering MBA in Forest Business Management starting from 2012-2013 academic year which is approved by the Academic Council of Tamil Nadu Agricultural University and this institution is in the process of getting recognition from AICTE, New Delhi for an all-round development by meeting infrastructural and manpower needs to meet the demands of forest based industries, plantation sectors and commercial banks.

The basic issue of sustainability, climate change and the new business paradigm are of utmost importance not only to the manufacturing firms, firms involved in sale of value added environmental goods and services but also to the educational institutions who are actively involved in imparting management education in India. The professional management educational institutions should give top-most priority in designing the novel management programs, its curriculum and syllabi by incorporating the need based courses and contents in the syllabi. It is not just for the survival of the institution or firm, but it poses a major challenge to the designer of such novel programs if we do not take appropriate action. As a management educationist if we plan, design and implement novel programs in management, not only it imparts linkages and feedback but also brings sound employment opportunities and pay packages which in turn facilitates product innovation, competitive prices and imparts customer satisfaction and desired feedback. Hence, designing a program so as to produce eco-products or eco-services needs a careful examination. At this backdrop, a program called "Forest Business Management" was planned and implemented in Forest College and Research Institute, a premier forestry institution in India functioning under the aegis of Tamil Nadu Agricultural University at Mettupalayam in the foothills of The Nilgiris.

Special Features of MBA in Forest Business Management

- Full Time Course
- National Syllabi with Adequate Practical Exposure (43.59 per cent)
- Combination of Management, Statistics, Economics and Technology Courses
- Industrial Tie up with Forest Based Industries
- Industry Sponsored Research and consultancy
- Internship and Inbuilt seminar courses (4.76 per cent)
- Research Projects as Partial Requirements (33.33 per cent)
- Good Placement with Forest Based Industries and Commercial Banks
- Cent Percent Job Opportunity for the position of Forest Range officers with B.Sc.(Forestry) as basic qualifications
- Opportunity for preparation and coaching for induction into Indian Forest Service

Curriculum Design and Development for Forest Business Management

A Syllabi Preparation Committee (SPC) was exclusively constituted for the purpose of identification of courses, designing the curriculum and syllabi under the chairmanship of Dr. P. Punniamoorthy, Professor and Head, Department of Management Studies, National Institute of Technology, Trichirappalli which included core courses, specialization and supporting courses. The course on Technology Management has been made mandatory for all the post graduate specialization degree programs in forestry which are being implemented during the academic year 2012-13 at Forest College and Research Institute, Mettupalayam

Table 6 Composition of Courses for MBA in Forest Business Management

SI. No	Courses	Credit Hours		Total
		Theory	Practical	
1.	Core Courses	12 (54.55)	10 (45.45)	22 (100.00)
2.	Specialization Courses	08 (66.67)	04 (33.33)	12 (100.00)
3.	Supporting Courses	02 (40.00)	03 (60.00)	05 (100.00)
	Total	22 (56.41)	17 (43.59)	39 (100.00)

(Figures in parentheses indicate percentage to total)

Table 6 reveals that the core courses comprised of theory and practical components which respectively accounted for 54.55 per cent and 45.45 per cent to the total credit load fixed by the Indian Council of Agricultural Research in its 4th Deans' Committee Meeting.

In respect of specialization courses, the credit load under theory and practical components were found to be 67 per cent and 33 per cent respectively. The supporting courses had 60 per cent practical exposure which included information, communication technologies and statistical applications in forest business management. With regard to theory alone, MBA in Forest Business Management had 54.55 per cent of credit load for core courses, 36.36 per cent of credit load for specialization courses and only 9 per cent of credits for supporting courses.

In a nutshell, the MBA in Forest Business Management Program gained a credit load of 56 per cent for theory part and 44 per cent credit load for practical to the total credit load earmarked for the program. The practical components consisted mainly of field visits to the forest based industries to gain exposure to the functional areas of management like planning, organizing and other associated functions besides attending some analytical aspects related to the courses.

Research Projects in Forest Business Management

In partial fulfillment of MBA in Forest Business Management degree program, the research credits comprise 20 credits for research project and one credit is exclusively meant for seminar in business related topics of current importance which infuses communication and compilation skill to the budding managers in forest business (Table 7). The research credits had a share of 95.24 per cent to the total research credits and 33.33 per cent to the total credit load earmarked for MBA in Forest Business Management.

Table 7 Details of Credit Load for Research Projects

SI. No	Code Number	Particulars	Credit Hours		Total
			Theory	Practical	
1	FOR 591	Seminar	00	01	01 (4.76)
2	FOR 591	Research	00	20	20 (95.24)
		Total	00	21	21 (100.00)

(Figures in parentheses indicate percentage to total)

Put together, the 60 credit hours were fixed as credit load for the award of MBA in Forest Business Management degree program. This work load is normally not prevailing in any other management program. MBA in Agribusiness Management offered in the Department of Agricultural and Rural Management has fixed the credit load of only 55 credit hours. The title chosen for the seminar and the research project should not be in the same area as envisaged in the university regulations. The research topic chosen by the candidate will be permitted for presentation before the combined sitting of management scientists and other biological scientists for validation of topic, methodology and tools of analysis. The approved topic will be permitted under the technical guidance of a Professor or Associate Professor in Management as Chairman, Advisory Committee with the members of advisory committee, one from the same department and the other from line department. The advisory committee will continuously evaluate the student's performance and the research project will be approved with the comments of an external expert identified by the Dean, School of Post Graduate Studies of Tamil Nadu Agricultural University, drawn from other State University/State Agricultural University in India.

Career Options and Linkages

Globalization has brought about radical transformations in what organizations need to do to maintain their competitiveness (Tanuja, 2008). Industrial demand for managerial resources in India far exceeds the supply. According to one estimate, total number of entry level managers needed by corporate India every year stands at 2735. But the best business schools in India are able to produce only about 1740 managers in a given year. The demand–supply gap amounting to almost 36 per cent has resulted in competition for scarce managerial talent, high levels of attrition, and an increase in the compensation levels of managerial professionals (Tanuja, 2008).

In respect of Forest Business Management, Forest College and Research Institute had established a very good rapport with the forest-based industries like Seshasayee Paper and Boards, Erode; Tamil Nadu Newsprint and Papers Limited, Karur; Tamil Nadu Tea Plantation Corporation, Coonoor; Tamil Nadu Forest Plantation Corporation, Pudukottai; Arasu Rubber Corporation, Nagercoil; commercial banks in Tamil Nadu; and the nongovernmental organizations. Past experiences in Tamil Nadu Agricultural University placement record showed that the farm graduates and management graduates were drawing the pay package ranging between Rs. 3.50 lakhs and 6 lakhs per annum. The companies visiting TNAU for placement prefer more number of management graduates for their organization due to their good communication skills. When a company recruits for its HR position, it looks at the following traits in a candidate.

- Excellent communication skills;
- The students' attitude;
- The number of guest lectures and the projects that the university offers;
- The number of lecturers in the university;
- Special programs conducted in the course;
- The number of courses/electives offered for HR (example).

The most important skill required is negotiating skill, which is more important for HR. This program is offering all the facilities including training and HR development to the students of forest business management.

FUTURE PROGRAMS IN MANAGEMENT

Considering the industrial need in respect of management graduates, the Department of Forest Resource Management of Forest College and Research Institute, Mettupalayam has planned to introduce certain promising programs in the areas of management in near future. The details of future programs planned to be introduced are delineated in Table 8.

Table 8 Future Programs in Management Education

SI. No	Program Planned	Areas of Specialization
01	MBA	Eco-Business Management
02	MBA	Energy and Environmental Management
3	MBA	Technology Management
4	PhD (Forestry)	Forest Economics
5	PhD (Forestry)	Development Management

On completion of MBA in Forest Business Management, the students had every option of doing Ph.D. in Agribusiness Management and Development Management. Especially this institution has planned to introduce such programs in consultation with and on approval by Tamil Nadu Agricultural University, Coimbatore shortly.

CONCLUSIONS AND POLICY IMPLICATIONS

In India, only one general management education was adopted and practiced during pre-independence era, but now it has been branched out nearly to more than 60 areas of specialized management education including the one, Forest Business Management started by Forest College and Research Institute, Tamil Nadu Agricultural University, Mettupalayam of Tamil Nadu. Diversification in management education to that level is mainly due to the need of specialization based expertise to the managers of technical departments. Lack of technical and managerial skill among the management graduate even after imparting specialization is mainly due to wide variation in the curriculum, method of teaching and syllabi for management education. Hence homogeneity in syllabi for management education, mode of teaching and the teaching methods should be very well defined, monitored and controlled under the umbrella organization of All India Council for Technical Education, New Delhi. It is noticed from the above discussions that there exists wide variation in designation of a degree in management education/diploma in management education and honoring such degree/diploma will not allow uniformity and build caliber among the candidates. Hence uniform award of MBA degree irrespective of institution need to be ordered and the same should be monitored by the respective universities and the AICTE, New Delhi through the validated national level syllabi for management education.

References

- Bhattacharyya, B. (2004) ***'Mechanism of Promoting Research in B-Schools: Interface with***
- ***Industry,'*** Paper presented at National Workshop on Targeting Business Research for Excellence in Management Education: A Business Schools Perspective, 24-25 September, New Delhi.
- Dayal,(2002) Developing Management Education in India. ***Journal of Management*** **Research,** 2(2), 98-113.
- Government of India (2012). "Approval Process Hand Book – 2011-12" (All India Council for Technical Education: New Delhi) pp. 01-186
- Porter, M. (1998) 'Creating Tomorrow's Advantages *in* Gibson, R. (eds), ***Rethinking The***
- ***Future,*** London, Nicholas Brealey Publishing, pp. 48-61.
- Pillana,K.R. (2007) Knowledge Management For Indian business schools. ***Journal of Services Research,*** 7(2), 183-192.
- Pylee,M.V.(1967). "Management Education in India", ***Management Science,*** 13(10), pp. 209-C217
- Tanuja, A. (2008). "Factors Influencing Career Choice of Management Students in India" ***Career Development International,*** 13 (4): 2008, pp. 362-376.

SECTION – II

PRODUCTION FOR AGRIBUSINESS

Chapter 11

Role of Farmers' Organization in Promotion of Agribusiness

Flowers have their own importance since ancient times and are being used for decoration, worshiping as well as in satisfying the aesthetic feelings. A language of flowers were nurtured and developed in England during 18th and 19th centuries. Flower speaks a language, which could be universally understood by the people across the world and are provided as the best ambassadors of good will. In recent years, the floriculture is emerging as a market demand driven activity throughout the world and considered as a potential money spinner. Under WTO regime, the floricultural activities are being recognized as a lucrative profession with much higher margin of returns per unit area when compared to the field and other horticultural crops.

Among horticultural crops, cut flowers and ornamentals perhaps do fetch higher value. The words of John Keats, "A thing of beauty is a joy forever", connotes fresh flowers have been one of the best medium of expression of greetings, regards, love and care and respect in enhancing aesthetics of daily décor in human habitations. It is being rightly said "flowers speak a millions of unspoken words..." They fit all occasions, be it at birth, marriage or death ceremonies. Hence, floriculture is fast emerging as a major commercial venture in the world scene. (Vanishree, 2007)

The social and economic aspects of flower growing were, however, recognized much later. The offering and exchange of flowers on all social occasions, in places of worship, their use for adornment of hair by women and for home decoration have become an integral part of human living. With changing life styles and increased urban affluence, floriculture has assumed a definite commercial status in recent times and particularly during the past 2 to 3 decades. Appreciation of the potential of commercial floriculture has resulted in the blossoming of this field into a viable agri-business option. Availability of natural resources like diverse agro-climatic conditions helped in producing wide range of temperate and tropical flowers, almost all through the year in some part of the country or other. Improved communication facilities have increased their availability in every part of the country. The commercial activity of production and marketing of floriculture products is also a source of gainful and quality employment to section of people.

The demand for flowers and ornamental plants for different needs like religious, official ceremonies, parties, house decoration, weddings, funerals, etc., is on the rise. This demand for fresh flowers and plants is increasing world-wide over the coming years. The recent liberalization policy of the Government of India has given fillip to commercialized agriculture particularly horticultural crops. Growing of flowers is in vogue in India since long time. Nevertheless, growing of cut-flowers has emerged as an important industry mainly to cater to the needs of the demand in the overseas market. It is being viewed as a high growth industry in our economy. There is a tremendous transformation in our floriculture sector mainly due to the entry of corporate who are producing cut-flowers to meet the emerging demand in the developed countries for floricultural products. The Government of India has also identified floriculture as a niche area with vast potential for export. There are many incentives given by the government for setting up of floricultural units as Export Oriented Units (EOUs).

Tamil Nadu is a leading state in area and production of flowers in the country. A large number of flowers like jasmine, tuberose, rose, chrysanthemum, marigold, crossandra, barleria, lily, limonium, alstroemeria, liatris, freesia, iris, lisianthus, calla, carnation, gerbera and anthurium are commercially cultivated in the state. Many hi-tech units with export tie-ups are there in the state. There are several commercial tissue culture laboratories. The daily average trade of cut-flower is over Rs. 2 lakh and loose flower over Rs. 5 lakh in Tamil Nadu itself (Sudhagar, 2013).

Jasmine is one of the major commercial flower crops of India. It has its special place in South Indian flower crops. It is mainly used for extraction of scented oil. India exports this oil mainly to England, United States of America, Holland, Sweden, Japan, Norway and European Union.

Flower and concrete yield in jasmines vary considerably according to the species and cultivars and management practices. The flower yield and concrete recovery in three important species of jasmine range between 4.33 tonnes and 10.14 tonnes per ha. It is a wide variation in the yield that stressed the need for undertaking a study on assessing the production and marketing of jasmine in Tamil Nadu.

PROBLEM FOCUS

In Sathyamangalam of Erode District, the jasmine is cultivated in around 40000 acres and the marketing of jasmine is taken over by the Farmers Flower Growers' Association which is not prevalent in any part of Tamil Nadu or even in South India. The farmers have themselves established an association to protect their produce from offers of low price by the traders during 2007 and the practice of marketing and distribution of jasmine is taken up by the association. In this context, to document the practice of production and marketing of jasmine by the farmers flowers growers' association, an effort has been taken to assess the same by adopting a case study approach with the specific objective of assessing the area, production and productivity of flowers with special reference to jasmine in the case farms of Tamil Nadu in comparison with India.

DESIGN OF THE STUDY

The main objective of this paper is to document the area, production, productivity and export of flowers in India and to document the supporting infrastructure available in the state of Tamil Nadu. To document the same, the data were collected from the secondary sources of both central and state government reports. The reports taken into account for collecting the details on area, production, productivity and export of flowers are furnished as follows.

- Agricultural Statistics at a Glance 2012 of Directorate of Economics and Statistics, Ministry of Agriculture, Government of India, New Delhi
- Food and Agriculture Organization website

- Indian Horticulture database 2011 of National Horticulture Board, Government of India, New Delhi
- Hort-Stat – 2008 of Government of Tamil Nadu, Chennai

The data gathered for different periods with respect to the area production and productivity of flowers, export of flowers, assembling and trading centers for flowers and the importing institutions and states involved in flower export were gathered from the above reports and the data were analyzed using descriptive statistics like mean. The current change in the data in respect of area, production and productivity of flowers over the base period was also documented to assess the technological and extension impact due to the government intervention through its planned programs in horticulture. The results of this analysis will be better utilized by the policy makers and the planners in horticulture and agriculture for further development of these sectors.

The productivity of flowers in the state was found to be marginally low when compared to the National Average Productivity and hence a comparison made with the intensely managed jasmine gardens of Sathyamangalam Farmers Flower Growers' Head Association of Erode district which was established during the year 2007 and their experiences were also incorporated in this study as a case study approach and as an illustration for easy implementation by other jasmine garden owners.

PRIMARY DATA

The farmers who are cultivating flowers in the Sathyamangalam taluk formed the base for the study. In the first stage, the Sathyamangalam taluk was selected as it has higher area under flowers cultivated by the farmers. The intention of conducting the survey among the flower growing farmers have been conveyed to the President and Secretary of the Sathyamangalam Farmers Flower Growers' Head Association for the project work on production and marketing of flowers. The President of the society has arranged the contact with the farmers from the selected villages. The number of farmers and the list of villages selected are presented in Table 1.

Table 1 Number of Villages Selected for the Study

Sl. No	Name of the Village	Area under Flowers in Ha	Percentage to Total	Number of Farmers Selected	Percentage to Total
01	Nanjappakavundan Pudhur	023.00	05.87	01	05.00
02	Kenjanoor	028.00	07.14	02	10.00
03	Kembanaickenpalayam	030.00	07.65	02	10.00
04	Phandampalayam	026.00	06.63	01	05.00
05	Akkarai Negamum	032.00	08.16	02	10.00
06	Chickarasampalayam	102.00	26.02	03	15.00
07	Bahuthampalayam	052.00	13.27	03	15.00
08	Badhramangalam	038.00	09.69	02	10.00
09	Chenbagapudhur	014.00	03.58	01	05.00
10	Kothamangalam	019.00	04.85	01	05.00
11	Raj Nagar	028.00	07.14	02	10.00
	Total Area under Flowers	392.00	100.00	20	100.00

From the farmer, the details of land use, crop management for its production and productivity, marketing of flowers, cost of production and the constraints involved are assessed using the structured and pretested questionnaire exclusively designed for the purpose and the same were analyzed and documented with respect to the objectives enshrined in the study by adopting a descriptive statistics like mean and percentage analysis.

The project work on production, marketing and trade aspects of flowers with special reference to jasmine in Tamil Nadu is an important topic today as most of the farmers have become knowledgeable and established farmers' organization themselves and took the practice of marketing of their own produce produced in the farm. The aspects of involvement of farmers' organization in direct marketing of their farm produce have not been much explored at national and regional level and hence a special effort has been taken to assess the production potentials, marketing and distribution made by the producers at micro level and the export of flowers at macro level and the details are analyzed and discussed under the following major heads.

- Area Production and Productivity of Flowers
- Area Production and Productivity of Jasmine
- The Production Potentials of Flowers in the Sample Farms

AREA PRODUCTION AND PRODUCTIVITY OF FLOWERS

The state-wise area production and productivity is analyzed with respect to flowers at national and regional level and the same is analyzed and discussed under the following heads for providing clear picture for further decisions.

Area Production and Productivity of Flowers in India

The details on area production and productivity of flowers in India over the years are analyzed and the results are discussed in Table 2. It provides an overall understanding on whether the area is increasing or decreasing at macro level and the magnitude of production over years provide the information to the policy makers on the extent of adoption of technology and its efficacy and the need based efforts to enhance the productivity of flowers to meet the requirements of both domestic and international markets.

Table 2 State wise Area Production and Productivity of Flowers in India during 2010-11

Sl.No	States	Area in Thousand Ha	Production of Loose Flowers in Thousand Tonnes	Pro-duction of Cut Flowers in Lakhs	Produc-tivity in Tonnes per Ha
01	Andaman and Nicobar	04.20	004.70	000.00	01.12
02	Andhra Pradesh	21.80	133.70	6202.00	06.13
03	Arunachal Pradesh	01.20	00.00	2860.00	00.00
04	Assam	00.00	00.00	000.00	00.00
05	Bihar	00.20	02.30	011.00	11.50
06	Chandigarh	00.00	00.00	000.00	00.00
07	Chhattisgarh	06.90	27.10	000.00	03.93
08	Dadra and Nagar Haveli	00.00	00.00	000.00	00.00
09	Daman and Diu	00.00	00.00	000.00	00.00
10	Delhi	05.50	05.70	1038.00	01.04
11	Goa	00.00	00.00	000.00	00.00
12	Gujarat	12.50	49.50	5063.00	03.96
13	Haryana	06.20	60.30	1084.00	09.73
14	Himachal Pradesh	00.70	00.60	0605.00	00.86
15	Jammu and Kashmir	00.10	00.20	0066.30	02.00
16	Jharkhand	01.60	22.00	1711.30	13.75
17	Karnataka	27.00	203.90	5860.00	07.55
18	Kerala	00.00	00.00	000.00	00.00
19	Lakshadweep	00.00	00.00	000.00	00.00

Sl.No	States	Area in Thousand Ha	Production of Loose Flowers in Thousand Tonnes	Pro-duction of Cut Flowers in Lakhs	Produc-tivity in Tonnes per Ha
20	Madhya Pradesh	07.70	06.00	000.00	00.78
21	Maharashtra	17.50	91.10	7914.00	05.21
22	Manipur	00.00	00.00	000.00	00.00
23	Meghalaya	00.00	00.00	000.00	00.00
24	Mizoram	00.10	00.00	162.00	00.00
25	Nagaland	00.08	00.00	017.00	00.00
26	Odhisha	07.40	03.70	5911.00	00.50
27	Pondicherry	00.30	02.40	000.00	08.00
28	Punjab	01.70	08.20	000.00	04.82
29	Rajasthan	05.40	09.60	000.00	01.78
30	Sikkim	00.20	00.00	0230.00	00.00
31	Tamil Nadu	32.00	247.30	000.00	07.73
32	Tripura	00.00	00.00	000.00	00.00
33	Uttar Pradesh	10.40	17.60	2958.00	01.69
34	Uttaranchal	01.30	02.30	3416.00	01.77
35	West Bengal	23.10	59.20	23919.00	02.56
	Total	195.08	957.40	69027.60	04.91

(Source: Agricultural Statistics at a Glance 2012)

Table 2 reveals the details of state wise area under flowers, production and productivity in India. Total area under flowers in India is arrived at 1.95 lakh ha. Of which, Tamil Nadu is the front runner with an area of 0.32 lakh ha which accounts for 16.40 per cent to the total area under flowers in India. The second highest area under flowers in India is the Karnataka state which has 13.84 per cent to the total area. The absolute area under flowers with Karnataka is 0.27 lakh ha. Third place goes to West Bengal in respect of area under flowers which has 0.23 lakh ha. Among the four constituent states of South India, the Andhra Pradesh, Karnataka and Tamil Nadu put together have the share of 41.41 per cent to the total area under flowers in India, revealing that south India is the leader in area and production of flowers in India.

In respect of production of flowers in India, the South Indian states had the highest share in flower production which accounted for 61 per cent to the total production of flowers. Tamil Nadu is the front runner in production of flowers followed by Karnataka and Andhra Pradesh which respectively accounted for 25 per cent, 21 per cent and 14 per cent to the total production of flowers in the country. It might be due to the adoption of technology, conducive climate and higher returns from flowers that motivated them to involve actively in establishing flower crops in their holding. In respect of cut-flower production, West Bengal is the front runner followed by Maharashtra. West Bengal alone accounted for 34 per cent of total cut-flower production in the country revealing that it is a potentially sound state in delivering the requirements of cut-flower needs of the nation.

In respect of productivity of flowers, the scenario is totally different. Jharkhand is the state in India capable of realizing around 14 tonnes per ha followed by Bihar achieving the productivity of 11.50 tonnes of flowers. Tamil Nadu is the state capable of realizing the productivity to the level of only 7.73 tonnes per ha.

To sum up, in respect of area and production of flowers, South India is the leader, particularly Tamil Nadu state is the leader. When one could compare the productivity of flowers, Jharkhand state is capable of achieving the highest productivity in India. In respect of cut-flower production, West Bengal is leading the nation to the tune of 34 per cent to the total cut-flower production and hence Tamil Nadu should learn the lessons of Jharkhand and West Bengal and should examine the ways and means of increasing the productivity of cut-flowers.

Area Production and Productivity of Horticultural Crops in Tamil Nadu

The horticultural crops are contributing higher returns to the farmer and hence increased production and productivity will definitely contribute to the upliftment of rural economy. In this context, the area production and productivity of horticultural crops in Tamil Nadu is analyzed and the results are presented in Table 3.

Table 3 Area Production and Productivity of Horticultural Crops in Tamil Nadu during 2012-13 (Estimated)

Sl. No	Name of the Crops	Area in Lakh Ha	Production in Lakh Tonnes	Productivity in Tonnes / Ha	Percentage to Total
01	Fruits	03.32	85.35	25.71	29.96
02	Vegetables	02.84	90.52	31.87	25.63
03	Spices and Condiments	01.73	10.87	06.28	15.61
04	Plantation Crops	02.75	11.99	04.36	24.82
05	Flowers	00.32	03.23	10.09	02.89
06	Medicinal & Aromatic Plants	00.12	00.68	05.67	01.09
Total		11.08	202.64	18.29	100.00

Source: agritech.tnau.ac.in/govt_schemes_services/pdf/tn_horti1213.pdf

Table 3 reveals that the area under horticultural crops in Tamil Nadu is about 11.08 lakh ha which is capable of producing the horticultural output to the tune of 202.64 lakh tonnes leaving the average productivity of 18.29 tonnes per ha. Among the horticultural outputs, the share of flowers is found to be only marginal which is about three lakh tonnes with a productivity of 10.09 tonnes per ha which accounted for only around 3 per cent to the total and hence a detailed analysis was made to highlight the area production and productivity of flowers in Tamil Nadu.

Area Production and Productivity of Flowers in Tamil Nadu

In the preceding section, we have seen the area under horticultural crops. The ensuing section deals with area production and productivity of flowers district-wise specifically to have clear information on the districts that has rich potential to cultivate and harvest flowers in a large scale. The details are analyzed and the results are presented in Table 4.

Table 4 Area Production and Productivity of Flowers in Tamil Nadu during 2011-2012

Sl.No	Name of the District	Area in Ha	Production of Flowers in Tonnes	Productivity in Tonnes per Ha
01	Coimbatore	1394	11821	08.48
02	Cuddalore	1018	7565	07.43
03	Dharmapuri	2389	17671	07.40
04	Dindigul	4795	38598	08.05
05	Erode	1763	14015	07.95
06	Kancheepuram	0430	3286	07.63
07	Kanyakumari	0253	1734	06.86
08	Karur	0747	6756	09.04
09	Krishnagiri	3222	25891	08.03
10	Madurai	2135	17025	07.97
11	Nagapattinam	0360	2782	07.73
12	Namakkal	0162	1297	07.99
13	Perambalur	0188	1650	08.76
14	Pudukkottai	0375	2895	07.71
15	Ramanathapuram	0124	0961	07.75
16	Salem	2382	19627	08.24
17	Sivagangai	0020	0160	08.02
18	Thanjavur	0352	2719	07.72
19	The Nilgiris	0052	0504	09.68
20	Theni	0470	3772	08.03
21	Thiruchirappalli	1447	11575	08.00
22	Tirunelveli	2466	20247	08.21
23	Thiruvallur	1876	14460	07.71
24	Thiruvannamalai	3175	27321	08.61
25	Thiruvarur	0034	0230	06.76

Sl.No	Name of the District	Area in Ha	Production of Flowers in Tonnes	Productivity in Tonnes per Ha
26	Thoothukkudi	0665	5104	07.68
27	Vellore	3879	31172	08.04
28	Villuppuram	0623	4350	06.98
29	Virudhunagar	0943	7386	07.83
	State Total	37739	302574	08.02

Table 4 reveals that the area under flowers in the state of Tamil Nadu is about 37739 ha. Among the 29 districts, the Dindigul district is the front runner in augmenting higher area under flowers which is about 4795 ha followed by the Vellore district in possession of 3879 ha which respectively account for 12.71 per cent and 10.28 per cent to the total area under flowers in the state of Tamil Nadu. Krishnagiri and Thiruvannamalai districts are having almost similar area under flowers which account for around 8.50 per cent of total flower growing areas of Tamil Nadu.

In respect of production of flowers in Tamil Nadu, Dindigul district is found to produce higher output of flowers which is about 38598 tonnes followed by Vellore district capable of producing 31172 tonnes of flowers which respectively accounted for 12.76 per cent and 10.30 per cent to the total output of different variety of flowers in Tamil Nadu. The third place is shared by Thiruvannamalai and Krishnagiri districts and their production is hovering around 26 thousand tonnes to 27 thousand tonnes. The percentage share of production of flowers to the total output of flowers is almost the same as to that prevailed in the area under flowers.

In respect of productivity of flowers in Tamil Nadu, the overall productivity is about 8.02 tonnes per ha. When we analyze the productivity across the state of Tamil Nadu, the Nilgiris district is found to be the topper and she had attained the productivity of 9.68 tonnes per ha, followed by Karur district which has achieved the productivity to the tune of 9.04 tonnes per ha. The third highest achiever in productivity is found to be the Perambalur district and the fourth place goes to Coimbatore district which respectively achieved the productivity level of 8.76 and 8.48 tonnes per ha.

To sum up, when we compare the area production and productivity of flowers in Tamil Nadu, area and production trend of flowers is almost same. In both the cases, Dindigul district is found to be the topper. But when we see the productivity, the Nilgiris district is found to produce higher productivity of flowers followed by Karur district. These details revealed that there exists potential to increase the area under flowers and to achieve higher productivity per unit area through sustained technological infusion among the target farmers.

AREA PRODUCTION AND PRODUCTIVITY OF JASMINE IN TAMIL NADU

So far the study had focused on the area production and productivity of flowers in India and Tamil Nadu which will be of immense usel to the researchers and the policy makers. However, our study is focusing especially on the production and marketing aspects of jasmine, the details of area production and productivity of jasmine in Tamil Nadu is analyzed and the results are presented in Table 5.

Table 5 Area Production and Productivity of Jasmine in Tamil Nadu During 2011-2012

Sl.No	Name of the District	Area in Ha	Production of Flowers in Tonnes	Productivity per Ha
01	Coimbatore	0296	2291	07.73
02	Cuddalore	0207	1602	07.73
03	Dharmapuri	0659	5104	07.74
04	Dindigul	0902	6991	07.50
05	Erode	1346	10433	07.75
06	Kancheepuram	0195	1507	07.72
07	Kanyakumari	0116	0902	07.77
08	Karur	0196	1519	07.75
09	Krishnagiri	1141	8843	07.75
10	Madurai	1610	12475	07.74
11	Nagapattinam	0357	2766	07.74
12	Namakkal	0083	0641	07.72

Sl.No	Name of the District	Area in Ha	Production of Flowers in Tonnes	Productivity per Ha
13	Perambalur	0086	0665	07.73
14	Pudukkottai	0253	1958	07.73
15	Ramanathapu-ram	0124	0961	07.75
16	Salem	0850	6588	07.75
17	Sivagangai	0017	0131	07.70
18	Thanjavur	0066	0510	07.72
19	The Nilgiris	0000	0000	00.00
20	Theni	0253	1958	07.73
21	Thiruchirappalli	0637	4938	07.75
22	Tirunelveli	1625	12594	07.75
23	Thiruvallur	1412	10944	07.75
24	Thiruvannama-lai	0666	5163	07.75
25	Thiruvarur	0011	0083	07.54
26	Thoothukkudi	0383	2967	07.74
27	Vellore	1118	8665	07.75
28	Villuppuram	0279	2160	07.74
29	Virudhunagar	0695	5389	07.75
	District Total	15583	120748	08.18

(Source: Hort - Stat 2008)

Table 5 reveals the details of area under jasmine and its production district wise. Among the 29 districts, Thiruvelveli district is found to have higher area under jasmine followed by the Madurai district. Both the districts have almost equal area. The difference is only marginal and the fourth place in respect of area under jasmine goes to Thiruvallur District which is very near to the capital of Tamil Nadu, which might be the only reason to augment higher area under jasmine as most of the garland users demand it. The extraction and supply of jasmine is very timely to the city because of its proximity to Chennai. The total area under jasmine in the state of Tamil Nadu is arrived at 15583 ha. Still the area under jasmine is very low when compared to the demand for the produce during the festive occasions and holy days.

With regard to the production of jasmine flowers in Tamil Nadu, Thirunelveli and Madurai districts are on par almost and they have produced the jasmine flowers to the tune of 12594 tonnes and 12475 tonnes respectively sharing 10.43 per cent and 10.33 per cent to the total jasmine production in the state of Tamil Nadu. The third place goes to Thiruvallur District and the Erode District is coming under fourth place in jasmine production. As the study is focusing on Erode District, its productivity is hovering around 7.75 tonnes per ha. The productivity of jasmine in almost all the districts is found to be unique and it is hovering around 7.72 to 7.75 tonnes per ha advocating the Department of Horticulture and Plantation Crops of Tamil Nadu that the Department should have the support of officials of Directorate of Economics and Statistics, Government of Tamil Nadu in arriving at the estimated figures for future planning and forecasting the crop area and its production.

Area Expansion Program in Jasmine

The area expansion is carried out through National Horticulture Mission in each and every state of India. In respect of jasmine, the details of area expansion planned and implemented in potential districts of Tamil Nadu are analyzed and the details are presented in Table 6.

Table 6 Area Expansion Programs for Jasmine in Tamil Nadu

Sl. No	Name of the District	Area Expansion in Jasmine and Cut Flowers	Percentage to Total
1.	Coimbatore	120.00	14.12
2.	Dindigul	205.00	24.12
3.	Dharmapuri	150.00	17.65
4.	Erode	170.00	20.00
5.	Krishnagiri	190.00	22.35
6.	Salem	015.00	01.76
	Total	850.00	100.00

(Source: nhm.xic.in/action plan/action plan Tamil Nadu. Pdf.)

Table 6 reveals that in Tamil Nadu, the action plan has identified six important districts for expanding the area under jasmine. During the year 2012-2013, Dindigul District was blessed with higher area expansion followed by Krishnagiri District both of which accounted for 24.12 and 22.35 per cent respectively. Erode District is positioned in the third place in respect of area expansion in jasmine which is approximately 170 ha. In toto for the year 2012-2013, Tamil Nadu had an area of 850 ha under the area expansion of jasmine.

THE PRODUCTION POTENTIALS OF FLOWERS IN THE SAMPLE FARMS OF TAMIL NADU

In respect of area production and productivity of flowers, the Erode District holds the fourth place. The sample farms selected for the study belonged to Sathyamangalam Tehsil of Erode District. The farmers who are cultivating flowers have organized themselves and established a society for promoting the concept of direct marketing so as to gain higher share in the consumers' money. The society is titled as "Sathyamangalam Farmers Flower Growers' Head Association" which was registered during 2007 and functioning at Sathyamanagalam. Currently its strength rose to somewhere around 1500 members. The society has appointed the President and Secretary to govern the society through the process of election if there exists competition and the society is paying a lump sum amount as salary every month for their untiring service to the society. Before listening to the society and its routine functions, it is important to know how the farmers are being benefitted from cultivation and marketing of jasmine because of the society and hence the results are analyzed and the details are discussed as follows.

3.1. Details of Size of Holding Available with the Sample Farms

3.2. Cropping Pattern in Vogue with the Sample Farms

3.3. Area under Flowers in Erode District

3.4. Area Production and Productivity of Jasmine in the Sample Farms

3.5. Quantity of Jasmine Produced and Distributed to Different Parts of India and other Places

Details of Size of Holding Available with the Sample Farms

The size of holding available with the farmers is a measure to call them as marginal, small, medium and big farmer's category to know who are actively participating in the production and marketing aspects of jasmine in Erode District particularly among the members of the society. The details of size of holding are analyzed and the results are presented in Table 7.

Table 7 Details of Size of Holding Available in the Sample Farms

Sl.No	Particulars of Holding	Average Size of Holding in Acres	Percentage to Total
01	Owned land under cultivation	06.73	85.41
02	Land under non agricultural uses	00.42	05.33
03	Owned land (1+2)	07.15	90.74
04	Leased-in land	00.73	09.26
05	Total size of holding	07.88	100.00

Table 7 reveals that the size of holding available with the sample farms was arrived at 7.88 acres (3.15 Ha) which is categorized as medium size of holding. Normally the farmers of Coimbatore and Erode belt are having higher size of holding in their farms and are dynamic in establishing mostly annual horticultural crops. The owned land under cultivation is arrived at 6.73 acres which accounted for 85 per cent of the total size of holding per farm. Wherever possible, the farmers are also utilizing others land on lease which accounted for nine per cent to the total size of holding. In this circumstance, it is important to analyze the cropping pattern prevalent among the farmers of flower growers association at Sathyamangalam.

Cropping Pattern in Vogue with the Sample Farms

The details of horticultural and agricultural crops cultivated by the farmers of Sathyamangalam are analyzed and the details are presented in Table 8 as it assumes greater importance that how much area is dedicated exclusively for the jasmine cultivation by the farmers.

Table 8 Cropping Pattern in Vogue with the Sample Farms during 2012-13

Sl.No	Name of the Crops Cultivated	Area in Acres	Percentage to Total
01	Jasmine	00.92	11.22
02	Mullai	00.25	03.05
03	Mysore Malli	00.45	05.49
04	Sambangi	00.44	05.37
05	Banana	02.10	25.61
06	Turmeric	00.74	09.02
07	Sugarcane	02.46	30.00
08	Cholam	00.32	03.90
09	Other Crops	00.52	06.34
	Total Area under Crops	08.20	100.00
	Cropping Intensity	104.06	

Table 8 reveals that the total area under crops is at 8.20 acres per farm leading to the cropping intensity of 104.60 per cent in Sathyamangalam Taluk. Among the horticultural crops, the area under jasmine crop is arrived at 0.92 acres which accounted for 11.22 per cent to the total area under crops. The highest area is allocated to banana and sugarcane, which being commercial crops have ready market at all the times.

Around 25 per cent of the size of holding available with the farmers was used for cultivating various flower crops which included sambangi, jasmine, mullai and Mysore malli. Mysore malli is the crop which will flower abundantly during the winter season and hence the farmers are cultivating Mysore Malli to reap higher income during the winter season.

Area under Flowers in Erode District

Floriculture is picking up in Sathyamangalam and Bhavanisagar blocks in Erode district with more farmers coming forward to cultivate flower plants. The total acreage under floriculture in the district is growing at the rate of 10 to 15 per cent every year. Over 1,100 hectares are covered under floriculture in these two blocks this year so far. The two blocks have good water sources and suitable conditions for planting flower bushes. Besides, farmers in these two blocks could easily access the two major flower markets—Coimbatore and Mysore. These factors have motivated more farmers to go in for the cultivation of flower plants in these blocks. The details of flower crops cultivated are analyzed and the results are presented in Table 9.

Table 9 Area under Flowers Crops in Erode District during 2010 - 11

Sl.No	Name of the Flower Crops Cultivated	Area in Ha	Production in Tonnes	Productivity in Tonnes per Ha	Percentage to State Total Production
01	Jasmine	1061	9284	08.75	09.99
02	Mullai	0067	0570	08.51	02.42
03	Rose	0013	0094	07.23	00.67
04	Tuberose*	0091	1138	12.51	NA
05	Marigold*	0050	0730	14.60	NA
06	Total	1282	11816	09.22	05.20

Source: tnhorticulture-tn.gov.in/flowers; *The Hindu Dt. 26.10.2010

Table 9 reveals that the total area under flower crops in Erode District is arrived at 1282 ha which has yielded 11816 tonnes of flowers per annum with an average productivity of 14.60 tonnes per ha. Out of the total area under floriculture, jasmine bushes are grown in over 1061 hectares, tuberoses in 91 hectares, mullai in 67 ha, marigold in 50 ha and the rose is cultivated in 13 ha.

The Horticulture Department has taken up a series of efforts to promote flower cultivation as the government had identified floriculture as one of the thrust areas under the National Horticulture Mission. The department of horticultural crops is providing up to 50 per cent subsidy for taking up jasmine and tuberose cultivation. Subsidies are also being extended to the farmers under few other programmes.

Area Production and Productivity of Jasmine in the Sample Farms

Jasmine, mullai, kanagambaram and marigold were grown in vast tracts of land in Sathyamangalam and its surroundings. Jasmine and mullai alone were raised in over 4,000 and odd acres of land in Sathyamangalam,

Rajan Nagar, Kothamangalam, Puliamptty, Bhavanisagar and nearby areas. Over 40,000 farmers depend on flower cultivation. Every day at 5 a.m., the flowers were plucked, kept in a place and transported to the market, situated near the Municipal Bus stand, Sathyamangalam.

The area production and productivity of jasmine in Sathyamangalam area are analyzed and the details are presented in Table 10.

Table 10 Area Production and Productivity of Jasmine Flowers in the Sample Farms

Sl. No	Name of the Village	Area under Flowers in Ha	Production in Tonnes	Productivity in Tonnes/Ha	Percentage to Total
01	Nanjappakavundan Pudhur	023.00	106.72	04.64	05.87
02	Kenjanoor	028.00	145.88	05.21	07.14
03	Kembanaickenpalayam	030.00	159.60	05.32	07.65
04	Phandampalayam	026.00	153.92	05.92	06.63
05	Akkarai Negamum	032.00	192.32	06.01	08.16
06	Chickarasampalayam	120.00	592.80	04.94	30.61
07	Bahuthampalayam	062.00	327.98	05.29	15.82
08	Badhramangalam	038.00	235.60	06.20	09.69
09	Chenbagapudhur	014.00	081.20	05.80	03.57
10	Kothamangalam	019.00	097.28	05.12	04.86
11	Raj Nagar				
	Total Area under Flowers	392.00	2093.30	05.45	100.00

Table 10 reveals that the area production of jasmine cultivated in the sample villages was 392 ha which has realized an output level of 2093 tonnes leaving the average productivity of jasmine at 5.45 tonnes per ha. Chickarasampalayam is the village with a higher area and production which accounted for 31 per cent of the total area in the sample villages followed by Bahuthampalayam with a higher area under jasmine which had a share of around 16 per cent of the total area. Highest productivity of jasmine is visible in Badhramanagalam village of Erode District.

Details of Quantity of Jasmine Produced and Distributed by the Society

The farmers of Sathyamangalam farmers flower growers head association at Sathyamangalam produce different varieties of flowers. Among them, jasmine is found to have a higher share. The harvested flowers are brought to the market around 6 a.m. and the auction is being conducted from 6.30 a.m. onwards. No commission was collected from the farmers for selling their flowers market. But a nominal fee for occupying the place was levied. The president said the association was functioning for the benefit of the flower growers and the annual flower sale turnover touched Rs. 10 crore and hence the details are analyzed and the results are presented in Table 11.

Table 11 Quantity of Jasmine Flowers Produced and Distributed to Different Parts of India and Other Places per Day during 2012-13

Sl. No	Place	Quantity of Flowers Exported in Tonnes During		Total Quantity in Tonnes	Percent-age to Total
		Summer Months	Winter Months		
01	Georgia	03.000	01.200	04.200	05.83
02	Simoka	02.000	00.600	02.600	03.61
03	Mumbai	02.000	01.200	03.200	04.44
04	Bangalore	12.000	03.500	15.500	21.50
05	Hyderabad	03.000	00.600	03.600	04.99
06	Mysore	13.500	04.000	17.500	24.27

Sl. No	Place	Quantity of Flowers Exported in Tonnes During		Total Quantity in Tonnes	Percentage to Total
		Summer Months	Winter Months		
07	Delhi	02.500	01.000	03.500	04.85
08	Kerala	06.000	02.500	08.500	11.79
09	Coimbatore	07.500	02.000	09.500	13.18
10	Erode	01.000	00.500	01.500	02.08
11	Thiruppur	01.300	00.600	01.900	02.63
12	Sathy Local	00.400	00.200	00.600	00.83
Total Quantity of Flowers		54.200	17.900	72.100	100.00

Table 11 reveals that the flowers are distributed to around 12 places regularly. During the summer months, around 54 tonnes of jasmine are distributed to different markets every day and during the winter months, the same is reduced to less than one third distributed during the summer months which arrives at about 18 tonnes per day. Among the different destinations, Bangalore, Mysore, Coimbatore and Kerala are the potential markets which had respectively consumed 21.50 per cent, 24.27 per cent, 13.18 per cent and 11.79 per cent. It is reported that an amount of Rs 10 crore per annum is achieved through the marketing of flowers.

Georgia, Mumbai and Delhi are the other important markets for which regular supply from Sathyamangalam farmers is practiced and the traders are actively involved in procuring the same and distributing to the different consumer segments and keeping them satisfied with quality produce from Sathyamangalam. Export of flowers to Singapore is also in the offing. In this context, it is important to analyze the price that prevailed in the farmers' market for flowers during the study period and the same is documented in the following section.

Price of Jasmine that Prevailed in the Farmers' Flower Market

Jasmine grown in the Sathyamangalam area had more number of petals and pleasant fragrance than those from other areas. So the demand in Gulf countries was very high. The wholesale traders collect the flowers and transport them to Mumbai, from where it was airlifted. The President of Sathyamangalam Farmers' Flower Growers Association said that every day the flowers were collected from flower growers and dispatched to Mysore and Kochi specifically. From Kochi they were transported to Mumbai. He said if flight service to Mumbai from Coimbatore was introduced, more quantity of flowers may be sent to Mumbai. Such quality flowers are traded and daily price varied significantly and hence the price that prevailed on the day of survey was analyzed and the results are presented in Table 12.

Table 12 Price of Jasmine that Prevailed in the Farmers' Flower Market during the Study Period

Sl. No	Time of Auction A.M.	Price per Kg of Jasmine		Percentage Change
		08.11.2013	09.11.2013	
01	07.00	250.00	235.00	(-) 06.00
02	07.30	225.00	225.00	00.00
03	07.45	201.00	201.00	00.00
04	08.00	190.00	190.00	00.00
05	08.30	165.00	165.00	00.00
06	08.45	153.00	144.00	(-) 05.88
07	09.00	135.00	135.00	00.00
Average Price per Kg		188.43	185.00	(-) 01.82

On plucking from the farm, if the flowers reach the market at very early hours, it attracts higher price. If flower reached the farmers' market prior to 7 am, the flower is priced at Rs 250 per kg; this was the status on 8th November 2013, but the same was little reduced to Rs 235 per kg on the very next day. Table 12 reveals that the average price for the day on 8th November and 9th November 2013 is hovering around 185 to Rs 188 per kg. If time exceeds 9 a.m., the price starts reducing gradually. Occasionally it rose to higher price when number of traders visited the market. The only advantage in the farmers' market is that the traders visit the farmers themselves, participate in the open auction of flowers and get their share to meet their own consuming ends.

Normally, the farmers take their produce to the retailers or wholesalers and get the money from them for their produce. Here, the society is attracting the traders and fixing the minimum price after assessing the price in the other markets of capital cities. After learning on the market information, the president and secretary of the society force the traders against bargaining, make them participate in the open auction conducted by the society and the produce gets higher price per unit because of competition between the traders.

The adverse climatic conditions like heavy mist have resulted in a sharp decline in jasmine production in Sathyamanangalam and nearby areas and prices touching a peak of Rs 1,200 per kg. The price, which touched Rs 1,200 per kg on Saturday and Sunday, came down to Rs 200 on the week days. Similar is the situation in Nilakottai flower market of Dindigul District.

Price of Jasmine that Prevailed in Other Markets of Tamil Nadu Price of Jasmine has shot up during the festive occasions and during the holidays and holy days in different markets of Tamil Nadu. The details of prices that prevailed in different flower markets have been enquired and the details are presented in Table 13.

Table 13 Price of Jasmine that Prevailed in Other Markets of Tamil Nadu

Sl. No	Name of the Markets	Name of the District	Price that Prevailed per Kg of Jasmine During	
			05.01.2014	12.01.2014
01	Madurai	Madurai	600.00	1000.00
02	Chennai	Chennai	350.00	1200.00
03	Nilakkottai	Dindigul	600.00	1000.00
04	Coimbatore	Coimbatore	500.00	1000.00
05	Thiruppur	Thiruppur	600.00	1100.00
06	Sathyaman-galam	Erode	450.00	1000.00
07	Dindigul	Dindigul	500.00	1100.00
08	Tanjore	Thanjavoor	500.00	1000.00
09	Thirunelveli	Thirunelveli	450.00	1200.00
10	Thovalai	Kanyakumari	600.00	1200.00
Average Price Per Kg			515.00	1080.00

Table 13 reveals that the price of jasmine is found to increase during this week of reporting. The increase in the price of jasmine was arrived at 109.71 per cent when compared to the first week of January 2014. It is mainly due to the Pongal festival demand and the arrival of flowers to the various markets was found to be reduced due to the mist that prevailed in different places. Among the markets, Thirunelveli, Thovalai and Chennai had offered highest price of Rs 1200 per kg of jasmine on 12th January 2014. From this, one could understand that on the festive occasions and holy days the flower price had reached the peak of Rs 1200 per kg. This called for technological improvement to induce the plants to flower even during the offseason to meet the demand.

Thovalai flower market in Kanayakumari district is one of the biggest flower markets in the state. The market used to transact nearly 20 tonnes of flowers in a single day on Onam festival eve and nearly 5000 traders from Kerala visit the market. The market receives flowers from across the state as far as from Hosur in Krishnagiri District and Sathyamangalam in Erode District of Tamil Nadu and from more than 50 villages surrounding the market. More than 60 flower merchants and hundreds of flower farmers depend on the market and Onam is significant for them. But the farmers are keeping their fingers crossed as the region is yet to get sufficient rains favoring flower growth.

COST OF PRODUCTION OF JASMINE IN ERODE DISTRICT

Highly varying prices of jasmine on one day and on another day, as for example, it goes to Rs 40 or Rs 45 per kg during certain season, called for arriving at the cost of production of jasmine so as to assess whether the farmer producers are able to get a profit even on the lowest price received by them. In this circumstance, the cost of production of jasmine is discussed under the following sub heads, viz.,

- Cost of cultivation of jasmine in the sample farms
- The cost of production and returns realized out of jasmine production and
- The production constraints prevalent in the study environment

Cost of Cultivation of Jasmine in the Sample Farms

The cost of cultivation of jasmine takes the items such as the apportioned cost of establishment cost, the average maintenance cost and the harvesting costs. The details are analyzed and the results are presented in Table 14.

Table 14 Cost of Cultivation of Jasmine Flowers in the Sample Farms

S. No	Details of Expenditure	Number of Units	Price Per Unit	Expenditure / Acre
		Fixed Cost		
1.1	Rental Value of Owned Land per Acre	One Acre	17680.00	17680.00
1.2	Interest on Fixed Capital	12 Per cent	2121.60	2121.60
1.3	Land Cess and Taxes		0010.00	0010.00
1.4	Apportioned Establishment Cost	1:10	3456.00	3456.00
	Annual Fixed Cost		23267.60	23267.60
		II. Production Cost		
2.1	Weeding	12	200.00	2460
2.2	Pruning	02 + 03	400 +200	1465
2.3	Earthing Up	04	400.00	1543
2.4	Manures and Fertilizers	06 Bags	900.00	5780
2.5	Spraying of Chemicals	00	00	4570
2.6	Irrigation	00	00	1950
2.7	Harvesting	128	200	25750
2.8	Miscellaneous	00	00	2025
2.9	Production (Variable) Cost (II)			45543.00
		Post-Harvest Expenses		

S. No	Details of Expenditure	Number of Units	Price Per Unit	Expenditure / Acre
3.1	Commission Charges	00	00	1376.00
3.2	Transportation Charges to Farmers Flower Market	00	00	1200.00
3.3	Sub Total Cost (III)			2576.00
		Total Cost		
4.1	Annual Fixed Cost (I)			23267.60
4.2	Variable Cost (II)			45543.00
4.3	Post-Harvest Expenses			2576.00
4.4	Total Cost			71386.60
	Cost of Production of Jasmine			
5.1	Output Realized per Acre in Tonnes		130.00	369200.00
5.2	Cost of Production per Kg of Jasmine			25.14

Table 14 reveals that the total cost of cultivation of jasmine is about Rs 71386 per acre per annum. Among the total cost, the annual fixed cost was arrived at Rs 23267 per acre, the annual maintenance cost and the variable cost was arrived at Rs 45543 per acre and the post-harvest charges amounted to Rs 2576 per acre accounting for 32.59 per cent, 63.80 per cent and 3.61 per cent respectively of the total cost of cultivation incurred for establishing and maintaining jasmine garden per acre.

Cost of Production and Returns Realized from Jasmine Garden

Estimation on cost of production of jasmine per unit will be the right indicator to the farmers at what price the commodity should be priced or the minimum price should be declared. In this respect, the cost of production and the returns realized from the sale of jasmine per acre is analyzed and the results are presented in Table 15.

Table 15 Cost of Production and Revenue Realized in Jasmine Flower Per Acre

Sl. No	Particulars of Cost and Return	Units	Price per Unit	Total
01	Output Realized per Acre of Jasmine in Kgs	2840	130.00	369200.00
02	Total Cost incurred in Jasmine Cultivation			71386.00
03	Cost of Production per Kg of Jasmine			25.14
04	Net Return Realized per Annum			297814.00
05	Output - Input Ratio			05.17

Table 15 revealed that the yield realized per acre from the cultivation of jasmine is arrived at 2840 kgs which had generated a gross revenue of 3.69 lakhs. For realizing this out turn, the farmer in Sathyamangalam belt should be in a position to invest Rs 71386 per acre. The net return realized by the farmer through the cultivation of jasmine was arrived at Rs. 2.98 lakhs. The cost of production of jasmine per kg is arrived at Rs 25.14 which is also considered to be marginally high. Sometimes the flower is priced even between Rs 30 to Rs 40 which will not add many returns to the farming community. In this context, the society should have the price declared in the capital city markets to the incoming traders and accordingly the traders should fix the minimum price per kg of jasmine above Rs 50 so that the farmer producer will get benefitted. In addition, the society itself can explore the possibilities of running the Floral Concrete Extraction Unit at Sathyamangalam to benefit further due to abundant resource availability.

Production Constraints Faced by the Jasmine Growers

Any production activity will normally attract certain constraints. Timely identification and removal of the constraints will smoothen the process of production of any commodity. It is true with jasmine producers and they also experienced certain constraints which are analyzed and the results are delineated in Table 16.

Table 16 Production Constraints Faced by the Jasmine Growing Farmers in the Study Area

Sl. No	Production Constraints	Rank
01	Insufficient Labor for Harvesting and Weeding Operations in the Farm	I
02	Lack of Technology to induce flowers during the Offseason	II
03	Pest and Disease Outbreak in Jasmine	III
04	Water Scarcity to the Tail Ends and Certain Pockets	IV
05	Non Availability of Plant Protection Chemicals	V
06	Absence of Collection Points near to the Villages	VI

Table 16 reveals that the first and foremost constraint is the insufficient labor for harvesting or extraction of flowers and for the weeding operations in the farm and it is ranked as number one constraint. Lack of technology to induce flowers during the off season is found to be the second important constraint, the farmers have reported. However, the scientific institutions like the Horticultural College and Research Institute, of the Tamil Nadu Agricultural University, Periyakulam, can explore this problem and do the needful in this respect. The third important problem reported was found to be the pest and disease outbreak in the season and hence this problem is widespread and the farmers are adopting the practice of spray in an uncontrolled manner. To alleviate this problem, the society can explore the possibility of recruiting a horticultural graduate as manager who will coordinate the technology related problems and thus higher productivity can be realized by the society.

Among the constraints, the last one is found to be the absence of collection points near to the production site. The society can explore the ways and means of collecting the produce from the distant places through the pickup vans so that the produce will be reachable in the early hours and also it can fetch higher price per unit.

SUMMARY AND CONCLUSIONS

Recommendations

- Pest and disease outbreak in the season are found to be the biggest problem and hence the farmers are adopting the practice of spray in an uncontrolled manner. To alleviate this problem, the Sathyamangalam Farmers Flower Growers' Head Association (SaFGA) can explore the possibility of recruiting a horticultural graduate as manager who will coordinate the technology related problems and thus higher productivity can be realized by the society.
- Among the production constraints, the absence of collection points near to the production site is recognized. The Sathyamangalam Farmers Flower Growers' Head Association can explore the ways and means of collecting the produce from the distant places through the pickup vans so that the produce will be reachable in the early hours and it can also fetch higher price per unit.
- Currently, the SaFGA has rented a vacant land for farmers to the backside of the municipal bus stand on rental basis and the marketing of jasmine is being practiced from this land by the society. In this regard, the government through the Public Works Department or Revenue Department or the town panchayat should come forward to identify a land suitable for the Sathyamangalam Farmers' Flower Growers' Head Association (SaFGA) for their efficient functioning. Around 3000 farmers depend on the society for their livelihood and they are involved in direct marketing of flowers through SaFGA. This novel effort should be supported by the Government at the earliest by providing place and erecting an integrated market yard for the flower procurement and trade including parking and storage facilities.
- Fluctuation in price of jasmine flower buds called for uncertainty in the income of the jasmine producers. Since the Government of Tamil Nadu is poised for enhancing the income of the farmers by following a phrase "Doubling the Production and Tripling the Income of the Farmers", to fulfill this goal, the Government should come forward to erect an integrated farmers flower market at Sathyamangalam, Nilakottai, Thovalai, Madurai and Thirunelveli following the pattern of Koyambedu market at Chennai by getting some share from the farmers association and the rest borne by the Government for en-

hancing further trade development and if need be a Special Economic Zone of flowers may also be announced.

- Another important constraint is transportation infrastructure. Here what is needed is providing a direct flight service from Coimbatore to Mumbai will readily enhance the transportation of flower buds on time which will in turn facilitate the producer traders to explore new markets keeping Mumbai as the base.

Chapter 12

Evolution of A Mango Farmer onto an Agripreneur

INTRODUCTION

The horticulture sector has in the past few years, proved to be the engine of growth in agriculture for improving the productivity per unit area, source of generating employment, improving the economic well being of the farming community and the entrepreneurs involved in agriculture and horticulture and thereby enhancing exports and capable of earning considerable foreign exchange.

Fruits and vegetables are considered to be the protective food because it helps us to develop a defensive mechanism in human system. Mango is called as "The King of Fruits" preferred by all sections of people for its delicious taste, flavor, attractive color, nutritive value and superior fragrance. India is the topper in mango production in the world. The important mango producing states in India are Andhra Pradesh, Uttar Pradesh, Karnataka, Bihar, Gujarat, Maharashtra, Tamil Nadu, West Bengal, Kerala and Orissa. The productivity of mango is found to be declining over the years. The national average productivity is found to be as low as 10 tonnes per ha (Biswas and Lalitkumar, 2011).

It has been proven that the horticulture crops for which the Indian topography and agro-climatic conditions is well suited could be an ideal choice for achieving sustainability by small farmers with continued governmental support. However, the importance of horticulture is not only in improving the productivity of land but also providing nutritional security to the people (Singh 2010).

In India, several horticultural development initiatives are being taken up through the National Horticulture Mission through the State Department of Horticulture and Plantation Crops. Though NHM has initiated its efforts to develop the horticulture sector through the area expansion programs, existing level of production and productivity of fruits particularly the mango is yet to be documented at national level and the data is found to be dismal. This chapter has been compiled by the author to show that a sound data and analysis on area, production and productivity and export of mango will be a very good tool for the planners and policy makers in horticulture and agriculture to take up planned investment in a phased manner for sustainable development of horticulture.

DESIGN OF THE STUDY

Murugamalai Pudhur, a village of Periyakulam taluk of Theni district is blessed with mango orchards which are manned by progressive farmers of Periyakulam town. The area under mango plantation, production and productivity in Theni district is considerably high next to Krishnagiri District. Mr. O. Parasuraman is one of the progressive farmers who has established the mango plantations in an area of 131 acres in Murugamalai Pudhur village and managed the plantation in a different style. He is also directly involved in marketing and distribution of harvested mango to different parts of the country and has turned to be an innovative agripreneur in the village. Hence, we visited the farm along with the students of Third Year B.Sc. (Horticulture) degree program of Horticultural College and Research Institute, Tamil Nadu Agricultural University, Periyakulam to learn on the innovativeness and to transform those ideas and knowledge to the students.

The Agripreneur also delivered his lecture to the students of B.Sc. (Horticulture), regarding his experiences on production and distribution of mango which were found to be technologically advanced. His efficient management of the farm for higher productivity and following of strategic actions in distribution of mango by aligning the traders from different parts of the country commanded higher attention from his class and hence a case study approach is adopted to document his success story to the horticulturists in the state of Tamil Nadu and to the nation for follow up and implementation.

The details of land use, orchard management for its production and productivity, marketing of mango, cost of production and the constraints involved are analyzed and documented in this study by adopting a descriptive statistics like mean and percentage analysis.

PRICE SPREAD ANALYSIS

Price spread in general, is referred to as the difference between the price paid by the consumer and the price received by the producers per unit of the commodity. Price spread analysis would estimate the share of different market intermediaries in the consumer rupee and this would often facilitate the understanding of the relative efficiencies otherwise of alternative channels in marketing.

To analyze the price spread in the distribution of mango, the following formula is used.

$$Ps = (Pp/Cp) \times 100$$

Where,

Ps = Producer share in consumer money expressed in per cent,

Cp = Consumer's price per unit of mango in rupees,

Pp = Producer's price per unit of mango in rupees. Similar approach was used by Sekhar et.al. (1998).

The price spread analysis involved computation of different marketing cost and marketing margin at each stage and their expression as a percentage to the consumer's rupee. The various costs incurred in the marketing process were considered for each of the identified channels and separate price spread was worked out. The profit margin for each market functionaries in the different marketing channels was computed by subtracting the price paid and the marketing cost incurred by them from the price received by him on the sale of the commodity. Thus the marketing costs and marketing margins were distinguished.

ESTIMATION OF MARKETING EFFICIENCY

Marketing efficiency is the degree of market performance. The movement of goods from producer to the ultimate consumers at the lowest possible cost consistent with the provision of services desired by the consumer is termed as marketing efficiency or efficient marketing. To analyze the marketing efficiency, the formula adopted by Shephered is used.

$$ME = (V/I) - 1$$

Where,

ME = Marketing Efficiency Index,

V = Value added to the output in the marketing system. Here it is referred to as the consumers price per kg of mango,

I = Input used in the marketing process. Here it is referred to as the marketing costs incurred per kg of mango. Similar methodology was used by Sekhar et.al. during 1998.

The lower the value of marketing cost higher would be the efficiency.

Calkin's Index

Calkin's index of marketing efficiency was estimated using the following formula:

$$ME=1+[\text{Sum of profit or margin/Sum of marketing cost}]$$

The lower the value of the index, higher would be the efficiency.

Agripreneurial practices are prevalent among the progressive farmer of Murugamalai village who is actively engaged in hi-tech production and distribution of mango from his 140 acres of horticultural activities. The details of activities taken up with respect to production and marketing of fresh mango are discussed in the following sections.

- General characteristics of case farm
- Area, production and productivity of mango orchard
- Cost of production of fresh mango

- Marketing practices adopted in distribution of fresh mango
- Marketing efficiency
- Constraints faced by the case farm

GENERAL CHARACTERISTICS OF CASE FARM

Under the head, general characteristics of case farm, the age, education and expertise gained in production of mango by the farm functionaries as a whole are analyzed and presented in Table. 1.

Table: 1 Age and Educational Qualification of Human Resource Personnel with the Farm

S. No	Details of HR Personnel	Age in Years	Education in Years	Experience in Years
1.	Manager-in-Chief	52	11 (PUC)	32
2.	Line Managers	48	17 (PG)	20
3.	Supervisors	37	15 (UG)	08
	Mean	46	14.67 or 15	20

It revealed that the Manager-in-Chief is pre-university qualified aged about 52 years having an experience in the field of production and distribution of mango for 32 years in planning and executing the tasks carefully and sequentially. The line managers, who are brothers of the owner manager, support the owner in all the activities of maintaining the mango orchard. The average education of the human resource personnel who are caring for the plantation is graduate level and the experience in this field is arrived at 20 years. From this, one could understand that the educational qualification and expertise gained in the field instill confidence in their mind and taking appropriate technological interventions needed for the development of mango orchard.

Table 2 revealed the details of strength of human resources available with the farm in managing the plantations. There are 5 permanent staff and 70 labor force involved in various activities of the farm. There are two line managers who are the brothers of owner manager (Manager-in-Chief) and the two supervisors are his relatives in caring the plantation. With respect to labor force used for various activities, permanent labor included are 20 in numbers and the casual laborers are called for intensive activities with respect to harvesting and distribution of mango. In that laborers used for harvesting purposes are 40 in numbers who accounted for 57 per cent to the total labor force employed in the farm. The labor force used for grading, packing and forwarding of mango to different markets are 10 in numbers who accounted for 14 per cent to the total labor force.

Table 2 Strength of Human Resources in Managing the Mango Plantations

S. No	Category	Number of Persons	Percentage to Total
	I - Permanent Staff		
1	Manager-in- Chief	01	20.00
2.	Line Managers	02	40.00
3.	Supervisors	02	40.00
	Total	05	100.00
	II - Labor Force		
1	Permanent Laborers	20	0.29
2.	Laborers for Harvesting Purpose	40	0.57
3.	Laborers for Grading, Packing and Forwarding	10	0.14
	Total	70	100.00

To sum up, 3 managers are available at decision making level that are accounted for 60 per cent to the total staff employed permanently in the case farm. The supervising personnel employed in the farm are accounted for 40 per cent to the total staff on permanent category. The staff including the labor force employed at the farm for execution of day-to-day activities reveal an organizational hierarchy in an orderly manner permiting target based actions in the case farm.

AREA, PRODUCTION AND PRODUCTIVITY OF MANGO ORCHARD

The area, production and productivity of mango in the case farm are discussed in the following sub-heads. They are

- Age Class Distribution and Yield of Mango Orchard
- Area Production and Productivity of Different Varieties of Mango Orchard

Age Class Distribution and Yield of Mango Plantation

The mango plantation in the case farm is in different age groups. The number of trees available in the case farm is classified into three different class intervals and the details of age class interval is analyzed and presented in Table 3.

Table 3 Productivity of Mango Plantation in the Case Farm

S. No	Age of the Trees in Years	No. of Trees	Percentage of Trees	Yield per Tree in Kgs	Total Yield in Tonnes	Percent-age of Yield
1	05-10	0625	09.43	119	0074.00	03.84
2	11-20	1200	18.12	208	0250.00	12.97
3	21-30	4800	72.45	334	1603.00	83.19
	Total	6625	100.00	661	1927.00	100.00

Table 3 revealed that 9.43 per cent of mango trees in the plantation are in the age group I belonging to the age of 5 to 10 years; 18.12 per cent of mango trees are in 11 to 20 years of age class and designated as group II and 72.45 per cent of the mango trees are between the age group of 21 to 30 years and designated as group III. Group III trees in the age group of 21 to 30 years are capable of yielding 334 fruits per tree giving an aggregate yield of 1603 tonnes followed by class II (11 to 20 years) that are capable of yielding 208 fruits per tree thereby producing 250 tonnes and the age class I has yielded 74 tonnes of mango with an average fruit yield of 119 mango fruits per tree. In toto, all the 6625 trees in the 3 age classes are capable of yielding 1927 tonnes of fresh mangoes. In Group III, 72 per cent of trees are capable of generating an output of mango to the tune of 83 per cent revealed that as age advances in the mango plantation, the yield of mango also increased. On seeing this, one would be interested to know the area under mango, its production and productivity which are delineated in the ensuing section.

Area Production and Productivity of Different Varieties of Mango

The area, production and productivity of mango varieties cultivated in the case farm are delineated in Table 4. It revealed that the case farm has established 12 varieties of mango in which Neelam is the predominant variety in 90 acres accounting for 67.37 per cent of the total area under mango production followed by Thothapuri variety accounting for 13.47 per cent to the total varieties raised in the farm. Senthura occupied the third place in the farm accounting for 11.23 per cent area.

Table 4 Area Production and Productivity of Mango Varieties in the Case Farm

S. No	Mango Varieties	Area in Acres	Area in Per cent	Average Yield / Tree in kgs	Total No. of Trees	Total Yield in Tonnes
1.	Mulgoa	00.50	00.37	200	25	05.00
2.	Thothapuri	18.00	13.47	300	900	270.00
3.	Senthura	15.00	11.23	300	750	225.00
4.	Manoranjitham	01.00	00.75	50	50	02.50
5.	Neelam	90.00	67.37	300	4500	1350.00
6.	Kalappad	01.00	00.75	200	50	10.00
7.	Banganapalli	04.00	03.00	200	150	30.00
8.	Alphonsa	02.00	01.50	200	100	20.00
9.	Chakkarakutty	00.50	00.37	200	25	05.00

S. No	Mango Varieties	Area in Acres	Area in Per cent	Average Yield / Tree in kgs	Total No. of Trees	Total Yield in Tonnes
10.	Rumani	00.10	00.07	100	05	00.50
11.	Imampas-anth	01.00	00.75	100	50	05.00
12.	Mallika	00.50	00.37	200	20	04.00
	Total	133.60	100.00	2350	6625	1927.00
	Mean	11.13		196	50	36.06

Banganapalli, Alphonso, Manoranjitham and Imampasanth are the other select sweet varieties of mango established respectively under 3.00 per cent, 1.50 percent, 0.75 per cent and 0.75 per cent to the total area under mango plantation established in the farm. Other varieties cultivated are Mallika, Mulgoa, Kalappad, Chakkarakutty and Rumani in 1.93 per cent to the total area under mango orchard. Put together, the area under mango orchard established in the farm is arrived at 53.44 ha (133.60 acres).

Total number of mango trees in the case farm is arrived at 6625 numbers in 53.44 ha and the population per ha is arrived at 125 trees and the number of mango trees per acre is arrived at 50 in numbers. In respect of productivity, 36.06 tonnes could be generated in the case farm which is 218 per cent higher than the national productivity level of mango. When we compare the productivity of mango in the case farm with the Theni District mango productivity, it arrived at 171 per cent higher revealing that the farmer is practicing productivity of mango by applying good agronomic practices which are needed for the mango orchards on time. Kaltar application, watering, manure application, three protective chemical sprays at the time of flowering and bud initiation stages and providing pole support to the branches of mango trees are the special activities initiated in the farm. Besides, trench cutting, basin formation for water harvesting are the soil and water conservation measures additionally taken up to keep the orchard evergreen. In addition to these measures, developing newer varieties of mango and initiation of offseason flowering in consultation with the co-author Dr. M. Selvarajan, Professor, Department of Fruits of Horticultural College and Research Institute, Periyakulam are also taken up in the case farm.

COST OF PRODUCTION OF FRESH MANGO

An enterprise involved in production will always incur the need based costs for producing the quality output. The need based costs are categorized into fixed costs and the variable costs. Fixed costs are the costs incurred for establishing infrastructural support facilities and the variable costs incurred are exclusively for producing the quality output and the related post harvest expenses. In the case farm, these costs are discussed in Table 5 and 6 separately.

Table 5 revealed that the progressive farmer cum agripreneur is incurring the following items of cost under the head of fixed costs. They included sinking of bore wells, mini truck for collecting the horticultural produce from the inner area of the plantation, farm equipments and tools, plastic trays for carrying the produce from the plantation site to the marketing site, the buildings for laborers and officer room are the major items.

Table 5 Details of Fixed Cost Incurred in the Case Farm

S. No	Particulars of Asset	Expected Life in Years	Actual Cost Incurred in Lakhs	Annual Fixed Cost in Rs/Acre
1	Sinking of bore wells 3 nos @ Rupees one lakh per well	30	03.00 (11.11)	74.85
2	Mini truck @ Rs 8.00 lakhs	10	08.00 (29.63)	598.80
3	Farm equipments and tools @ 2.00 lakhs	05	02.00 (07.41)	299.40
4	Plastic trays, 1000 nos @ Rs 500/ tray	05	05.00 (18.52)	748.50
5	Building for laborers and office room	30	06.00 (22.22)	149.70
6	Erection of four check dams	10	03.00 (11.11)	224.55

S. No	Particulars of Asset	Expected Life in Years	Actual Cost Incurred in Lakhs	Annual Fixed Cost in Rs/Acre
7	Interest on fixed capital @ 12%			2198.47
Total fixed cost in Rs		---	27.00 (100.00)	4294.27
Annual fixed cost / ton		---	---	297.72

(Figures in Parentheses Indicate Percentage to Total Fixed Cost Incurred in the Case Farm)

Among these costs, the mini truck and the plastic trays consumed higher costs which accounted for about 48 per cent to the total fixed cost incurred in the case farm. The buildings for the laborers and the office room expenses accounted for around 22 per cent of the total fixed cost. To arrive at the cost of production of mango per ton, the fixed costs incurred in the respective items were amortized based on the life expectancy of these assets and the annual fixed costs were worked out. The total annual fixed cost incurred per acre was arrived at Rs. 4294.27 and the same for per ton of mango was arrived at Rs. 297.72 and the variable costs incurred in producing mango is presented in Table 6.

Table 6 revealed the details of variable costs incurred in producing a ton of mango in the case farm under study and the post harvest expenditure incurred in transportation of mango from the production site to the distribution centres. The production cost or variable cost incurred were land preparation, basin formation, watering, spraying of chemicals, manuring, harvesting and watch and ward expenses.

Table 6 Cost of Production of Mango Cultivation

S. No	Details of Expenditure	Expenditure/ Acre in Rs.	Expenditure / Ton	Total Exp. In Lakhs
I	Annual Fixed Cost	4294.27	297.72	05.74
		Production Cost		
1	Basin formation once in a year @ Rs. 3.00 lakhs per year	2245.51	155.72	03.00

S. No	Details of Expenditure	Expenditure/ Acre in Rs.	Expenditure / Ton	Total Exp. In Lakhs
2	Clearing of bushes, pruning and maintenance @ Rs. 5 lakhs	3742.52	259.54	05.00
3	Electricity and watering charges @ Rs. 6 lakhs per annum	4491.02	311.44	06.00
4	Cost of spraying of plant protection chemicals and nutrients @ Rs. 5.80 lakhs	4341.32	301.06	05.80
5	Watch and ward and other miscellaneous expenditures @ Rs. 1.5 lakhs	1122.75	77.86	01.50
6	Cost of manuring @ Rs. 2 lakhs per annum	1497.00	103.81	02.00
7	Harvesting cost of mango @ Rs. 8 lakhs	5988.02	415.26	08.00
	Sub Total (II)	23428.14	1624.69	31.30
	III - Post Harvest Expenditure			
1	Loading & transportation cost @ Rs. 17.50 lakhs for 2242 tonnes	12500.00	780.55	17.50
2	Cost of 315 tonnes of mango procured from other sources	---	---	34.65
	Sub Total (III)	12500.00	780.55	52.15

S. No	Details of Expenditure	Expenditure/ Acre in Rs.	Expenditure / Ton	Total Exp. In Lakhs
	Grand Total (I+II+III)	40222.41	2702.96	89.19
	Gross income generated in the farm	02.77	---	365.90
	Net profit in lakhs	02.37	---	276.71

These expenses in toto are arrived at Rs. 1624.69 per ton of mango which is accounted for 60 per cent to the total cost of producing a ton of mango and the post harvest expenditures including transportation cost is arrived at Rs 780.55 per ton of mango which is accounted for 29 per cent to the total cost of producing a ton of mango. Put together, the cost of production per ton of mango is arrived at Rs. 2703/-. The annual fixed cost items accounted only for 11 per cent which is meagre in mango production. Hence, it is a handy information for those who are interested in establishing a mango orchard in a large scale.

MARKETING AND DISTRIBUTION OF FRESH MANGO

Production of such huge tonnage of mango is a possible means, the cash farm would have a planned way of distribution of mango from deterioration and hence an effort has been taken to analyze the marketing practices prevalent and the method of distribution of mango and the same are presented in the following sub-heads.

- Distribution of mango to different markets in India.
- Details of mango sold and revenue realized per annum.
- Price spread in distribution of mango.

Distribution Of Mango To Different Markets In India

The quantity of mango produced in the case farm is arrived at 1927 tonnes (Table 4). Besides this, the farmer is also procuring around 315 tonnes from nearby plantations and all the same is distributed to different consumer markets in India. The details are analyzed and presented in Table 7.

Table 7 Distribution of Mango to Different Markets in India

S. No	Name of the Market	Quantity Exported in Tonnes	Percentage of Quantity Exported	No. of Loads	Tonnes per Trip
1	Kerala	0650	28.99	130	05.00
2	Karnataka	1000	44.60	100	10.00
3	Maharashtra	0320	14.27	020	16.00
4	Jammu & Kashmir	0192	08.57	012	16.00
5	Jabalpur	0048	02.14	003	16.00
6	Nasik	0032	01.43	002	16.00
	Total/Mean	2242	100.00	267	11.50

Table 7 reveals that the quantity of mango exported is arrived at 2242 tonnes which are sent to different markets through 267 lorry loads with an average capacity of 11.50 tonnes per load. The potential markets for the mango are Kerala, Karnataka, Maharashtra, Jammu and Kashmir, Jabalpur and Nasik. Among the total tonnage of mango, 44.60 per cent were distribution to Karnataka markets followed by Kerala and Maharashtra markets which are respectively accounted for 28.99 per cent and 14.27 per cent.

Jammu and Kashmir is the state capable of absorbing 8.57 percent (192 tonnes) of mango farm the case farm of Tamil Nadu and the Jabalpur is able to receive 48 tonnes from the case farm which accounted for 2.14 percent of the total mango distributed to different markets in India.

Details Of Revenue Realized In The Farm

The details of mango sold and the revenue realized per annum are outlined in Table 8. It revealed that 2242 tonnes of mango are distributed at an average price per ton of Rs.17250 and 20.52 per cent (460 tonnes) of mango are distributed at the rate of Rs.15000 per ton followed by 410 tonnes. Through sale of mango at an average price of Rs.17250 per ton, the case farm could generate a gross revenue of Rs.365.90 lakhs per annum revealing that the farm is generating a revenue of Rupees One Lakh in a day which is an appreciable tax free income from horticulture.

Table 8 Details of Mango Sold and Revenue Realized per Annum

S. No	Quantity Disposed in Tonnes	Average Price/ Tonne	Total Revenue Earned in Lakhs
1	205	22000	45.10
2.	460	15000	69.00
3.	410	12000	49.20
4.	360	15000	54.00
5.	340	18000	61.20
6.	220	21000	46.20
7.	110	25000	27.50
8.	137	10000	13.70
Total/Mean	2242	17250	365.90

Such progressiveness should be identified by the Department of Horticulture and Plantation Crops of Government of Tamil Nadu and the farm should be awarded with cash prize for further motivation. Besides, such success stories should be videographed and presented before the small and marginal farmers of horticulture, so that it may motivation them for area expansion and productivity enhancement.

Price Spread In Distribution of Mango

Though marketing of mango is effected through different channels, the case farm at Periyakulam is practicing six important channels for delivery and distribution of mangoes. They are:

I. Producer à Wholesaler (Kerala) –> Wholesaler-cum-retailer –> Consumers.

II. Producer à Wholesaler (Madhya Pradesh) –> Wholesaler-cum-retailer (MP) à Consumers.

III. Producer à Wholesalers (Karnataka) –> Retailers à Consumers.

IV. Producer à Wholesalers (Jammu) àRetailers à Consumers.

V. Producer à Wholesalers of central market (Jabalpur) —> Retailers –> Consumers.

VI. Producer à Wholesalers of central market (Nasik) –> Retailers à Consumers.

In all the channels, the produce is sent to only the identified wholesaler, the consignment delivered at wholesaler's desk and the same are further distributed to the consumers. Such details are not known to the authors because those channels are not traceable. However, based on the telephonic conversation made by the case firm, the following two channels and their marketing cost could be identified and hence the price spread was worked out for the following channels, viz.,

I. Producer → Wholesales (Kerala) →Wholesaler-cum-retailer → Consumers.

II. Producer → Wholesaler (Madhya Pradesh) →Wholesaer-cum-retailer (MP) → Consumers.

Price spread is the difference between the producers' price and the consumers' price. It is also called as marketing margin which contains marketing cost and net margin to the intermediaries. In respect of channel I, the cost of marketing of mango incurred by the producer is arrived at Rs.7.30 which accounted for 12.17 per cent to the consumers' rupee and the marketing margin gained by the producer is arrived at Rs.5.55 which accounted for 09.25 per cent. Both marketing margin and producers' price per kg of mango arrived at Rs.22.80 which accounted for 38 per cent to the consumers' rupee paid for a kg of mango.

The wholesaler at Kerala (Kochin) central market has incurred Rs.10.10 per kg of mango which accounted for 16.83 per cent of consumers' rupee and he has gained the marketing margin of Rs.5.20 per kg which accounted for 8.67 percent of the consumers' rupee. The consumers' rupee in Kerala market arrived at Rs.60 per kg.

From the wholesaler of Kerala market, wholesales-cum-retailers of different markets started procuring the mango at a cost of Rs.45.40 and the same was sold to the consumers at Rs.60 per kg after incurring marketing cost of Rs.8.50 which accounted for 14.17 per cent of the consumers' rupee after gaining the marketing margin of Rs. 6.10 per kg of mango which accounted for 10.17 percent of the consumers rupee. It is understood that the wholesaler and the wholesaler-cum-retailers are incurring higher transportation cost followed by loading and unloading charges. The absolute value of marketing margin varies from commodity to commodity, channel to channel, market to market and time to time (Biradhar, 1996).

Table 9 Price Spread in Distribution of Mango through the Channel – I Producer → Wholesaler (Kerala) → Wholesaler-cum-Retailer → Consumer

Sl. No	Particulars of Cost	Amount (Rs/Kg)	Per Cent
01	Producer		
01	Producers' Price	17.25	28.75
1.1	Marketing Cost		
1.1.1	Grading and Packing Charges	01.20	
1.1.2	Loading Charges	02.30	
1.1.3	Transportation Cost	02.90	
1.1.4	Miscellaneous Expenses	00.90	
	Marketing Cost at Producers' Level	07.30	11.67
1.2	Marketing Margin	05.55	09.25
02	Wholesaler		
2.1	Procurement Price by the Wholesaler	30.10	50.17
2.2	Marketing Cost		
2.2.1	Unloading, Sorting and Reloading Charges	03.30	
2.2.2	Transportation Cost	04.20	
2.2.3	Weighing Charges	00.60	
2.2.4	Watch and Ward Charges	00.80	
2.2.5	Miscellaneous Charges	01.20	
	Marketing Cost at Wholesalers' Level	10.10	16.83
2.3	Marketing Margin to the Wholesaler	05.20	08.67
03	Wholesaler cum Retailer		
3.1	Purchase Price Incurred by the Wholesaler cum Retailer	45.40	75.67
3.2	Marketing Cost		
3.2.1	Loading and Unloading Charges	02.10	
3.2.2	Transportation Charges	04.10	
3.2.3	Watch and Ward Charges	00.70	

Sl. No	Particulars of Cost	Amount (Rs/Kg)	Per Cent
3.2.4	Miscellaneous Cost	01.60	
	Marketing Cost to the Wholesaler cum Retailer	08.50	14.17
3.3	Marketing Margin to the Wholesaler cum Retailer	06.10	10.17
3.4	Sale Price	60.00	100.00
04	Consumer		
4.1	Procurement Price by the Consumer	60.00	100.00

In respect of channel II, the producers' price is same as in channel I but the percentage share is reduced because of higher consumers' price per kg of mango. Here, the producer has incurred a marketing cost of Rs.9.60 which accounted for 13.71 percent of consumers' rupee and the marketing margin is arrived at Rs.5.90 which accounted for 8.43 per cent of consumers' rupee. Put together, the producer is able to receive Rs 23.15 per kg of mango which accounted for 33 per cent of the consumers' rupee which arrived at Rs.70 per kg of mango.

The wholesaler of Madhya Pradesh has received the produce at a cost of Rs.32.75 per kg and incurred marketing cost of Rs.13.50 at wholesalers' level which accounted for 19.29 per cent of consumers' rupee. Wholesaler of Madhya Pradesh has gained the margin of Rs.6.25 per kg which accounted for 8.93 per cent to the total consumers' rupee. Here also the transportation cost consumed was found to be high.

The wholesaler-cum-retailer of Madhya Pradesh markets procured the mango at the price of Rs.52.50 per kg and incurred the marketing cost of Rs.10.95 and marketing margin of Rs.6.55 respectively which accounted for 15.64 per cent and 9.36 per cent of consumers' rupee.

In these two channels, the producer is playing an active role in indentifying the wholesalers and sends the consignment directly and earns higher percentage share of consumers' rupee. If these progressive farmers together joined and promoted an interest group at higher level, the procurement and distribution under the experienced personnel of the case firm, higher percentage share of consumers' rupee could be generated by all the farmers of commodity interest group which permits collective bargaining.

Table 10 Price Spread in Distribution of Mango through the Channel – II Producer → Wholesaler (Madhya Pradesh) → Wholesaler-cum-Retailer (MP) → Consumer

Sl. No	Particulars of Cost	Amount (Rs/Kg)	Per Cent
01	Producer		
01	Producers' Price	17.25	24.63
1.1	Marketing Cost		
1.1.1	Grading and Packing Charges	01.22	
1.1.2	Loading Charges	02.40	
1.1.3	Transportation Cost	04.80	
1.1.4	Miscellaneous Expenses	01.18	
	Marketing Cost at Producers' Level	09.60	04.29
1.2	Marketing Margin	05.90	08.43
02	Wholesaler		
Sl. No	Particulars of Cost	Amount (Rs/Kg)	Per Cent
2.1	Procurement Price by the Wholesaler	32.75	46.79
2.2	Marketing Cost		
2.2.1	Unloading, Sorting and Reloading Charges	04.30	
2.2.2	Transportation Cost	05.60	
2.2.3	Weighing Charges	00.90	
2.2.4	Watch and Ward Charges	01.10	
2.2.5	Miscellaneous Charges	01.60	
	Marketing Cost at Wholesalers' Level	13.50	19.29
2.3	Marketing Margin to the Wholesaler	06.25	08.93
03	Wholesaler cum Retailer		
3.1	Purchase Price Incurred by the Wholesaler cum Retailer	52.50	75.00
3.2	Marketing Cost		
3.2.1	Loading and Unloading Charges	01.90	

Sl. No	Particulars of Cost	Amount (Rs/Kg)	Per Cent
3.2.2	Transportation Charges	06.35	
3.2.3	Watch and Ward Charges	01.25	
3.2.4	Miscellaneous Cost	01.45	
	Marketing Cost to the Wholesaler cum Retailer	10.95	15.64
3.3	Marketing Margin to the Wholesaler cum Retailer	06.55	09.36
3.4	Sale Price	70.00	100.00
04	Consumer		
4.1	Procurement Price by the Consumer	70.00	100.00

MARKETING EFFICIENCY

It refers to the market performance. The movement of goods from producer to the ultimate consumers at the lowest possible cost consistent with the provision of services desired by the consumer is termed as marketing efficiency or efficient marketing. To arrive at the marketing efficiency, as stated in the design of the study, the Shepherd's formula (Acharya and Agarwal) and the Calkin's Index were adopted and the results are presented in Table 11.

Table 11 Marketing Efficiency in Distribution of Mango

Marketing Channel	Value of Output (Consumers' Price)	Input (Marketing Cost)	Marketing Efficiency Ratio (O/I)
Shepherd Method			
I	60.00	25.90	01 .32
II	70.00	34.05	01.06
Kalkins Index			
Marketing Channel	Marketing Margin	Marketing Cost	Marketing Efficiency

Marketing Channel	Value of Output (Consumers' Price)	Input (Marketing Cost)	Marketing Efficiency Ratio (O/I)
I	16.45	25.90	01.64
II	18.70	34.05	01.55

Table 11 reveals that channel II is found to be the efficient channel which has yielded the lowest value in both the methods. Marketing efficiency arrived at using the Shepherd method is 1.06 in respect of channel II. But the Channel I had the score of 1.32 in which the marketing cost is lower. Kalkins Index revealed that channel II is efficient by indicating 1.55 as its score while the channel I had the score of 1.64. From this, one could infer that channel II is capable of fulfilling different intermediaries, resulting in the lowest score of marketing efficiency and ranked as an efficient channel.

CONSTRAINTS FACED BY THE CASE FARM

The main problems faced by the mango orchard owners are delineated as follows. They are:

- Delayed payment for the produce shifted from the production site.
- Damage of trees at the time of harvesting. It is not a serious problem. However, it will be regulated with canopy management.
- Inadequate storage facilities locally. This is a serious problem. If these facilities are installed, storage becomes easy.
- Non-availability of pre-cooling facilities.
- Poor infrastructure and transportation facilities. The road facilities need to be improved for easy transportation and access to the plantation site.

The traders who are making delayed payment are avoided by the producer in future supply. They are given instructions clearly that the time bound payment alone will ease the supply.

The town panchayat can improve the roads and provide storage and ripening chambers in Periyakulam with the support of the farmers and their contribution to make the industry sustainable.

CONCLUSIONS AND POLICY IMPLICATIONS

Agripreneurial practices are prevalent among the progressive farmer of Murugamalai village of Theni District actively engaged in hi-tech production and distribution of mango to different parts of the country from his 134 acres (53.44 ha) of horticultural initiatives and hence a case study approach is adopted to document the success story of the farmer. The farm has managerial, supervisory and execution level of employees to achieve its mandate of production and sale following the concept of span of management. Though the farm had the trees of different age classes, the group III age class accounted for 72 per cent, capable of producing 83 per cent of output of mango revealing that as age advances, the yield of mango also increased in the orchard. Among the twelve varieties established in the orchard, Neelam is the single variety that accounted for 68.39 per cent to the total area under mango cultivation. The total yield of mango obtained arrived at 1927 tonnes with an average productivity of 36 tonnes per ha which is 216 per cent higher productivity than the national average productivity of mango. Kerala and Karnataka alone have received 73 per cent of mango produced in the case farm and the rest goes to Maharashtra, Jammu and Kashmir, Jabalpur and Nasik and the farmer is able to gain 29 per cent share of consumers' rupee through channel I and 25 per cent share of consumers' rupee in Channel II. The farmer-cum-agripreneur stated that pre-cooling and storage facilities are almost nil in the zone which is a big impediment to the farmer and hence the following recommendations and its implementations will help the horticultural development in a big way.

RECOMMENDATIONS

Road infrastructure to the plantation site is very poor which commands immediate execution by the town panchayat of Periyakulam to ease transportation of horticultural produce. Store house to store the largely produced horticultural output becomes an impediment to the producers and hence pre-cooling and appropriate storage facilities will serve the immediate needs of producers of horticultural produce.

References

- **Biswas, B.C.** and Lalitkumar (2011). "Revolution in Mango Production – Success Stories of Some Farmers" ***Fertilizer Marketing News***, March 2011. Pp. 1-23
- Biradhar, R.D., (1996). "Marketing Costs, Margins and Price Spread of Selected Agricultural Commodities in Kolhapur District, Maharashtra", Agricultural Marketing 38(4):Pp. 21-29
- Karthick, V. , K. Mani and A. Anbarasan (2013). "Mango Pulp Processing Industry in Tamil Nadu – An Economic Analysis" ***American International Journal of Research in Humanities, Arts and Social Sciences***, 2(1): March- May 2013, Pp. 48-52
- Sekhar, C., R.C. Panda, R.S.Vinaya Rai and P. Devaraj (1998). "Marketing of Forest Produce from Farm Land – A Case Study in Palani, Tamil Nadu, India", Journal of Tropical Forest Products, 4(1): Pp. 68-73
- Singh S.K. (2010). "More Resource Allocation for Research Necessary" ***The Hindu Survey of Indian Agriculture 2010***, pp. 40-43
- Subodh Varma (2013). " How Mangoes Became 'AAM' " ***Times of India, August 25, 2013***. P. 8
- Viswanatha Reddy K and Pramod Kumar (2010). "An Economic Analysis of Mango Processing Plants of Chittoor District of Andhra Pradesh, ***Indian Journal of Agricultural Economics***, 65(2): 2010, pp. 277-297

Chapter 13

Commercial Nursery as an Agribusiness Venture

INTRODUCTION

Nursery is defined as a place where plants are grown, nurtured and sold out. Generally, various commercial crop growers require good quality saplings or grafts of genuine type (NIOS, 2013; Web Reference 2). Tree nurseries are a key success factor in many forestry, horticulture and agricultural development interventions. The purpose of the nurseries has varied from commercial biomass production, to land rehabilitation and forest conservation, to local capacity building and livelihood enhancement. Partners involved with operating those nurseries have included farmers, entrepreneurs, commercial firms, non-government organizations (NGOs), communities, projects, and government agencies. Exceptional individual nurseries are known to produce over 10,00,000 seedlings (Roshetko et.al. 2010).

A tree nursery is a managed site, designed to produce tree seedlings grown under favorable conditions until they are ready for planting. It can be an informal, small-scale arrangement or a large commercial enterprise. Nurseries vary in size, facilities (supplies, tools, equipment, etc.), types of seedlings produced, and operations. They also differ significantly in quality and quantity of planting stocks produced. However, all nurseries primarily aim to produce sufficient quantities of high quality seedlings to satisfy the needs of seedling users. Users include the nursery operator themselves, individuals, community organizations, farmer groups, government agencies, non-government organizations, corporate or private customers (Web Reference 1).

Nurseries may often provide income generating opportunities for the operators and enhance the social capital, technical capacity and leadership skills of communities. Despite the great variety, there are four major nursery types:

- Institutional Nurseries,
- Project Nurseries,
- Group Nurseries, and
- Individual Nurseries.

Institutional nurseries are further comprised of three sub-categories: research nurseries, government nurseries, and industrial nurseries (Roshetko et.al 2010). Let us focus on the individual nurseries for better understanding of young agripreneur.

INDIVIDUAL COMMERCIAL NURSERIES

Individual commercial nurseries are also known as household nurseries or backyard nurseries and are generally established and managed by individual farmers and/or their families. Although of limited size, individual nurseries often produce a broad variety of species. Individual nurseries usually employ simple and often substandard technologies as operators generally have meagre resources. Families almost exclusively manage individual nurseries, with all members contributing. However, nursery establishment and management are constrained by a lack of family labor and resources. Many large commercial oriented individual nurseries employ workers to help with nursery operations. Operators of individual nurseries can be effective farmer-to-farmer extension agents (farmer specialists). Their nurseries and farms may serve as examples for small-scale farmers. Commercial-oriented individual nurseries evolve as self-funded enterprise through seedling sales and the provision of technical services.

The nursery enterprise is an important source of employment and income generation. It is important for the entrepreneur to make an estimate of expenditure and probable income during the operation of the nursery. Similarly, it is also beneficial for the entrepreneur to make a nursery production plan of work, manpower and raw material requirement for successful operation of the nursery. The anticipation of income is also an area which requires planning for proper financial appraisal in the nursery business. Proper financial planning of the income, expenditure statements is an essential factor in the success of the nursery business. Entrepreneurs can use business analysis as a guide for important decisions such as business expansions, financing, marketing strategies, operations planning, and product selection. When applied properly, this information can increase nursery business profitability, control costs, reduce the risk of business failure, boost employee productivity and job satisfaction, enhance physical efficiency, and improve management professionalism. It can also support the evaluation of costs and returns for individual species to determine a more profitable species/product mixture (Web Reference 6).

The advantages of individual nurseries are: independent decision-making in all aspects of nursery management; income generation from seedling sales; produce technical innovations that are appropriate for small-scale operating conditions; and appropriate venue for farmer-to-farmer training and extension. The limitations of individual nurseries are: seedling quality may be poor due to operators' limited resources and knowledge; poor access to assistance from research, technical and development agencies; and working alone results in limited access to new ideas. The above facts are true to many of the nurseries operated by individual farmers. But the farmer-turned-agripreneur has very good technical background in seed processing; employing labor force in a big way; procuring seeds from different sources; producing quality and unmatchable seedlings and distributing to different parts of the country and above all earning profits for his quality services in this sector motivated the researcher to document his experiences and the incentives associated with the commercial individual nursery so that the young agripreneur minds will explore this as one of the agribusiness venture as an alternative to employment seeking behavior. Hence the above ideas resulted in the making of this chapter.

DESIGN OF THE STUDY

The Krishi Vigyan Kendra (KVK), an organ of Gandhigram Rural Institute - Deemed University, Gandhigram, Dindigul, Tamil Nadu is one of the dynamic institution playing a vital role in dissemination of technologies to the poor farmers, empowering them through Transfer of Technologies and building confidence in their mind to establish agribusiness ventures. In and around the Gandhigram Rural Institute, there are more than 100 commercial forest nurseries established in an area ranging from one acre to more than 15 acres. Arjun nursery, one of the successful nursery established at Narasingapuram, near Gandhigram is a specialized one in sandal nursery, in addition it also produces seedlings of other species in larger quantities and distributes them to different parts of India. This nursery was purposively chosen for a case study approach.

The owner of the nursery is Mr. Murugan of Narasingapuram. He had established 6 nurseries being maintained under sylvan environs of Gandhigram foothills in an area of around 16 acres concentrating mainly on sandal nursery which is of heavy demand in the Forest Departments of different states of India and he is supplying regularly in different size poly bag containers. The seedlings are hale and healthy with attractive lush green color which too attracted the researcher to go deep into the various aspects of nursery establishment like area and production of seedlings, varieties of seedlings, quantity produced and distributed to different consuming markets, price per unit of seedlings and the cost of production of seedlings. Analyzing this information will give an idea to the young minds who wanted to become entrepreneurs with little investment. The study used the descriptive statistics alone as tools of analysis using the mean and percentage analysis and the information delivered in this chapter will be of immense use to the emerging agripreneur to take this as an agribusiness venture to empower themselves.

Commercial forest nursery as an agribusiness venture commanding simple technology usage, little investment at the time of establishment and little experience in the field of running the nursery to infuse confidence in establishing the nursery on their own are the pre-requisites in establishment of nursery. Being an ancestral business venture, the owner manager of the case nursery has gained much expertise in developing the healthy seedlings in a hygienic sylvan environment commanding interest to explore the study and hence a detailed analysis was made and the results are presented in the following sub-heads:

1. Land Use Pattern in the Commercial Nursery
2. Area and Production of Seedlings in the Commercial Nursery
3. Distribution of Seedlings to Different Markets
4. Quantity of Seed Purchased from Different Sources

5. Cost of Production of Seedlings
6. Quantity Sold and Revenue Realized from Nursery

Land Use Pattern in the Commercial Nursery

The agripreneur or the owner manager of the commercial nursery is producing the seedlings of commercial tree species and horticultural importance to meet the greenery requirement of the nation. How effectively and efficiently he is allocating the land resource between the competing species (product) is very important and hence the details are analyzed and the results are presented in Table 1.

Table 1 Land Use Pattern in the Commercial Nursery

Sl. No	Type of Land Use	Area in Acres	Percentage to Total
01	Forest Nursery	10.60	66.24
02	Horticulture Nursery	02.75	17.19
03	Naked Seedlings	00.75	04.69
04	Roads	01.20	07.50
05	Buildings and Dump Yards	00.70	04.38
	Total	16.00	100.00

Table 1 reveals that the owner manager of the commercial nursery is blessed with 16 acres (06.40 ha) in the locale of Narasingapuram Village of Dindigul District with an allocation of 66 per cent of the land under commercial forest nursery and 17 per cent of the land under the nursery of horticultural importance. These two nurseries had occupied 83 per cent of the total land area. In addition to these, another 5 per cent of the land area is allotted for bulk nursery where teak is raised and the stumps are distributed to the consuming ends. The nursery also had allotted area for road facility and dump yard for transportation of raw materials required for the establishment of nursery which respectively accounted for 7.50 per cent and 4.38 per cent to the total land area owned by the agripreneur. From this, one could understand that the allocation of land is planned and prioritized among different competing ends for easing the business revealing that the owner manager has clear vision to take the nursery in a big way.

Area and Production of Seedlings in the Commercial Nursery

The components of commercial nursery included are tree species of forestry importance and the horticultural importance. The area under different species of tree seedlings and the quantity produced are analyzed and the results are presented in Table 2.

Table 2 Area and Production of Seedlings in the Commercial Nursery

Sl. No	Name of the Nursery	Area Under Nursery (Acres)	Quantity of Seedlings Produced	Percentage to Total
		Forest Nursery		
01	Sandal	04.50	1200000	52.86
02	Neem	00.75	60000	02.64
03	Teak	00.30	40000	01.77
04	Gmelina	00.20	30000	01.32
05	Pungam	00.50	60000	02.64
06	Basia	00.50	50000	02.20
07	Casuarina	00.20	30000	01.32
08	Silver Oak	00.30	40000	01.77
09	Mantharai	00.25	35000	01.54
10	Melia	00.25	35000	01.54
11	Ailanthus	01.30	140000	06.17
12	T.Bellerica	00.20	30000	01.32
13	T.Kadappa	00.25	35000	01.54
14	Thespesia	00.30	40000	01.77
15	Raintree	00.25	35000	01.54
16	Gulmohar	00.30	40000	01.77
17	Red Sanders	00.25	35000	01.54
Total Forest Nursery		**10.60**	**1935000**	**85.25**
		B. Horticulture Nursery		
18	Chenbagam	00.25	35000	01.54

Sl. No	Name of the Nursery	Area Under Nursery (Acres)	Quantity of Seedlings Produced	Percentage to Total
19	Amla	00.20	30000	01.32
20	Tamarind	00.20	30000	01.32
21	Pomegranate	01.30	140000	06.17
22	Citrus	00.50	60000	02.64
23	Naval	00.30	40000	01.76
	Total Horticulture Nursery	02.75	435000	14.75
Grand Total (A+B)		13.35	2270000	100.00

Table 2 reveals that the forestry species of commercial importance numbering 17 are established in an area of 10.60 acres (4.24 ha). The total quantity of forestry seedlings produced in the nursery is arrived at 19.35 lakhs in which the sandal seedlings alone arrived at 12 lakhs which accounted for 52.86 per cent to the total forestry seedlings produced in the nursery. The next important seedlings produced in the highest order is Ailanthus which arrived at 1.40 lakhs that accounted for 6.17 per cent followed by Pungam and Basia (Iluppai) which accounted for 2.64 per cent and 2.20 per cent respectively to the total. Total quantity of forestry seedlings produced accounted for 85 per cent to the total seedlings produced in the case nursery.

With respect to horticulture seedlings, the major items produced are Chenbagam, Amla, Tamarind, Pomegranate, Citrus and Naval which put together accounted for 15 per cent to the total seedlings produced in the commercial nursery. Among the seedlings of horticultural importance, pomegranate seedlings alone arrived at 1.40 lakhs which accounted for 6 per cent to the total seedlings produced in the case nursery.

To sum up, Table 2 reveals that the case nursery had produced 22.70 lakh seedlings in which 85 per cent of the seedlings belonged to forestry importance and the 15 per cent of the seedlings produced were of horticultural importance. According to Roshetko et.al., 2010, the individual commercial nurseries could produce 10 lakh seedlings on an exceptional case. But in the case nursery, one single species (sandal) alone was produced to the tune of 12 lakhs revealing that Tamil Nadu is the pioneer and the leader in producing seedlings and establishment of individual commercial nurseries. The only one dominant species produced and sold in the nursery is sandal seedlings which accounted for 52.86 per cent to the total seedlings produced in the nursery outlines the fact that potential scope exists for the sandal seedlings which is of high demand in the Forest Departments of different states. The details are analyzed and presented in the ensuing sections.

Distribution of Seedlings to Different Markets

The seedlings which are produced in the commercial nursery had good markets in different states of South India. The details of the same are analyzed and presented in Table 3.

Table 3 Distribution of Sandal and Other Seedlings to Different Markets

Sl. No	Name of the Markets	Quantity Sold	
		Sandal	Other Seedlings
01	Andhra Pradesh	500000 (41.67)	450000 (42.06)
02	Karnataka	400000 (33.33)	250000 (23.36)
03	Kerala	250000 (20.83)	150000 (14.02)
04	Tamil Nadu	50000 (04.17)	220000 (20.56)
	Total	1200000 (100.00)	1070000 (100.00)

(Figures in Parentheses Indicate Percentage to Total)

Among the potential markets, Andhra Pradesh was found to be the leader in receiving large amount of seedlings from the case nursery of Tamil Nadu. Andhra Pradesh received sandal and other forest seedlings at the rate of 5 lakhs and 4.5 lakhs per annum which respectively accounted for 42 per cent each. The main recipient of seedlings in Andhra Pradesh is found to be the Forest Department for afforestation related activities followed by Karnataka which had received 4 lakhs seedlings of sandal and 2.50 lakhs of other forest seedlings which respectively accounted for 33 per cent and 23 per cent to the total seedlings distributed from the commercial case nursery. Kerala state had received sandal and other forest seedlings from the case nursery to the tune of 2.50 lakhs and 1.50 lakhs which respectively accounted for 21 per cent and 14 per cent to the total seedlings distributed. Tamil Nadu was found to receive the sandal seedlings very poorly, this is because sandal is the product of their own forest nursery. However, they are receiving around 2.20 lakh seedlings of other forest species from the case nursery which accounted for 21 per cent to the total seedlings distributed from the case nursery. Other forest seedlings received by Tamil Nadu were found to be on the higher side due to the higher quantity demanded for the birth day event of the Honorable Chief Minister of Tamil Nadu and the mass planting was taken up on that day.

To sum up, Andhra Pradesh was found to be the good market for the sandal seedlings followed by Karnataka and Kerala. Tamil Nadu was found to be the poor taker of sandal seedlings from private nurseries as they have their own interest in establishing sandal nursery in select pockets like Mettupalayam of Coimbatore District and Sathyamangalam of Erode District. In toto, sandal seedlings are produced and distributed to the tune of 12 lakhs and other forest seedlings to the tune of 10.70 lakhs from the case nursery of Tamil Nadu. To promote further, certification of forest nursery can be made. A certificate signifies that the nursery has met all requirements of the Plant Act (Smith 2011) if any and the stock is apparently free of injurious pests and plant diseases. Issuance of a certificate is based upon periodic inspection of nursery stock by an official with the Department of Certification. To produce such a huge stock, the basic raw material of seed availability and its mobilization becomes more important which are also analyzed and presented in Table 4.

Quantity of Seed Purchased from Different Sources

The owner manager of commercial nursery is procuring seed from different sources for the production of seedlings. The details of procurement of seeds are analyzed and are presented in Table 4.

Table 4 Quantity of Seed Purchased from Different Sources

Sl. No	Name of the Seed Procured	Place of Purchase	Quantity Purchased in Tonnes / Numbers	Percentage to Total
01	Sandal Seed	Forest Department – Bangalore	01.600	16.33

Sl. No	Name of the Seed Procured	Place of Purchase	Quantity Purchased in Tonnes / Numbers	Percentage to Total
02	Melia Seed	Forest Department – Bangalore	05.000	51.02
03	Melia – Wildlings (Numbers)	Anthiyur – Tamil Nadu	20000	00.00
04	Teak Seeds	Forest Department – Coimbatore	00.200	02.04
05	Other Seeds	Local Collections	03.000	30.61
Total			09.800	100.00

The Forest Department of Bangalore city, Karnataka is found to be the major supplier of seeds of both sandal and Melia to the commercial nursery which respectively accounted for 16.33 per cent and 51 per cent of the total supply of seeds. Bulk supply of Melia is obtained from the Bangalore city which arrived at 5 tonnes per annum as it has poor germination. Teak seeds are purchased to the tune of 1.60 tonnes from the Forest Department of Bangalore city. Other forest seeds are obtained from the local forest areas and from the Forest Department of Government of Tamil Nadu which arrived at 3 tonnes that accounted for 30.61 per cent to the total receipt of seeds from different sources. Since Melia is found to have poor germination percentage, the wildlings are collected from Anthiyur forest areas and are immediately planted in poly bags for ensuring good survival and then they are transferred to the main nursery for establishment.

Cost of Production of Seedlings

In respect of commercial nursery and its successful running, they always attract the young Agripreneurs/Entrepreneurs in learning the aspects of investment and its associated benefits. On learning the intensity of cost involvement and the returns realized, a confidence will kindle the young entrepreneurs to think about similar investment. To ease their thinking, this study has focused on furnishing the details of costs involved in establishment of nursery.

The costs involved in the production of nursery seedlings are categorized into fixed and variable costs. Under the category of fixed costs, the rental value of land, erection of bore wells, duties and taxes and the miscellaneous costs incurred towards fixed investment are furnished in Table 5. Rental value of land alone accounted for 21 per cent to the total cost of establishment of nursery which arrived at Rs 15 lakhs. Erection of 5 number of bore wells has consumed an amount of Rs 3.75 lakhs which accounted for 5.25 per cent to the total cost. These two major items of fixed cost shared around 26 per cent of the total cost incurred in the establishment and production of nursery seedlings.

Table 5 also reveals the details of variable expenses incurred in the production activities of important and different species of forestry and horticultural seedlings. Among the variable expenses, wages paid to the employment of 52 laborers in the commercial nursery arrived at Rs. 39 lakhs which accounted for nearly 55 per cent of the total cost of establishment and production of nursery seedlings. The Chattanooga region (Tennessee State, USA) employed only about 154 persons in the nursery/ greenhouse industry. An additional 49 persons were employed through indirect and induced effects (Jamey Menard, 2000) of nursery. The indirect aspects of employment are not quantified and discussed in the study. The next important variable expenses are cost of poly bags which arrived at 7.35 lakhs which accounted for 10 per cent to the total cost of producing the nursery seedlings.

The total variable expenses arrived at Rs 51.53 lakhs and the total fixed costs arrived at Rs.19.90 lakhs, both of which accounted for 72 per cent and 28 per cent respectively to the total cost of producing the nursery seedlings which arrived at 71.43 lakhs.

With the investment of 71.43 lakhs per annum, the owner manager is producing 22.70 lakh seedlings and the cost of production of one seedling is arrived at Rs. 3.15 and the output–input ratio in the production of seedlings from commercial nursery is arrived at 3.29 revealing that for a rupee of investment in the commercial nursery, the owner manager could reap the gross return of Rs. 3.29 and a net return of Rs. 2.29 revealing that the commercial nursery is highly profitable, if it is manned with proper technology, timely distribution to the potential market at an appropriate time and by identifying the demand driven months for the nursery. All the above said measures will bring laurel to the owner manager of the commercial nursery.

Table 5 Cost of Production of Seedlings in the Commercial Nursery

Sl. No	Details of Cost	Amount Incurred (Rs)	Percentage to Total Cost
	Annual Fixed Cost		
01	Rental Value of Owned Land @ Rs 100000 per acre per annum	1500000	21.00
02	Erection of Bore Wells @ Rs 75000 per Well	375000	05.25
03	Duties and Taxes	15000	00.21
04	Miscellaneous Items including Cess	100000	01.40
	Total Annual Fixed Cost (A)	1990000	27.86

Sl. No	Details of Cost	Amount Incurred (Rs)	Percentage to Total Cost
	Variable Expenses in Commercial Nursery		
01	Cost of Sandal Seed	960	00.01
02	Cost of Wildlings of Melia	20000	00.28
03	Cost of Forest Seeds	1200	00.02
04	Cost of Processing of Seed (Seed Treatment)	10000	00.14
05	Cost of Horticulture Seeds / Crafts @Re 1 per craft /Sucker	295000	04.13

Sl. No	Details of Cost	Amount Incurred (Rs)	Percentage to Total Cost
06	Cost of Teak Seeds @ Rs 80 per kg	16000	00.22
07	Cost of Poly Bags @ Rs. 175 per Kg	735000	10.29
08	Cost of Red Earth @ Rs. 1000 per Load	20000	00.28
09	Cost of Coir Pith Waste @ Rs 750 per Load	15000	00.21
10	Wages Paid @ Rs 250 per Labor per Day for 300 women day per annum for 52 Women for filling of bags, planting, processing, watering, shifting and loading etc	3900000	54.55
11	Electricity Charges per annum @ Rs 5000 per Well	15000	00.21
12	Miscellaneous Charges	125000	01.75
	Total Variable Cost (B)	5153160	72.14
	Total Cost (A+B)	7143160	100.00
Details of Output Produced			
01	Quantity of Seedlings Produced	2270000	
02	Gross Return Generated on Sale of Seedlings of 22.70 Lakh numbers	23545000	
03	Cost of Production per Seedlings	03.15	

Sl. No	Details of Cost	Amount Incurred (Rs)	Percentage to Total Cost
04	Net Return Generated per annum	16401840	
05	Output – Input Ratio	03.29	

Quantity Sold and Revenue Realized from the Nursery

Quantity sold and revenue realization is the backbone to the survival and sustainable running of commercial nursery. Identifying the monsoon times and keeping the plant ready for sale to the user groups, maintaining the disease-free healthy seedlings are the important factors that decide the offering of price by the user agencies. Since the owner manager of commercial nursery was maintaining very healthy seedlings, he could reap higher benefit of returns from the commercial nursery and the details are analyzed and presented in the Table 6.

Table 6 Quantity Sold and Revenue Realized From Commercial Nursery

Sl. No	Name of the Seedlings	Price Per Seedling (INR)	Cost of Production (INR)	Quantity Sold in Numbers	Gross Return (INR)
01	Sandal Seedlings	15.00	03.15	1200000	18000000
02	Ailanthus	07.00	03.15	140000	980000
03	Other Forest Seedlings	04.00	03.15	595000	2380000
04	Horticulture Seedlings	07.00	04.15	295000	2065000
05	Teak Stumps	03.00	00.82	40000	120000
		Total		2270000	23545000

Table 6 reveals that the production of higher volume of sandal seedlings and distribution of the same to the Forest Department attracted a gross return of Rs 180 lakhs on sale of 12 lakh sandal seedlings at the rate of Rs 15 per saplings.

The cost of production of saplings on an average consumed only Rs.3.15 per plant yielding a net return of Rs.11.85 per sapling. On sale of Ailanthus seedlings, an amount of Rs 9.80 lakhs could be realized as gross return and the sale price was arrived at Rs 7 per plant yielding 3.85 as net return per plant.

The sale of other forest seedlings yielded an amount of 23.80 lakhs as gross return by sale of 5.95 lakh seedlings at the rate of Rs 4 per plant and the horticultural seedlings had yielded a gross revenue of Rs. 20.65 lakhs on sale of 2.95 lakh seedlings at the rate of Rs. 7 per plant. The gross revenue earned from the commercial nursery was arrived at Rs.235.45 lakhs per annum from the sale of 22.70 lakh seedlings and the net return from the commercial nursery per annum was arrived at Rs 164 lakhs yielding an output–input ratio of 3.29 revealing that for a rupee of investment in the commercial nursery, a farmer could reap Rs 2.29 as net return. However, the farmer-cum-agripreneur is facing the problem of investment initially on learning the demand for seedlings. Similar problem experienced in New Jersey nursery producers too, are production costs, market demand, and their own managerial expertise (Web Reference 5). The owner manager of case nursery had rich expertise in managing the nursery as it is an inherited family business.

On seeing the cost and return statements presented in the Tables 5 and 6, it is clearly understood that the commercial nursery is capable of generating higher profits per annum. One should learn that the lessons of time bound investment, appropriate technology usage in seed processing and ensuring survival percentage, applying plant protection on the appropriate hours, keeping liaison with the user agency regarding disposal of the nursery seedlings and entering into a pre-agreed contract with user group will ensure higher profit and impart loyalty to the producer and user groups. On ensuring this, the business will become sustainable and at the same time, one should keep vigil on the threats from competitors too. One should know the strategies of managing the competitors too when they compete for your output and take away your profits.

An important element in the success of a nursery enterprise is to consistently produce and market high quality products. The market for nursery stock is expected to increase in size due to the following factors:

- Home owners with higher income levels and greater disposable incomes
- Greater numbers of hobby gardeners

- Greater public interest in gardening and landscaping
- Increased environmental awareness and the knowledge that trees contribute to a clean environment
- New markets such as industrial landscaping.
- An increasing awareness of the therapeutic value of gardening (Government of Alberta, 2004) (Web Reference 4). Though these factors influence much in developed nations, Indian consumers are not resorting too much due to the movement of greenery is with Government and the real estate practitioners are resorting very little to these.

CONCLUSION AND POLICY IMPLICATIONS

The commercial case nursery had produced 22.70 lakh seedlings in which 85 per cent of the seedlings belonged to forestry importance and the 15 per cent of the seedlings produced were of horticultural importance. The only one dominant species produced and sold in the nursery is sandal seedlings which accounted for 52.86 per cent to the total seedlings produced in the nursery outlines the potential that exists for the sandal seedlings that is of high demand in the Forest Departments of different states. Andhra Pradesh is found to be the good market for the sandal seedlings followed by Karnataka and Kerala. Tamil Nadu is found to be the poor taker of sandal seedlings from private nurseries as they have their own interest in establishing sandal nursery in select pockets like Mettupalayam of Coimbatore District and Sathyamangalam of Erode District. In toto, sandal seedlings are produced and distributed to the tune of 12 lakhs and other forest seedlings to the tune of 10.70 lakhs from the case nursery of Tamil Nadu. The Forest Department of Bangalore city, Karnataka is found to be the major supplier of seeds of both sandal and Melia to the commercial nursery. With the investment of Rs. 71.43 lakhs per annum (US $ 115210), the owner manager is producing 22.70 lakh seedlings (2.27 million) and the cost of production of one seedling is arrived at Rs. 3.15 and the output-input ratio from the enterprising commercial nursery is arrived at 3.29 revealing that for a rupee of investment in the commercial nursery, the owner manager could reap the gross return of Rs. 3.29 and a net return of Rs. 2.29 revealing that the commercial nursery is highly profitable and hence the young agripreneur can enter into similar investment on a small scale.

RECOMMENDATIONS

- Since the owner manager is producing good quality seedlings, particularly in sandal, called for Certification of Nursery by approaching the Department of Certification, Government of Tamil Nadu. Certification of Forest Nursery will aid in getting a Brand Name for the nursery.
- The owner manager of commercial nursery can explore the possibilities of getting a government approval for production of seedlings and distributing the same for user agencies in a big way through institutional tie-up or linkage to earn a brand name of "Government Approved Nursery" so that more demand could be brought in.
- The educational institutions who taught commercial nursery as a course in their curriculum should make use of the rich expertise available with the owner manager of Arjun Nursery. So that much agripreneur development could be visualized and achieved.

References

- **James M. Roshetko**, Enrique L. Tolentino, Jr., Wilfredo M. Carandang, Manuel Bertomeu, Alexander Tabbada, Gerhard E.S. Manurung and Calixto E. Yao (2010). "***Tree Nursery Sourcebook - Options in Support of Sustainable Development***" (World Agroforestry Centre: Bogor, Indonesia) pp. 01-52
- Smith C.H. and Campbell, J.G. (2011). "***Directory of Mississippi Certified Nurseries and Nursery Dealers***" (Mississippi Department of Agriculture and Commerce: Mississippi) 2011-12. P.1
- Jamey Menard, Kim Jensen, and Burton English (2000). "***Economic Impacts of Construction on the Nursery/Greenhouse and Landscape Services Sectors in Tennessee***" (Institute of Agriculture, The University of Tennessee: Knoxville) p.6
- Government of Alberta (2004). " ***Commercial Tree Nursery Industry***" (Agriculture and Rural Development: Alberta)

Web References

1. http://www.worldagroforestry.org/downloads/publications/PDFS/MN16651.PDF
2. http://oer.nios.ac.in/wiki/index.php/Nursery_and_Nursery_Management 2013
3. http://www.mdac.state.ms.us/publications_and_forms/publications/pdf/bpi_nurserybook.pdf
4. http://www1.agric.gov.ab.ca/$department/deptdocs.nsf/all/agdex1370
5. http://aesop.rutgers.edu/~farmmgmt/ghpubs/usda_speech.pdf
6. http://agropedialabs.iitk.ac.in/agrilore/sites/default/files/HNM-book.pdf

Chapter 14

Sheep Production as A Business

INTRODUCTION

In India, about 64 per cent of the population is engaged in agriculture and rearing livestock, a subsidiary occupation to agriculture. The livestock plays an important role, specifically in the rural economy. It provides a stable and well-distributed income throughout the year. Sheep husbandry is an important part of the people of arid, semi-arid and hilly region of the country. The resource poor people in the lower strata of society rear the sheep. It is the most important species of livestock for utilization of available sparse vegetation and owing to their multifaceted utility (wool, meat, skin, milk and manure) constitute an important part of rural economy. The total population of sheep in India was 61.5 million that was 6.0 per cent of the world's sheep (1024.0 million) population. India ranks fourth in global sheep production (Suresh et.al. 2008).

India possesses one of the largest livestock populations in the world. In 2004, the country had 61 million sheep, while Tamil Nadu had 5.59 million sheep. The sheep population in the state had always registered a positive growth rate except during 1989–94 and 1994–97 periods. In 2006-07, sheep farming contributed to 45 million kg of wool and 2.3 million tonnes of meat (Economic Survey, 2007-08). The meat sector contributed about 73 per cent of total livestock exports from India during the year 2007. The huge expected increase in the demand for meat in developing countries especially in South-East Asia in the next 20 years presents an excellent opportunity for enhancing export of live goats and chevon from India (Selvakumar et.al. 2012). Sheep farming provides regular employment and income to the rural farmers and is little influenced by the vagaries of nature. Hence, it becomes imperative to study the economics of sheep production from time to time to exploit this growing sector more efficiently (Prabhu et.al. 2009).

THE ADVANTAGES OF SHEEP FARMING

Sheep can use practically all types of available forage including crop residues and even the grass available in the polluted canals and banks. Abundance of forage is key to sheep production. The successful producer also must have a genuine interest in sheep farming, managing the sheep farm efficiently, and the labor to care for the sheep health and hygiene. The main advantages of sheep production are:

- Sheep are easy to handle and generally require little input
- Sheep production does not require elaborate facilities and equipments
- Sheep consumes the available roughage as their primary feed
- Sheep help control weeds in the farm
- Sheep provide cash income from wool, mutton and pellets
- Sheep requires minimal amount of supplemental feeding
- Sheep feeds on roughages and gains weight and provides quick return on investment

DISADVANTAGES OF SHEEP PRODUCTION

- Sheep farming commands utmost care during severe drought and severe rain
- Sheep are subject to predation by panther, tiger, lion, bears, wild and domestic dogs
- Sheep requires better fencing and care
- Parasitic attacks can create health problems when sheep grazed on irrigated pastures and the pastures raised using polluted waters
- Consumer preference towards sheep meat is low when compared to the goat meat

Although the economic contribution of sheep seems to be quite substantial in the agricultural economy as well as in the national economy, the sheep farmers are yet ignorant about scientific management practices. If their breeding, feeding and management aspects are effectively supervised, the country can become one of the leading wool producing nations of the world, therefore, improvement in existing sheep population is essential. Considering the vitality of above stated facts, the present case study was attempted to carry out with the specific objective of determining the management practices followed in sheep production for making it as a successful commercial agribusiness venture.

VISIT TO A SUCCESSFUL CASE FARM

Periyakulam Taluk of Theni District in the foothills of Kodaikanal has not only attracted the farmers for its rich vegetation, good climate and normal rainfall but also has motivated them to practice cultivation of fruits and vegetables under higher area. Because of rich vegetation, it has also attracted the owners of sheep herd from Ramanathapuram District which is mainly a drought hit zone unable to do the grazing of sheep and hence many of the owners of sheep herd had migrated to the Kodaikanal foothills and are guiding the sheep flocks for grazing. Many of the studies and feasibility reports advocated sheep rearing in pens by providing supplemental feeding with feed concentrate mix for gaining weight in a short span of time. Mathis and Tom Ross (2005) in their report also advised the producers' of sheep in knowing the nutritional requirements of animals during different phases of production, viz., nutrient requirements during reproduction, gestation, lactation, etc. But the farmers who are guiding the sheep flocks for grazing in common lands and in the forest floor are maintaining the sheep flock in good health and commanding higher revenue on sale of major and minor products realized from sheep farming without providing any additional feed concentrate. The owner-manager of the case farm, Mr.Sakthivel is guiding the sheep flocks and rearing successfully in a vision based approach which has motivated the author to take the students for a learning purpose. This chapter is the outcome of taking the students of third year B.Sc. (Horticulture) of Horticultural College and Research Institute, Tamil Nadu Agricultural University, Periyakulam for the course entitled "Production Economics and Farm Management". One would be appreciated on learning the treasure of outcome of this publication.

GUIDANCE FOR A SUCCESSFUL AGRIBUSINESS VENTURE

Sheep has got special importance among all other livestock species due to their multifaceted utility. India possesses one of the largest livestock populations in the world. Since sheep farming provides regular employment and income to the rural farmers and is little influenced by the vagaries of nature, it becomes imperative to study the economics of sheep production from time to time to exploit this growing sector more efficiently. Being the course teacher, it is the duty of the author to expose the students to the successful agribusiness venture which are not exposed to them in the right sense and hence a detailed analysis of sheep farming is made as a case study and the findings are presented in the following different subheads to ease the students, academicians, planners, bankers and the researchers as a guidance tool in entering into this small agribusiness venture.

1. General Characteristics of the Sheep Farm
2. Details of Animal Population in the Case Farm
3. Details of Grazing Land Available to the Case Farm
4. Costs and Returns in Sheep Rearing Venture
5. Constraints Faced by the Agribusiness Firm

GENERAL CHARACTERISTICS OF THE SHEEP FARM

The single enterprise formed the backbone of the study since it is considered as a successful venture in sheep rearing. Hence, the results under the general characteristics are analyzed and are discussed in detail under the following subheads:

1.1 Size of the Family of the Case Farm
1.2 Size of Holding Available with the Case Farm
1.3 Land Use Pattern Prevalent in the Case Farm

Size of the Family of the Case Farm

The discussion on the size of the family is very important in this context. Because, it provides the opportunities for utilizing the family labor and clearly spell out the possibilities of whether to hire the labor or to run the farm with the available labor. Hence, an analysis was made and the results are presented in Table 1.

Table 1 Size of the Family of the Case Sheep Farm

Sl. No	Particulars of Family	Number of Persons	Number of Years in Education	Percentage to Total
01	Earners	02	05	33.33
02	Dependent for Education	03	10	50.00
03	Dependent for Livelihood	01	00	16.67
04	Attached Farm Labor	01	05	00.00
	Size of the Family	06	05	100.00

Table 1 reveals that the size of the family with the case farm is found to be six in numbers. Out of the six members, two are earners who accounted for 33 per cent of the total. The earners are the family laborers comprising both husband and wife looking after the production and management activities of sheep farm. Three are the dependent children for education and are studying in eleventh, ninth and seventh standards in government schools of Periyakulam. Importance has been given for the one who has completed the tenth standard. He has scored state first in tenth standard examination and got the Chief Minister' Award for scoring the highest mark in the Secondary School Examination conducted by the State Board of Secondary Education, Government of Tamil Nadu. The owner-manager of the case farm revealed that the elder son is going to pursue veterinary education for running the enterprise in a scientific and technical way and also for managing the same to elevate it to number one position in Tamil Nadu. Another dependent is the mother of the owner manager caring the agricultural crops to be grown in the vacant land with the support of one attached farm servant.

On examining the size of the family, the case farm utilizes every individual human resource in a proper direction with a vision to educate, guide and direct them to take the farm to new heights of achievment in the agribusiness.

Size of Holding Available with the Case Farm

Analyzing the details on size of holding is equally important to know how effectively the resources are being used and the products and byproducts are generated for various end uses. The details are analyzed and the results are presented in Table 2.

Table 2 Size of Holding Available with the Sheep Farm

Sl. No	Details of Land Use	Size of Holding (Acres)	Percentage to Total Holding
01	Crop Cultivation	02.00	88.89
02	Pen for Sheep Flocks	00.20	08.89
03	Residential Use	00.05	02.22
	Total Land Use	02.25	100.00

Table 2 reveals that the owner manager of the case sheep farm is in possession of 2.25 acres, out of which two acres is earmarked for crop cultivation exercise that accounted for 89 per cent to the total size of holding with the farmer. For accommodating the sheep flocks, only 8.89 per cent of land is allotted and the pen was erected to protect the animals. During the rainy season, a temporary measure of providing plastic tarpaulin above the pen is given to protect the animals from the rain. The residential space for habitation of family of owner-manager is accounted to be only 2.22 per cent of the total size of holding.

From the details of size of holding, one could understand that strict allocation of space and the output realization is carefully planned for each entity. However, the land use pattern delineates the details of what was grown and what was realized and the details are presented in Table 3.

Land Use Pattern Prevalent in the Case Farm

Under the head of crop cultivation as deli eated in Table 2, the area earmarked is only 2 acres, out of which paddy and cotton are cultivated in 0.70 acres each under partly irrigated condition which respectively accounted for 25.93 per cent to the total area under crops followed by chillies raised in an area of 0.60 acres which accounted for 22.22 per cent to the total area under crops. After harvesting the paddy crop, pulses and gingelly are raised in an area of 0.40 acres and 0.30 acres respectively which accounted for 14.81 per cent and 11.11 per cent to the total area under crops.

Table 3 Land Use Pattern in Practice with the Sheep Farm

Sl. No	Details of Crops Cultivated	Area Under Crops (Acres)	Quantity of Output (Quintals)	Gross Return Realized (Rs)
01	Paddy	00.70 (25.93)	12.60	10080
02	Chillies	00.60 (22.22)	07.00	14000
03	Cotton	00.70 (25.93)	06.00	12000
04	Pulses*	00.40 (14.81)	03.00	06000
05	Gingelly*	00.30 (11.11)	01.20	04800
Total		02.70 (100.00)	29.80	46880
Cropping Intensity in Per Cent*		135.00*		

(Figures in Parentheses Indicate Percentage to Total)

On examining the cropping intensity practiced in the farm, it is found to be of 135 per cent which revealed that the resources, even though available scarcely, are effectively utilized to reap the benefits to sustain the livelihood through farming and the farming in turn helps providing roughages to the sheep, often it will serve as hideouts and act as emergency feed to sheep.

To sum up, under the head of general characteristics of the farm, the human resources are effectively utilized and the resources are directed to achieve vision based actions to transform the farm into a more specialty enterprise. The land resources are properly allocated to the competing needs of food, shelter and revenue making by mitigating the land resource for repeated usage and the cropping intensity raised to the level of 135 per cent which is above the district and the state average revealing that the private farms are efficient and the successful model should be emulated periodically by government organizations and the emerging agripreneurs to make the farm enterprise a successful one which could contribute to the growth of Indian economy.

DETAILS OF ANIMAL POPULATION IN THE CASE FARM

The first section dealt with crop cultivation and the general land use details in a fit-in- manner. The second part deals with details of animal population raised in the farm and how these animals are manned effectively is analyzed and the results are presented in Table 4.

Table 4 Details of Animal Population in the Sheep Farm

Sl. No	Details of Animals	Numbers	Name of the Breed	Percentage
01	Male Goat	01	Thallasseri	00.47
02	Female Goat	10	Thallasseri	04.69
03	Male Sheep	05	Meacheri	02.35
04	Female Sheep	75	Meacheri	35.21
05	Female Sheep	120	Mylambadi	56.34
06	Pilot Dogs	02	Rajapalayam	00.94
	Total	213		100.00

Table 4 reveals that the farm has a mixture of animals which arrived at 213 in numbers, out of which goat population alone accounted for 5 per cent and the sheep population arrived at 94 per cent. To protect these goat and sheep population from poaching and to safeguard them from the wild animals, the farmer-cum-agripreneur has trained two dogs of Rajapalayam breed which guarded the sheep and goat herd and are serving the role of pilots as in human VIP escorts. The dog number one proceeds to the target place for grazing and the other dog shapes the herd to lead in a unidirectional pattern. The owner manager and his wife are the followers on other side and back to escort the entire team of sheep herd. The pilot alone accounted for around one per cent to the total animal population raised in the case farm.

It is interesting to note that the role played by the pilot dogs is commendable as they are communicating differently to the animals when they face threat due to wild animals or the transportation threats when they have to cross the roads which are not visible to humans. The human folk have learnt violation on many events. But the animals have communication which shapes them and protects them from threats.

Birth and Death of Sheep Kids in the Case Farm

The section two dealt with number of animals available in the farm. In continuation of their active existence, how many animals are produced on reproduction and added to the total population and how many died due to multifarious issues are analyzed and the details are presented in Table 5.

Table 5 Birth and Death of Sheep Kids in the Case Farm

Sl. No	Particulars of Birth/Death/ Theft of Animals	Number of Animals	Percentage to Total
01	Sheep Kids Delivered	195	91.55
02	Goat Kids Delivered	018	08.45
	Total Animals Delivered	213	**100.00**
03	Died After Birth	011	05.16
04	Theft of Animals	005	02.35
05	Lifting of Sheep by Wild Animals	008	03.76
06	Died due to Disease Attack	009	04.23
	Total Number of Animals Lost	033	15.50

Table 5 reveals that the total animals delivered and added to the existing sheep flock arrived at 213 in numbers, out of which, the goat kids delivered accounted for 8.45 per cent to the total kids. The number of sheep kids delivered in the case farm arrived at 195 which accounted for 91.55 per cent to the total kids delivered in the farm. Out of the delivered kids, 5.16 per cent died immediately after birth followed by lifting of sheep by wild animals that accounted for 3.76 per cent while grazing in the forest floor. While penning the goats, theft could also be observed out of careful watch and ward service extended which also accounted for 2.35 per cent to the total number of animals lost.

During the rainy season, an outbreak of blue tongue disease and foot and mouth disease also caused death of animals which accounted for 4.23 per cent to the total number of animals lost. A brief on the blue tongue disease is presented for better understanding which is widespread during the rainy season particularly in the southern part of Tamil Nadu.

Blue Tongue Disease

Blue tongue is an insect-borne, viral, non-contagious disease that occurs in many parts of Tamil Nadu. Mathis and Tim Ross (2005) have also reported that such disease is prevalent in New Mexico. It is transmitted from infected animals to susceptible sheep by the bites of a small insect commonly referred to as a gnat. The disease normally occurs from midsummer until frost. Early symptoms usually include excessive salivation, reddening of the lips and mouth, and progressive darkening of the vascular areas of the mouth. Furthermore, the muzzle, lips, tongue, throat, and sometimes the ears and neck become swollen. Occasionally, sheep suffer from severe lameness as well. Not all signs of blue tongue appear in a single sheep or even in a single outbreak (Mathis and Tim Ross, 2005).

Best control methods involve controlling the gnat. Since this insect breeds in the mud along the edges of slow-moving streams or water tank overflow, try to eliminate these breeding sites. Breeding sites also can be sprayed with insecticides. A modified live-virus type of vaccine is available, but it is estimated that six to seven different viruses cause blue tongue. Occasionally, the vaccine may cause a reaction that is nearly as bad as the disease itself. Pregnant ewes, particularly in the first 50 days of gestation, should not be vaccinated. No satisfactory medical treatment has been found for animals with blue tongue. Generally, with proper care, most animals recover naturally within 14 days, although severely affected animals may recover more slowly. Isolate affected animals in a shaded area with palatable feed and fresh water. Antibiotics are of no value in the treatment of blue tongue, but they are helpful in preventing secondary infections (Mathis and Tom Ross, 2005).

DETAILS OF LAND AVAILABLE FOR GRAZING

Cage method of farming permits energy saving and fattening cycle becomes easier and the productivity per animal becomes increased. But it is not prevalent in grazing method by permitting the animals to graze in the open land. But the health of the animal could be maintained to certain extent. Hence, the details of resources available for open grazing is analyzed and the results are presented in Table 6.

Table 6 Details of Grazing Lands Available to Sheep Farm

Sl. No	Particulars of Grazing Land Available	Number of Days Per Annum	Amount Paid Per Annum	Percentage to Total
01	Private Patta Lands + Common Grazing Lands	014	4200.00	62.69
02	Common Grazing Lands alone	062	00.00	00.00
03	Forest Floor + Common Grazing Lands	289	2500.00	37.31
	Total	365	6700.00	100.00

Table 6 reveals that the private patta lands plus the common grazing lands like foreshore area of the tank, ponds, streams, etc., permit the animals for grazing only for 14 days for which an amount of Rs 4200 is to be paid by the owner manager of the case farm. This money accounted for 62.69 per cent to the total amount paid per annum for grazing of sheep flocks. The common grazing lands alone permitted the sheep flocks to graze for 62 days mainly during the rainy season. The forest floor plus common grazing lands permitted the sheep to graze for 289 days per annum which accounted for 37.31 per cent to the total amount paid per annum. The grazing in forest floor attracted a license amount of Rs. 2500 per annum which is to be obtained from the Forest Range Officer of Periyakulam, Theni District. However, in recent days, the forest officials are not demanding any license or permits for grazing of sheep flocks but they deny any goats to graze the forest floor. Hence, careful segregation is needed while permitting the sheep to graze the forest floor.

To sum up, for open grazing, the common lands and the forest floor alone contributed to the maximum number of days for grazing of sheep. Otherwise supply of additional roughages to the sheep flocks attracts costs that are always not possible from the side of farmer-cum-agripreneur.

COSTS AND RETURNS IN SHEEP REARING

One has to assume here that every business firm has a certain state of technology or know how. That is, how the firm is producing various levels of outputs with the given prices of inputs and outputs. With data on those variables, one can understand the relationships between costs and output levels. Basically we require data on output, fixed costs, variable costs and the prices of inputs and outputs. In sheep and goat rearing activity of the case farm, the costs are discussed under the head of fixed costs and variable costs separately and the details are analyzed and presented in Tables 7 and 8 respectively.

Table 7 Fixed Investment in Sheep Farming (Investment for a unit of 200 Sheep Consists of 195 Female and 05 Male)

Sl. No	Details of Fixed Expenses	Life of the Asset (Years)	Amount Incurred (Rs)	Annual Fixed Cost (Rs)
01	Cost of 195 Female Sheep weighing 20 Kg per animal @ Rs 250 per Kg	07	975000	139286 (60.61)
02	Cost of 5 Male Sheep Weighing 20 Kg per animal @ Rs 250 per Kg	07	25000	3572 (01.55)
03	Cost of 5 Goats Weighing 20 Kg per Animal @ Rs 250 per Kg	07	25000	3572 (01.55)
04	Cost of Pen for 195 adult female Sheep + and 5 Adult Female Goats @ 10 Sq. feet per animal @ Rs 200 per Sq. Feet	10	400000	40000 (17.40)
05	Cost of Pen for 5 Adult Male Sheep + 1 Adult Male Goat @ 20 Sq. Feet per Animal @ Rs 200 per Sq. Feet	10	24000	2400 (01.04)
06	Pen for 200 Kids @ 5 Sq. Feet per Kid @ Rs 200 per Sq.Feet	10	200000	20000 (08.70)
07	Cost of Equipments like Pails, Buckets and Ropes 10 Numbers @ Rs 500 per unit	02	5000	2500 (01.09)

Sl. No	Details of Fixed Expenses	Life of the Asset (Years)	Amount Incurred (Rs)	Annual Fixed Cost (Rs)
08	Erection of one Bore Well Unit @ Rs.75000	10	75000	7500 (03.26)
09	Erection of a Water Tank for Water arrangements for the 200 Animals	10	10000	1000 (00.44)
10	Miscellaneous Cost under Fixed Expenses	01	10000	10000 (04.36)
Total Fixed Cost (Rs.)			1749000	229830 (100.00)

Table 7 reveals that the Total Fixed Cost (TFC) incurred towards establishment of sheep farm arrived at Rs 17.49 lakhs which included cost of animals, cost incurred towards erection of pen for the animals, one bore well unit and the overhead water tank. The Annual Fixed Cost (AFC) worked out for incorporation in the cost analysis arrived at Rs. 2.30 lakhs. Out of which, the cost of animals alone accounted for 60.61 per cent to the total annual fixed cost followed by cost of pen which accounted for 17.40 per cent to the total annual fixed cost revealing that these two are the major investments under the head of fixed cost for establishing a sheep farm. The life term of the animal is fixed at seven years and the cost incurred towards purchase of sheep was arrived at 1.39 lakhs as annual fixed cost.

The miscellaneous expenses under fixed one time expenses towards consultation and other charges arrived at Rs 10000 per annum which accounted for 4.36 per cent to the total annual fixed cost followed by the erection of bore well for drinking and other farm management activities which accounted for 3.26 per cent to the total annual fixed cost. From this, one could understand that the cost of sheep for sheep rearing alone is the major item of cost followed by pen making charges and the annual fixed cost per animal is arrived at Rs 1079 and let us analyze the variable expenses incurred towards maintenance of sheep farm which is presented in Table 8.

Table 8 reveals that the imputed value of family labor spent towards maintenance of sheep farm was arrived at Rs 1.82 lakhs per annum which accounted for 69.60 per cent to the total variable expenses followed by insurance charges for the animals which arrived at 15 per cent to the total variable expenses.

Table 8 Details of Variable Expenses in Sheep Rearing

Sl. No	Details of Variable Cost	Amount (Rs in Lakhs)	Percentage to Total
01	Grazing Charges for 10 Days Per Month @ 500 per day in Private Farm Lands	15000	05.72
02	License Charges for Grazing Permits in the Forest Floor for 250 Animals per Annum	2500	00.95
03	Feeding of Male and Female Goats for 3 Months @ 300 grams per day per Animal @ Rs 10 per Kg of Feed	1350	00.51
04	Feeding of Kids for 3 Months @ 100 grams per day per Animal @ Rs 10 per Kg of Feed	900	00.35
05	Cost of Veterinary Medicine and Health Care to Animals 10 Visits @ Rs. 1000 per visit including Medicines	10000	03.81
06	Insurance Premium Charges @ 4 per cent	40000	15.25
07	Imputed Value of Family Labor @ Rs 300 for Men and 200 for Women Days	182500	69.60
07	Miscellaneous Expenses under the Head of Variable Cost	10000	03.81
	Total Variable Cost	262250	100.00

Grazing charges incurred towards grazing of sheep in private farms accounted for 5.72 per cent to the total variable expenses. The total variable cost (TVC) arrived at Rs. 2.62 lakhs and the maintenance cost per animal arrived at Rs 1231. Put together the total cost spent towards rearing of one sheep arrived at Rs 2310 per annum and the returns realized from the sheep farm is analyzed and the results are presented in Table 9.

Table 9 Returns Realized From Sheep Production

Sl. No	Details of Revenue Earned	Amount (Rs in Lakhs)	Percentage to Total
01	Sale of Animal by Live Weight @ Rs. 300 per Kg @ 20 Kg per Animal (180 Numbers)	1080000	61.28
02	Sale of Animal by Live Weight @ Rs 300 per Kg @ 12 Kg per Animal (90 Numbers)	324000	18.38
03	Insurance Claim for 14 Animals @ Rs 2000 Per Animal	28000	01.59
04	Sale of Manure realized from Sheep and Goat Farm @ 1.50 Kg of Pellets per day per Animal for 213 Animals @ Rs 2 per Kg of Manure	233235	13.23
05	Sale of Manure realized from Sheep and Goat Farm @ 1.50 Kg of Pellets per day per Animal for 180 Animals @ Rs 2 per Kg of Manure	97200	05.52
06	Gross Return Realized from Sheep and Goat Rearing (213 Animals)	**1762435**	**100.00**
07	Total Variable Cost Incurred in Sheep and Goat Rearing	262250	53.29
08	Annual Fixed Cost incurred in Sheep and Goat Rearing	229830	46.71
09	Total Cost (Total Variable Cost + Annual Fixed Cost)	492080	100.00
10	Net Return Realized From Goat Farming	1270355	72.08
11	Cost of Production of Meat per Kg	105.15	
12	Output – Input Ratio	03.58	

Table 9 reveals that the gross return realized from a goat and sheep unit with the population of 213 animals was arrived at 17.62 lakhs per annum in that the net return realized was arrived at 72 per cent to the gross income generated by the sheep farm. Sale of meat from the animals alone accounted for 80 per cent of revenue and the manure drawn from those animals was accounted for 13 per cent of gross revenue after deducting the animals lost due to theft, lifting of sheep by wild animals, etc. A study conducted by Prabhu et.al. during 2009 revealed that the total cost incurred in sheep farming per farm with imputed value of family labor per farm was the lowest in small farms and the highest in marginal farms. The overall picture showed that the gross income obtained per farm per annum from sheep flock was highest in marginal farms followed by landless farmers, small farmers and was the lowest in large farms.

The total cost of production of 213 animals was arrived at 4.92 lakhs. Out of which the Total Variable Cost alone accounted for 53.29 per cent and the Annual Fixed Cost accounted for 46.71 per cent to the Total Cost incurred in the sheep farm. The cost of production of meat per kg was arrived at Rs. 105 and the output–input ratio was arrived at 3.58 revealing that for a rupee of investment in the sheep farm, it could generate Rs 3.58 as gross income which is highly profitable. To sum up, the total variable cost is found to be on the higher side in sheep rearing activity in which the imputed value of family labor is accounted to be the major cost. While examining the fixed investment, the cost of animals procured for rearing consumed higher cost and the analysis revealed that the cost of production of meat per kg is only Rs 105 and the selling price is arrived at Rs 300 per kg live weight and hence there is good scope for establishing sheep farm in the arid and semi-arid regions of Tamil Nadu which will yield profitability and hence the school drop-outs and the unemployed can choose this agribusiness venture by availing the loan from the commercial bank for getting sustainable income.

CONSTRAINTS FACED BY THE AGRIBUSINESS FIRM

There are few constraints faced by the owner-manager of the case farm which are considered to be the vital factors in running the sheep farm. They are:

- A general threat from the private land owners that the sheep would have entered into their farm and hence often the restrictions infused by them that the goat should not come by this way causes problem in maintaining the grazing time for the animals.

- Theft among the common public is more visible in this zone and hence the owner- manager of the case farm should infuse more vigil in caring the animals and hence it consumes much of his sleep and permits sleeplessness.
- Timely availability of insurance and the formalities for availing is a big problem and hence the farmer has withdrawn the insurance policy and takes personal care of animals.
- The governmental intervention through developmental programs is not percolated to the sheep flock owners and hence the Department of Animal Husbandry should launch special schemes to protect their interest in establishing and maintenance of sheep flocks.

SUMMARY AND CONCLUSIONS

In the case farm, the human resources are effectively utilized and the resources are directed to achieve vision based actions to make the farm into a more specialty enterprise. The land resources are properly allocated to the competing needs of food, shelter and revenue making by mitigating the land resource for repeated usage and the cropping intensity raised to the level of 135 per cent which is above the district average reveals that the private farms are efficient. It is also interesting to note that the role played by the pilot dogs is commendable as they are communicating differently to the animals and provide them safety when they face threat due to wild animals or the transportation threats while they have to cross the roads. With respect to open grazing, the common lands and the forest floor alone contributed to the maximum number of days for grazing of sheep. Otherwise supply of additional roughages to the sheep flocks attracts costs that are always not possible from the side of farmer-cum-agripreneur. The total cost of production of 213 animals was arrived at 4.92 lakhs. Out of which the Total Variable Cost alone accounted for 53.29 per cent and the Annual Fixed Cost accounted for 46.71 per cent to the Total Cost incurred in the sheep farm. The cost of production of meat per kg was arrived at Rs. 105 and the output–input ratio was arrived at 3.58 revealing that for a rupee of investment in the sheep farm, it could generate Rs 3.58 as gross income which is highly profitable and hence the young agripreneur can take up this as their agribusiness venture for profit making.

STRATEGIES AND RECOMMENDATIONS

- The governmental intervention through developmental programs is not percolated to the sheep flock owners and hence the Department of Animal Husbandry should launch special schemes to protect their interest in establishing and maintenance of sheep flocks.
- The problems of availing insurance and fulfilling the related procedures discouraged them from investing in insurance related measures and hence a personal vigil is imposed on the herd to protect from theft, lifting of sheep by wild animals, etc. Hence, a campaign can be arranged by the Department of Animal Husbandry along with the insurance agencies so that the grievances of sheep flock owners could be solved through need based amendments.
- One of the children of the case farm owner has got the state first rank in the examinations conducted by Board of Secondary Education of Government of Tamil Nadu and got the appreciation certificate and medal from the Honorable Chief Minister of Tamil Nadu. Hence the child's higher education may be made more professional as the father of the children is aiming to make him a veterinarian and hence allocation of seats to the children of sheep flock owners may be arranged on the lines of Farmers Quota. On fulfilling these recommendations, the sheep flock owners and their survival could be sustained.
- The successful model of the case farm dealing with agribusiness should be emulated periodically by government organizations and the emerging agripreneurs to make the farm enterprise a successful one which could contribute to the growth of Indian economy.

References

- Government of India (2008). "**Economic Survey 2007-08**" (Ministry of Finance: New Delhi)
- Mathis C.P. and Tim Ross (2005). " ***Sheep Production and Management***" (New Mexico State University: New Mexico) pp. 1-44
- Prabhu, M., K.N. Selvakumar, A.Serma Saravana Pandian and N. Meganathan (2009). "Economic Analysis of Sheep Breeding in Tamil Nadu", ***Indian Journal of Small Ruminants***, 15 (2): pp. 224-230.
- Selvakumar, K.N., M. Prabhu, A. Serma Saravana Pandian, G. Senthilkumar and B. Jaya Varathan (2012). "Livestock / Poultry Business Prospects in Tamil Nadu", ***In: Proceedings of the Seminar on Agribusiness Potential in Tamil Nadu*** (Madras Institute of Development Studies: Chennai and Indian Society of Agricultural Marketing: Nagpur), pp. 33-45
- Suresh A., Gupta D.C. and J.S. Mann (2008). "Returns and Economic Efficiency of Sheep Farming in Semi Arid Regions-A Study in Rajasthan", ***Agricultural Economics Research Review***, 21(2): pp. 227-234
- Verma, R.K., and N.K. Sharma (2009). "Knowledge of Sheep Farmers about Improved Sheep Production Technologies" ***Indian Journal of Animal Research*** 43(4):pp.275-278.

Chapter 15

Garlic Production – A Potential Business

INTRODUCTION TO SPICES

Spices are the '*Pearls of Developing Countries*'. Spices are in great demand in the industrialized nations. Since India is blessed with varied agro-climatic zones, it could produce a variety of spices and exports the same to 120 countries. At present, the production of different spices is around 3.20 million tonnes valued at approximately 4 billion US$ and holds a prominent position in World Spice Production.

A spice is a dried seed, fruit, root, bark or vegetative substance primarily used for flavoring, coloring or preserving the food. Spices are distinguished from herbs which are parts of leafy green plants also used for flavoring. Many spices have antimicrobial properties. A spice may have an extra use usually medicinal, for religious/ritual purposes/perfume production, or as a vegetable/cosmetic (Sajith Mohan et.al. 2013). In turmeric, the roots are consumed or the rhizomes are consumed. But in garlic, it is used as vegetable and serves as antibiotic. However, the details of uses of different spices particularly the medicinal uses are delineated in Table 1.

Table 1 Medicinal Properties of Indian Spices

Sl. No	Name of the Spice	Medicinal Properties of Spices
01	Allspice	Stimulant, digestive, carminative, anodyne against rheumatism and neuralgia
02	Aniseed	Mild expectorant, stimulating, carminative, diuretic, diaphoretic, in asthma powders, in veterinary medicine
03	Bay Leaves (Laurel)	Stimulant in sprains, narcotic and in veterinary medicine
04	Bishop's Weed (Ajowan)	Digestive, antispasmodic, stimulant, carminative, expectorant
05	Capsicum	Digestive, thermogenic, carminative, stimulant, cardiotonic, antipyretic, serdorific, rubefacient and sialogogue
06	Cardamom (Small)	Stimulant, tonic, diuretic, carminative, digestive, expectorant, cardiotonic and used in several pharmaceutical preparations
07	Cardamom (Large)	Hypnotic, appetizer, astringent to bowels, tonic to heart and liver
08	Cinnamon	Astringent, diuretic, carminative, aphrodisiac, deodorant, expectorant, febrifuge, stomachic
09	Clove	Refrigerant, ophthalmic, digestive, carminative, stomachic, stimulant, antispasmodic, antibacterial, expectorant, rubefacient, aphrodisiac, appetizer, emollient
10	Coriander	Carminative, diuretic, tonic, stimulant, stomachic, refrigerant, aphrodisiac, analgesic, anti-inflammatory
11	Cumin	Digestive, carminative, astringent, anti-inflammatory, constipating, diuretic, repulsive, galactogogue, uterine and nerve stimulant
12	Curry Leaf	Astringent, anthelmintic, febrifuge, stomachic, appetizing, carminative, constipating, anti-inflammatory, antiseptic, used in skin diseases, used in diarrhea, ulcers

Sl. No	Name of the Spice	Medicinal Properties of Spices
13	Garlic	Anti-cholesterol, antifungal, tonic, rubefacient, stimulant, thermogenic, aphrodisiac, used in cough, asthma, cardiopathy
14	Ginger	Digestive, carminative, emollient, appetizer, stomachic, rubefacient, anodyne, expectorant, anthelmintic, stimulant
15	Mint	Stimulant, stomachic, carminative, antiseptic, digestive, antispasmodic, contraceptive, used in vomiting, skin diseases, amenorrhea, dental caries
16	Mustard	Thermogenic, anodyne, anti-inflammatory, carminative, digestive, anthelmintic, sudorphic, tonic, emetic, used in vomiting, abdominal colic, dyspepsia, flatulence, skin diseases
17	Pepper	Anthelmintic, carminative, alterant, antiperiodic, diuretic, digestive, emmenogogue, rubefacient, stimulant, stomachic, used in fever, asthma, cough, dyspepsia, flatulence, arthritis
18	Tamarind	Refrigerant, digestive, carminative, laxative, antiscerbutic, febrifuge, ophthalmic, useful in gastropathy, datura poisoning, alcoholic intoxication, scabies, constipation
19	Turmeric	Thermogenic, emollient, anodyne, anti-inflammatory, vulnerary, depurative, antiseptic, appetizer, carminative, expectorant, stomachic, anthelmintic, stimulant, ophthalmic, tonic, used in skin diseases, dyspepsia, asthma, cough, bronchitis, inflammations, ulcers, worms, skin discolouration
20	Vanilla	Aphrodisiac

Table 1.1 reveals that the garlic has some important medicinal properties which are summarized as follows.

- ***Anti-cholesterol***: The regular consumption of garlic will lower the blood pressure by reducing the cholesterol level.

- ***Antibiotic in Nature***: Acts as an antibiotic and also serves as a tonic to the human being.
- ***Stimulant***: It helps in nerve-building.
- ***Thermogenic***: Regulator of body temperature.
- ***Aphrodisiac***: Garlic enhances the production of sperms and enhances the percentage of their survival.
- ***Cough Medicine***: It acts as a good medicine for cough in human being.
- ***Cardiopathy***: Garlic is capable of controlling cardiac diseases. Regular intake of garlic will eliminate the blocs in the blood vessel.
- ***Asthma Controller***: Controls asthma disease in human being and hence regular intake of garlic should be practiced in the diet for their well being.

AN INTRODUCTION TO GARLIC

Garlic is one of the important commercial spice or a condiment crop, second among the *Allium* species grown throughout India, in almost every home, kitchen garden for seasoning and flavouring the food as it has got manifold uses including many medicinal properties. It is also an important foreign exchange earner apart from meeting the domestic requirement of the country.

Chemical Composition of Garlic

Garlic is consumed by people in almost all walks of life and who take onion in every meal. It has higher nutritive value than other *Allium* species. It is rich in proteins, phosphorus, potash, calcium, magnesium and carbohydrates. Ascorbic acid is very high in green garlic. Nutritive composition of fresh peeled garlic clove and dehydrated garlic powder is given in Table 2. The uninjured bulb contains a colourless, odourless water soluble amino acid allin. On crushing garlic bulb the enzyme allinase breaks down allin to produce allicin of which the principle ingredient is the odoriferous diallyl disulfide. Garlic contains about 0.1 per cent volatile oil. The chief constituent of the oil are diallyl disulfide (60 per cent), diallyl trisulfide (20 per cent), allyl propyl disulfide (6 per cent), small quantity of diethyl disulfide and probably diallyl polysulfide. Diallyl disulfide is said to possess the true garlic odour.

Table 2 The Uses of Garlic and its Chemical Composition

Sl. No	Particulars	Unit	Fresh peeled Garlic Cloves	Dehydrated Garlic Powder
01	Moisture	Per Cent	62.80	05.20
02	Protein	Per Cent	06.30	17.50
03	Fat	Per Cent	00.10	00.60
04	Mineral Matter	Per Cent	01.00	03.20
05	Fiber	Per Cent	00.80	01.90
06	Carbohydrate	Per Cent	29.00	71.40
07	Calcium	Per Cent	00.03	00.10
08	Phosphorus	Per Cent	00.30	00.42
09	Potassium	Per Cent	-	00.70
10	Iron	Per Cent	0.0001	0.004
11	Niacin	Per Cent	-	00.70
12	Sodium	Per Cent	-	00.01
13	Vitamin-A	IU	-	17500
14	Nicotinic Acid	Mg/100g	00.40	-
15	Vitamin C	Mgl100g	1300	1200
16	Vitamin B	Mg/100g	-	00.68
17	Vitamin B2	Mg/100g	-	00.68

The yield of bulbs varies between 100 and 200 q/ha depending upon the variety and the regions where they are grown. Longer the day length given to the variety, higher is the yield. Agrifound Parvathi which is grown only in northern hills is the highest yielder.

The knowledge about cost of production is very necessary as garlic growing is considered now as business in view of follow up of improved agro-techniques and increase in the cost of cultivation of garlic. Farmers therefore, will have to take due care in harvesting maximum with optimum level of inputs including labour. In this situation, documenting the cost of production of garlic becomes very important to assess the profitability to the farmer.

In India, several horticultural developmental initiatives are being taken up through the National Horticultural Mission through the state department of horticulture and plantation crops. Different states in India are showing different levels of productivity in respect of spices and plantation crops especially in garlic. This has created a gloomy picture. Though NHM has initiated its efforts to develop the horticulture sector through the area expansion programs, existing level of production and productivity of spices particularly the garlic is yet to be documented clearly at national and regional level and the data found to be dismal in this regard. A sound data and its analysis on area, production and productivity, cost of production and export of garlic will be a very good tool for the planners and policy makers in horticulture to take up planned investment in a phased manner for sustainable development of horticulture and hence this study is aimed at with the following specific objectives.

OBJECTIVES OF THE STUDY

- To assess the area, production and productivity of garlic in Tamil Nadu and India.
- To assess the area, production and productivity of garlic in the sample farms.
- To document the constraints associated with garlic production and to suggest appropriate strategies for the same.

DESIGN OF THE STUDY

The objective of this chapter is to document the area, production, productivity of garlic in India and to document the supporting infrastructure available in the state of Tamil Nadu. To document the same, the data were collected from the secondary sources of both central and state government reports. The reports taken into account for collecting the details on area, production, productivity of garlic are furnished as follows.

- Agricultural Statistics at a Glance 2012 of Directorate of Economics and Statistics, Ministry of Agriculture, Government of India, New Delhi

- Food and Agriculture Organization Website
- Indian Horticulture Data Base 2011 of National Horticulture Board, Government of India, New Delhi
- Hort-Stat – 2008 of Government of Tamil Nadu, Chennai

The data gathered for different periods with respect to the area production and productivity of garlic, and states involved in garlic export and import were gathered from the above reports and the data were analyzed using descriptive statistics like mean and the current change in the data in respect of area, production and productivity of garlic over the base period was also documented to assess the technological and extension impact due to the government intervention through its planned programs in horticulture. The results of this analysis will be better utilized by the policy makers and the planners in horticulture and agriculture for further development of these sectors.

The productivity of garlic in the state was found to be marginally low when compared to the national average productivity and hence a comparison made with the intensely managed garlic farms of Kodaikanal taluk of Dindigul district and their experiences were also incorporated as a case study approach as an illustration for easy implementation by other plantation owners.

PRIMARY DATA

The farmers who are cultivating garlic in the Kodaikanal taluk formed the base for the study. In the first stage, the Kodaikanal taluk was selected as it has higher area under garlic cultivated by the farmers. The intention of conducting the survey among the garlic farmers have been conveyed to the Professor and Head, Horticultural Research Station, Kodaikanal for the project work on garlic production and marketing. The Professor and Head, has arranged the contact with the farmers from the selected villages. The number of farmers and the list of villages selected are presented in Table 3.

Table 3 Number of Villages Selected for the Study

Sl. No	Name of the Village	Area under Garlic in Ha	Percentage to Total	Number of Farmers Selected	Percentage to Total
01	Poomparai	112.00	21.46	03	21.42
02	Poondi	074.00	14.18	02	14.29
03	Kavunji	064.00	12.27	02	14.29
04	Kilavarai	054.00	10.34	01	07.13
05	Pallangi	030.00	05.76	02	14.29
06	Kodaikanal	010.00	01.90	02	14.29
07	Senbaganur	010.00	01.90	02	14.29
08	Other Villages	168.00	32.19	00	00.00
09	Total Area under Garlic	522.00	100.00	14	100.00

From the farmer, the details of land use, crop management for its production and productivity, marketing of garlic, cost of production and the constraints involved are assessed using the structured and pretested questionnaire exclusively designed for the purpose and the same were analyzed and documented in this study with respect to the objectives enshrined in the study by adopting a descriptive statistics like mean and percentage analysis.

DETAILS OF THE PROJECT ANALYSIS

The project work on production, marketing and trade aspects of garlic in India with special reference to Tamil Nadu is an important topic today as most of the researchers have not much explored on this aspect at national and regional level and hence a special effort has been taken to assess the production potentials, marketing and distribution made by the producers at micro level and the export of garlic at macro level and the details are analyzed and discussed under the following major heads:

- Area Production and Productivity of Spices
- Area Production and Productivity of Garlic

- The Production Potentials of Garlic Farms

AREA PRODUCTION AND PRODUCTIVITY OF SPICES

The area production and productivity is analyzed in respect of spices at national and regional level and the same is analyzed and discussed under the following heads for providing clear picture for further decisions.

Area Production and Productivity of Spices in India

The details on area production and productivity of spices in India over the years are analyzed and the results are discussed in Table 4. It provides an overall understanding on whether the area is increasing or decreasing at macro level and the magnitude of production over years provide the information to the policy makers on the extent of adoption of technology and its efficacy and the need based efforts to enhance the productivity of spices to meet the requirements of both domestic and international markets.

Table 4 Area Production and Productivity of Spices in India

Sl. No	Year	Area in Thousand Ha	Production in Thousand Tonnes	Productivity in Tonnes per Ha
01	1991-92	2005	1900	00.95
02	2001-02	3220	3765	01.17
03	2002-03	3220	3765	01.17
04	2003-04	5155	5113	00.99
05	2004-05	5909	8051	01.36
06	2005-06	2366	3705	01.57
07	2006-07	2448	3953	01.61
08	2007-08	2617	4357	01.66
09	2008-09	2629	4145	01.58
10	2009-10	2464	4016	01.63

Sl. No	Year	Area in Thousand Ha	Production in Thousand Tonnes	Productivity in Tonnes per Ha
11	2010-11	2940	5351	01.82
12	Percentage Change During 2010-11 Over 2001-02	(-) 08.70	(+) 42.12	(+) 38.56

(Source: Indian Horticulture Data Base – 2011)

Table 4 reveals that the area under spices between 2001-02 and 2010-11 was found to be reduced to the tune of nine per cent. In a decade of time, the reduction in area is found to be only marginal and the policy makers should not let the same to reduce further and hence the policy intervention is needed right now to augment the area under spices in India through some special motivational efforts to restore horticulture area and the spices area in particular. When we compare the production figures between the period, the production is found to be increased to the tune of 42 per cent and the productivity arrived at 39 per cent increase in a decade of time.

The increase in production is mainly due to the enhanced productivity of spices. During the period 2001-02, the productivity of spices was arrived at 0.95 tonnes per ha and the same was increased to 1.85 tonnes per ha during the year 2010-11. The productivity has almost doubled due to concerted effort taken by the officials of horticulture at central and state governments and the spices board through the different schemes to boost spices production. To be specific, the productivity of spices was found to increase steadily from 2005. It might be due to the National Horticulture Mission and its efforts to boost the productivity with the technological support of Spices Board, Kochi and hence it is important to analyze the role played by the related institutions like Spices Board.

The Role of Spices Board The Spices Board was constituted as a Statutory Body on 26th February, 1987 under Section (3) of the Spices Board Act, 1986. The Board is headed by a Chairman with its head office at Kochi and is responsible for the development of cardamom industry and promoting the export of all the 52 spices listed in the Spices Board Act, 1986. The primary functions of the Board include increasing the production and productivity of small and large cardamom; development, promotion and regulation of export of spices; granting certificate for export of spices;

undertaking programmes and projects for promotion of export of spices (like setting up of spices parks, support of infrastructure improvement in spices processing, etc.); assisting and encouraging studies and research for improvement of processing, grading and packaging of spices; striving towards stabilization of prices of spices for export and controlling and upgrading quality for export (including setting up of regional quality evaluation labs and training centers). With regard to cardamom, the Board also provides financial and other assistance for cultivation and processing of cardamom; monitoring prices; increasing domestic consumption; improving marketing; registering and licensing brokers (including auctioneers), e-auction system; undertaking, assisting or encouraging scientific, technological and economic research and improving quality. The Board also implements programmes for development of spices in North-East (NE) region and organic spices in the country; it also supports programmes aimed at better post-harvest practices.

Area Production and Productivity of Spices in Tamil Nadu

The area production and productivity of spices in Tamil Nadu is analyzed and the results are presented district wise in Table 5. The district wise details will give better understanding on the status of spices area and production so as to plan further at micro level to meet the requirements at macro level.

Table 5 Area Production and Productivity of Spices and Condiments in Tamil Nadu during 2010-11

Sl. No	Districts	Area in Ha	Area in Per Cent	Production in Tonnes	Productivity in Tonnes per Ha
01	Coimbatore	12300	06.13	419495	34.10
02	Cuddalore	1644	00.82	5271	03.21
03	Dharmapuri	8008	03.99	27405	03.42
04	Dindigul	10181	05.07	31681	03.11
05	Erode	14683	07.31	109572	07.46
06	Kancheepuram	0304	00.15	3579	11.79
07	Kanyakumari	3135	01.56	5894	01.88
08	Karur	1929	00.96	8917	04.62

Sl. No	Districts	Area in Ha	Area in Per Cent	Production in Tonnes	Productivity in Tonnes per Ha
09	Krishnagiri	4680	02.33	15872	03.39
10	Madurai	3981	01.98	10946	02.75
11	Nagapattinam	0438	00.22	1249	02.85
12	Namakkal	3993	01.99	16922	04.24
13	Perambalur	5834	02.91	13232	02.27
14	Pudukottai	0614	00.31	1236	02.01
15	Ramanathapuram	30991	15.44	21057	00.68
16	Salem	9850	04.91	109879	11.16
17	Sivagangai	6095	03.04	4954	00.81
18	Thanjavur	0333	00.17	1383	04.15
19	The Nilgiris	5484	02.73	17543	03.20
20	Theni	6351	03.15	38240	06.02
21	Thiruchirappalli	6070	03.02	13073	02.15
22	Tirunelveli	5570	02.77	22114	03.97
23	Thiruvallur	1296	00.65	1651	01.27
24	Thiruvannamalai	1280	00.64	11322	08.85
25	Thiruvarur	0209	00.10	0496	02.37
26	Thoothukkudi	36370	18.12	133458	03.67
27	Vellore	2525	01.26	19062	07.55
28	Villuppuram	4315	02.15	30349	07.03
29	Virudhunagar	12306	06.12	31557	02.56
30	State Total	200769	100.00	1127409	05.71

Hort Stat – 2008; Government of Tamil Nadu

Table 5 reveals that the area under spices in Tamil Nadu is about 2.01 lakh ha which is capable of producing the spices and condiments to the tune of 11.27 lakh tonnes leaving the average productivity of 5.71 tonnes per ha. This cannot be compared with the Indian productivity of spices as the figures of Tamil Nadu have incorporated the condiments too in the list of spices and hence higher productivity is accomplished. Paucity of data for spices alone for Tamil Nadu forced the researcher to use this data for area related analysis. When we compare the area under spices in Tamil Nadu, it accounted for only 6.83 per cent to the total area under spices in India. Within Tamil Nadu, among the spices dominated belts, Thoothukkudi is found to have higher area which accounted for 18 per cent followed by Ramanathapuram, Erode, Coimbatore and Virudhunagar accounting for 15 per cent, 7.30 per cent, 6.13 per cent and 6.12 per cent respectively. Dindigul district had only 5 per cent of the total area under spices in Tamil Nadu. Thoothukkudi, Ramanathapuram and Virudhunagar districts dominate in chilli spice whereas, the Coimbatore and Erode districts dominate in turmeric spice and the Dindigul district is blessed with chilli. The Government of Tamil Nadu has to take concerted efforts to augment the area under spices through some action programs like area expansion under NHM.

The comparative picture in respect of area production and productivity of spices and the garlic in particular is analyzed and the details are presented in Table 6. The details presented are an indicator for the state government to assess their position in area production and productivity of spices and the garlic in particular.

Table 6 Comparative Analysis of Area Production and Productivity of Spices

Sl. No	Particulars	Spices	Garlic	Percentage
		All India		
01	Area in Ha	2940000	209340	07.12
02	Production in Tonnes	5351000	1264690	23.63
03	Productivity in Tonnes per Ha	01.82	06.04	
		Tamil Nadu State		
01	Area in Ha	200769	0740	00.37
02	Production in Tonnes	1127409	4410	00.39
03	Productivity in Tonnes per Ha	05.62	05.96	
Percentage of Production in Tamil Nadu in India		21.07	00.35	

Table 6 reveals that only seven percent of the area is under garlic compared to the total area under spices of the nation. In respect of production of garlic, India had produced around 24 per cent of spices to the total out turn of spices in the country. The All India Productivity of garlic is arrived at 6 tonnes per ha. The state of Tamil Nadu had also attained similar productivity as gained by the nation. In respect of area and production of garlic in Tamil Nadu, it is not even half per cent to the total area and production of the country revealing that the state of Tamil Nadu should concentrate on area expansion in spices.

AREA PRODUCTION AND PRODUCTIVITY OF GARLIC

In the preceding section we have discussed the area, production and productivity status in respect of spices and the this section addresses the area production and productivity of garlic in particular and the same was analyzed and the results are discussed in the following subheads.

- International Production of Fresh and Chilled Garlic
- Area Production and Productivity of Garlic in India and
- Area Production and Productivity of Garlic in Tamil Nadu

International Production of Fresh and Chilled Garlic

Since garlic has many medicinal properties including anti-cholesterol, many nations are producing the garlic in the world. But appreciable level of production of garlic is confined to very few countries. There were only 9 countries involved in production of garlic and their share in world production of garlic is analyzed and the results are presented in Table 7.

Table 7 International Production of Fresh and Chilled Garlic (in Tonnes during 2011)

Sl. No	Name of the Country	Production in Tonnes	Percentage To World Total
01	China	19219939.00	81.02
02	India	1057800.00	04.46
03	Egypt	295845.00	01.25
04	Republic of Korea	295002.00	01.24

Sl. No	Name of the Country	Production in Tonnes	Percentage To World Total
05	Russian Federation	233950.00	00.99
06	Myanmar	209308.00	00.88
07	Bangladesh	209153.00	00.88
08	United States of America	190690.00	00.80
09	Argentina	174363.00	00.74
10	Other Countries	1835396.00	07.74
	World Total	23721446.00	100.00

(Source: Food and Agriculture Organization)

Table 7 reveals that China is the leader in producing garlic and its share is accounted for 81 per cent of global garlic production. India is in the second place in garlic production and its share is accounted for only 4.50 per cent to thc global production of garlic which arrived at 23.72 million tonnes. The countries like Egypt, the Republic of Korea, Russian Federation, Myanmar, Bangladesh and United States of America are able to produce the garlic of meagre quantity. From this one could understand that China has to fulfill the requirements of garlic to many nations and act as monopoly producer and seller of garlic and one can say that China is the price taker. India is second in the ranking of garlic production and meets the domestic requirement of garlic besides little for export to its sister nations.

Area Production and Productivity of Garlic in India

Followed by the international production of garlic, the discussion on area production and productivity of garlic in India is ideal to compare the position of India. Hence, the details are analyzed state wise and the results are presented in Table 8.

Table 8 Area, Production and Productivity of Garlic in India (2010-2011)

Sl. No	Name of the State	Area '000 ha	Percentage to Total Area	Production '000 Tonnes	Productivity Tonnes/ha
01	Andhra Pradesh	00.40	00.19	04.00	10.000
02	Bihar	04.25	02.03	29.75	07.000
03	Chhattisgarh	03.90	01.86	21.10	05.410
04	Gujarat	40.00	19.11	275.00	06.880
05	Himachal Pradesh	03.60	01.72	44.70	12.420
06	Jammu & Kashmir	02.30	01.10	32.00	13.910
07	Karnataka	04.20	02.01	30.20	07.190
08	Madhya Pradesh	54.00	25.80	228.00	04.220
09	Maharashtra	03.50	01.67	34.10	09.740
10	Orissa	11.00	05.25	35.80	03.250
11	Others	13.29	06.35	104.74	07.880
12	Punjab	03.70	01.78	40.50	10.950
13	Rajasthan	25.00	11.94	150.00	06.000
14	Tamil Nadu	00.40	00.19	03.10	07.750
15	Uttar Pradesh	35.10	16.76	190.50	05.430
16	Uttaranchal	01.20	00.57	07.30	06.080
17	West Bengal	03.50	01.67	33.90	09.690
	Total	209.34	100.00	1264.69	06.040
	Average	12.31		74.41	06.040

Source: NHRDF, Nashik

Table 8 reveals that the area under garlic in India arrived at 2.09 lakh ha. From this area, the country could produce 12.65 lakh tonnes of garlic. The resultant productivity is arrived at 6.04 tonnes per ha at national level. In respect of state wise analysis, Madhya Pradesh, Gujarat, Uttar Pradesh,

Rajasthan and Orissa are found to have higher area under garlic which respectively accounted for 25 per cent, 19 per cent, 16.76 per cent, 11.94 per cent and 5 per cent to the total area under garlic in India. Madhya Pradesh is the single state that accounted for one fourth of the area under garlic cultivation of the country.

In respect of productivity of garlic in India, the state of Jammu and Kashmir stood first with the average productivity of 13.91 tonnes per ha followed by Himachal Pradesh and Punjab both of which accounted for 12.42 tonnes and 11 tonnes per ha of garlic respectively. The secret of achieving higher productivity of garlic in Jammu and Kashmir and the Himachal Pradesh might be mainly due to the soil quality and the climate factors which are conducive for achieving highest productivity when compared to all other states in the country. Though Madhya Pradesh had higher area under garlic, it could not perform well in terms of productivity due to poor soil and environmental conditions for cultivation of garlic. However, need based research alone could address the real constraints behind the Madhya Pradesh state in realization of higher output of garlic. The state of Tamil Nadu could earn the productivity of garlic to only around 50 per cent of the topper state of Jammu and Kashmir in India. Though area under garlic cultivation is poor in Andhra Pradesh, it could contribute higher output in terms of productivity which arrived at 10 tonnes per ha.

To sum up, when we compare the state wise area, production and productivity of garlic, Madhya Pradesh has higher area followed by Gujarat. But in terms of productivity of garlic, the state of Jammu and Kashmir and the Himachal Pradesh could be the front runners followed by Punjab and the Andhra Pradesh states. West Bengal and Maharashtra are the other two states that had higher potentials to realize higher productivity of garlic and hence the Spices Board should initiate appropriate efforts to enhance the area and productivity of garlic in the country by intensifying research and developmental efforts to make the country a self-sustained one in terms of garlic and spices.

Area Production and Productivity of Garlic in Tamil Nadu

Tamil Nadu is one of the states which had poor area under garlic. Though there were 32 districts in Tamil Nadu, it has only five districts that are able to produce garlic to some extent. The details of area production and productivity of garlic with respect to Tamil Nadu is analyzed and the results are presented in Table 9.

Table 9 Area Production and Productivity of Garlic in Tamil Nadu during 2010-11

Sl. No	Districts	Area in Ha	Area in Per Cent	Production in Tonnes	Productivity in Tonnes per Ha
01	Coimbatore	01	00.14	06	04.00
02	Cuddalore	80	10.81	475	05.93
03	Dharmapuri	01	00.14	08	06.00
04	Dindigul	479	64.73	2921	06.10
05	Erode	42	05.67	237	05.63
06	Kancheepuram	03	00.41	17	06.00
07	Kanyakumari	00	00.00	00	00.00
08	Karur	00	00.00	00	00.00
09	Krishnagiri	08	01.07	49	05.83
10	Madurai	00	00.00	00	00.00
11	Nagapattinam	00	00.00	00	00.00
12	Namakkal	01	00.14	03	02.00
13	Perambalur	00	00.00	00	00.00
14	Pudukottai	00	00.00	00	00.00
15	Ramanathapuram	00	00.00	00	00.00
16	Salem	01	00.14	08	06.00
17	Sivagangai	00	00.00	00	00.00
18	Thanjavur	00	00.00	00	00.00
19	The Nilgiris	124	16.75	686	05.55
20	Theni	00	00.00	00	00.00
21	Thiruchirappalli	00	00.00	00	00.00
22	Tirunelveli	00	00.00	00	00.00
23	Thiruvallur	00	00.00	00	00.00
24	Thiruvannamalai	00	00.00	00	00.00
25	Thiruvarur	00	00.00	00	00.00
26	Thoothukkudi	00	00.00	00	00.00
27	Vellore	00	00.00	00	00.00

Sl. No	Districts	Area in Ha	Area in Per Cent	Production in Tonnes	Productivity in Tonnes per Ha
28	Villuppuram	00	00.00	00	00.00
29	Virudhunagar	00	00.00	00	00.00
30	State Total	740	100.00	4410	05.96

Hort Stat -2008; Government of Tamil Nadu

Table 9 reveals that among the 30 districts, the potential districts capable of cultivating garlic are Dindigul, The Nilgiris, Cuddalore and Erode accounting for 65 per cent, 16.75 per cent, 11 per cent and 6 per cent respectively to the total area under garlic in Tamil Nadu. The average productivity of garlic in Tamil Nadu is arrived at 5.96 tonnes per ha which is almost equal to the productivity of garlic at national level. In Dindigul district, Kodaikanal is the place which is ideally suitable for cultivating garlic. The quality of the garlic output produced in Kodaikanal is well received in many parts of the country and it is being exported abroad for its quality.

The productivity of garlic per ha in India and Tamil Nadu is found to be almost similar and ranging to the level of 5.90 to 6.00 tonnes and hence a micro level analysis was held at Kodaikanal among the farmers spread over in different villages by selecting one farmer per village and the details are analyzed and the results are presented in the ensuing section.

The Production Potentials of Garlic Farms in Kodaikanal The production potentials available with garlic farms are important to compare the production efficiency in terms of productivity. One can specifically suggest the need for technology adoption to enhance productivity further. In this respect, the following aspects were analyzed from the case farms. They are:

- Size of Holding Available with the Sample Farms
- Cropping Pattern in Vogue with the Sample Farms
- Area Under Garlic in the Villages of Kodaikanal
- Cost of Production of Garlic in the Sample Farms
- Revenue Realized from Sale of Garlic in the Sample Farms
- Production Constraints Faced by the Garlic Farms

Size of Holding Available with the Sample Farms Before analyzing the productivity at micro level, analysis on the size of holding available with the sample farms is ideal and it will reveal the status of small or medium farms as it has relevance in productivity. Sekhar et.al (1994) revealed that the small farms are efficient in achieving higher productivity per unit area and hence this discussion is imperative and the results are analyzed and the details are presented in Table 10.

Table 10 Details of Size of Holding Available with the Sample Farms

Sl. No	Particulars of Holding	Average Size of Holding in Acres	Percentage to Total
01	Owned Land under Cultivation	02.00	61.54
02	Land under Non Agricultural Purpose	00.25	07.69
03	Owned Land (1+2)	02.25	69.23
04	Leased-in Land	01.00	30.77
05	Total (3+4)	03.25	100.00

Table 10 reveals that size of holding per farm is found to be 2.25 acres which accounted for 69 per cent to the total size of holding under cultivation. The remaining 31 per cent of land is under lease with the sample farms selected and the sample farm is coming under the category of marginal farms which is less than one ha and hence these farms should be of highly efficient one in terms of attaining the productivity because of adoption of intensive farming practices.

Cropping Pattern in Vogue with the Sample Farms Cropping pattern practiced by the sample farmers is very important in the context of analyzing the crop rotation practiced by the sample farmers and hence these details are analyzed and the results are delineated in Table 11.

Table 11 Cropping Pattern in Vogue with the Sample Farms during 2012-13

Sl. No	Name of the Crop	Area in Acres	Percentage to Total
01	Butter Beans	01.00	16.00
02	Cabbage	00.50	08.00
03	Carrot	00.50	08.00
04	Garlic	00.60	09.60

Sl. No	Name of the Crop	Area in Acres	Percentage to Total
05	Knool-Khol	00.25	04.00
06	Potato	02.00	32.00
07	Turnip	01.40	22.40
	Total	06.25	100.00
	Average Size of Holding	03.25	52.00
	Cropping Intensity in Per cent		192.31

Table 11 reveals that the cropping intensity among the sample farms is found to be 192 per cent. It has outlined that the farmers are hard working to raise repeated crops of vegetables and spices in the size of holding available with the farms. In respect of cropping pattern with the sample farms, potato is found to have higher area among the sample farms which has shared 32 per cent of the size of holding followed by turnip, butter beans and the garlic which accounted for 22 per cent, 16 per cent and 10 per cent respectively to the total size of holding. Carrot and cabbage are found to have equal area of importance after the potato crop.

Area under Garlic in the Sample Villages of Kodaikanal Our focus is centered around on the cultivation details of garlic and hence the area, production and productivity of garlic is analyzed and the details are outlined in Table 12. The details of area production and productivity of garlic in the sample villages at micro level are compared with the national and regional level productivity.

Table 12 Area under Garlic in the Villages of Kodaikanal Taluk

Sl. No	Name of the Village	Area under Garlic in Ha	Percentage to Total	Production in Tonnes	Productivity per Ha
01	Poomparai	112.00	21.46	941.000	08.400
02	Poondi	074.00	14.18	599.400	08.100
03	Kavunji	064.00	12.27	505.600	07.900
04	Kilavarai	054.00	10.34	388.800	07.200
05	Pallangi	030.00	05.76	207.000	06.900

Sl. No	Name of the Village	Area under Garlic in Ha	Percentage to Total	Production in Tonnes	Productivity per Ha
06	Kodaikanal	010.00	01.90	068.000	06.800
07	Senbaganur	010.00	01.90	076.000	07.600
08	Other Villages	168.00	32.19	1192.800	07.100
09	Total Area under Garlic	522.00	100.00	3978.600	07.600

Table 12 reveals that the total area under garlic in the sample farms is arrived at 522 ha. From that the farmers are able to produce around 3979 tonnes of garlic leaving the average productivity of 7.60 tonnes per ha which is 26.67 per cent higher than the regional and national productivity of garlic. Among the seven villages, the Poomparai, Poondi, Kavunji are the villages found to have high potentials in terms of soil quality and hence higher area under garlic in these villages. The out-turn of garlic from Kodaikanal villages are sent to the Vadugapatti wholesale market functioning exclusively for garlic in Theni district and hence effort has been taken to document the functioning of garlic market with respect to its receipt and disposal of the garlic and its linkages with the national and regional markets. Before that the details of cost of production per ton of garlic will reveal the status of earning profit to the producers.

Cost of Production of Garlic in the Sample Farms The farmers of Kodaikanal trust garlic as one of the cash crop which will offer higher price per unit of the produce harvested from their farm. They can store the produce if the price is low and when the market is attractive, slowly they will release the produce to the market. During the time of storage of the produce, the farmers used to fumigate for drying and preserving the produce. However, the market is showing fluctuations in price for the garlic and the same goes from Rs 50 to Rs 250 per kg. In this circumstance, it is important to know how much it costs the farmer to produce a kg of garlic so that one could assess the level of profit earned by the producer and hence the related information were gathered from the sample farms and the data were analyzed and the results are presented in Table 13.

Table 13 Cost of Production of Garlic in the Sample Farms

S. No	Details of Expenditure	Number of Units	Price Per Unit	Expenditure / Acre
		Annual Fixed Cost		
1.1	Annual Fixed Cost (Rental Value of Owned Land per Acre)	One Acre	43330.00	43330.00
1.2	Annual Fixed Cost per Crop	One Crop	21665.00	21665.00
1.3	Cost of Gunny Bags @ Rs 35 per Bag assuming the life of 5 years	30.00		0210.00
		Production Cost		
2.1	Ploughing of Land @ Rs 700 per Plough	03.00	700.00	2100.00
2.2	Clearing the Weeds and Grasses using the women	05.00	200.00	1000.00
2.3	Land Preparation and Channel Formation @ Rs 400 per Man Day	05.00	400.00	2000.00
2.4	Cost of Garlic Seed @ Rs 150 per Kg	70.00	150.00	10500.00
2.5	Planting of Garlic @ one women day for 10 Kgs of Seed	10.00	200.00	2000.00
2.6	Application of NPK Fertilizer @ Rs 600 per Bag	60.00	600.00	36000.00
2.7	Application of Cow Dung Milk @ Rs 4000 per Tractor Load	02.00	4000.00	8000.00
2.8	Weeding Two Times @ 10 Women Days per Weeding	20.00	200.00	4000.00

S. No	Details of Expenditure	Number of Units	Price Per Unit	Expenditure / Acre
2.9	Application of Rogor Spray on 15th Day after Sowing @ Rs 80 per Tank including Cost of Chemicals	05.00	080.00	400.00
2.10	Spray of Bavistin on 20th Day after Sowing @ Rs 85 per Tank including the Cost of Chemicals	05.00	085.00	425.00
2.11	Spray of Kavach on 40th Day after Sowing @ Rs 130 per Tank including the Cost of Chemicals	05.00	130.00	650.00
2.12	Spray of Biosal on 20th Day after Sowing @ Rs 100 per Tank including the Cost of Chemicals	05.00	100.00	500.00
2.13	Harvesting Cost @ 2 Man Days and 3 Women Days	05.00	280.00	1400.00
2.0	Production (Variable) Cost (II)			68975.00
	Post Harvest Expenses			
3.1	Cleaning and Packing Charges in Gunny Bags @ 2 Man Days	02.00	400.00	800.00
3.2	Transportation Charges to Vadugapatti Wholesale Market @ Rs 50 per Bag	70.00	50.00	3500.00
3.3	Sub Total Cost (III)			4300.00
	Total Cost			
4.1	Annual Fixed Cost (I)			21875.00
4.2	Variable Cost (II)			68975.00
4.3	Post Harvest Expenses			4300.00
4.0	Total Cost			95150.00

S. No	Details of Expenditure	Number of Units	Price Per Unit	Expenditure / Acre
	Cost of Production of Garlic			
5.1	Output Realized per Acre in Tonnes	03.04	100000.00	304000.00
5.2	Cost of Production per Tonne of Garlic			31299.00
5.3	Cost of Production per Kg of Garlic			31.30

Table 13 reveals the details of annual fixed cost, variable expenses involved in producing garlic per acre, the output realized per crop per acre and the cost of production. The annual fixed cost involved in realizing the output of garlic is arrived at Rs 21875 per acre and the variable expenses involved in producing the garlic crop per acre is arrived at Rs 68975. The post-harvest expenses incurred by the farmer is arrived at Rs 4300 including transportation. Put together the total cost of realizing the output of garlic per acre is arrived at Rs 95150/. In the sample farms, the average productivity of garlic per acre is arrived at 3.04 tonnes (7.60 tonnes per ha) capable of consuming Rs 31.30 as cost towards realizing one kg of garlic. From this, one could understand that the farmer is realizing considerable return from the crop of garlic and this will act as the vicissitude against the losses in other crops due to price failure.

Revenue Realized from Sale of Garlic in the Sample Farm Any production process will generate some revenue. But in agriculture and horticulture, the revenue generation is a cumbersome process. It is mainly due to high cost of labor, high cost of inputs like seed, fertilizer and other pesticides besides poor rainfall and water availability for crop production activities. In the case of production of garlic, the Kodaikanal farmers are mainly depending on the rainfall and they establish the garlic during the time of onset of monsoon and harvest the produce within four months. The details are analyzed and the results are presented in Table 14.

Table 14 Revenue Realized from Sale of Garlic in the Sample Farm

Sl. No	Details of Revenue Generated	Return in Rs Per Acre
01	Gross Return Realized from Sale of Garlic	304000.00
02	Total Investment Made per Crop per Acre	095150.00
03	Net Return Per Crop Per Acre	208850.00
04	Percentage of Profit Earned by the Producer	68.70

Table 14 reveals that the garlic crop could generate a gross return of Rs. 3.04 lakhs per acre. For that the farmer could spend an amount of Rs 0.95 lakh as cost and the net return per crop per acre is arrived at Rs. 2.09 lakhs. The percentage of profit earned by the farmer is arrived at 69 per cent to the gross return.

To sum up, the total cost of raising garlic crop per acre is arrived at Rs 0.95 lakh and the per acre productivity of garlic is arrived at 3.04 tonnes leaving the cost of production per kg at Rs 31. These details revealed that there exists scope for realizing higher returns from garlic crop in Kodaikanal due to conducive climate and good soil health.

Production Constraints Faced by the Garlic Farmers Any production activity commands naturally certain constraints. In the activity of cropping of garlic, the Kodaikanal farmers face lot of constraints and these details are analyzed and the results are presented in Table 15.

Table 15 Production Constraints Faced by the Garlic Farmers

Sl. No	Production Constraints	Rank
01	Damage by the Wild Animals	I
02	Yield Reduction over a Period of Time	II
03	Poor Rainfall Availability per Annum	VI
04	Shortage of Farm Labor for Farm Operations	III
05	Higher Wages to Labor Commands higher Cost of Production	IV
06	Subsidy for Leased-in Land from the Department of Agriculture is Poor	V

Table 15 reveals that there were six important constraints the farmers in Kodaikanal face regularly. These are damage by the wild animals, yield reduction over a period of time, poor rainfall availability per annum, shortage of farm labor for farm operations, higher wages to labor commands higher cost of production and the subsidy for leased-in land are not available to their respective owner from horticulture and agriculture departments. Among these constraints, the first and foremost constraints formed the damage by the wild animals followed by the yield reduction over a period of time. The yield reduction is mainly due to depletion of soil nutrients during the rainy season and this in turn forced the farmer to invest on higher nutrients in the form of farmyard manure and the fertilizers. The garlic commanded the cow dung solution for realizing higher productivity which is practiced by almost all the farmers raising garlic in the Kodaikanal fringes.

SUMMARY AND CONCLUSION

The area under spices was found to be reduced to the tune of nine per cent in a decade of time. When we compare the production figures, the production is found to be increased to the tune of 42 per cent and the productivity is arrived at 39 per cent increase. The All India Productivity of garlic is arrived at 6 tonnes per ha. The state of Tamil Nadu had also attained similar productivity as gained by the nation. In respect of area and production of garlic in Tamil Nadu, it is not even half per cent to the total area and production of the country revealing that the state of Tamil Nadu should concentrate on area expansion in spices. China is the leader in producing garlic and its share is accounted for 81 per cent of global garlic production. India is in the second place in garlic production and its share is accounted for only 4.50 per cent to the global production of garlic which arrived at 23.72 million tonnes. The countries like Egypt, the Republic of Korea, Russian Federation, Myanmar, Bangladesh and United States of America are able to produce garlic of meagre quantity. Madhya Pradesh, Gujarat, Uttar Pradesh, Rajasthan and Orissa are found to have higher area under garlic cultivation in India. Madhya Pradesh is the single state that had one fourth of the area under garlic cultivation of the country. The state of Jammu and Kashmir stood first with the average productivity of 13.91 tonnes per ha followed by Himachal Pradesh and Punjab both of which accounted for 12.42 tonnes and 11 tonnes per ha of garlic productivity per ha.

In the sample farms, the average productivity of garlic per acre is arrived at 3.04 tonnes capable of consuming Rs 31.30 as cost towards realizing one kg of Garlic. Among the constraints, the first and foremost constraints formed the damage by the wild animals followed by the yield reduction over a period of time. The yield reduction is mainly due to depletion of soil nutrients during the rainy season and forced the farmer to invest on higher nutrients in the form of farmyard manure and the fertilizers. The garlic commanded the cow dung solution for realizing higher productivity which is practiced by almost all the farmers raising garlic in the Kodaikanal fringes.

- The garlic area is found to be on the decline and the productivity per ha is also found to be declining over the years and hence appropriate research and development and extension effort need to be intensified in the hilly areas through area expansion, technology dissemination and the technology demonstration to fetch higher productivity in spices and garlic.
- Since hill agriculture and its output commands higher consumer preference, the special subsidy to ensure production and productivity may be mooted through some special schemes.
- The Farmers' Retail Market in the foothill town may be established for the producer-consumer advantage by establishing Commodity Interest Groups and the CIGs can run the retail outlet on their own for higher percentage share by minimizing the role of middlemen in the name of product specific outlet (Spices Retail Outlet). The fruits and vegetable stores on the lines of Pazhamudhir Nilayams can also be established in different cities which had been accepted by all sections of the consumers and hence the same may be practiced in the foothill towns.

References

- Sajith Mohan, Sheena S. Rajan and Unnikrishnan, G. "Marketing of Indian Spices as a Challenge in India", **International Journal of Business and Management Invention,** 2(2): February 2013, pp. 26-31
- Government of Tamil Nadu (2008). "***Hort-Stat 2008***" (Directorate of Horticulture and Plantation Crops: Chennai) pp. 1-493

- Government of Tamil Nadu, "Season and Crop Report of Tamil Nadu – 2010" (Directorate of Economics and Statistics: Chennai)
- Government of India, "Agricultural Statistics at a Glance – 2011-12" (Directorate of Economics and Statistics: New Delhi)
- Government of India (2011). ***Horticulture Data Base 2011*** (National Horticulture Board: New Delhi) pp. 01-278
- Government of India (2012). Spice India (Spices Board, Kochi) July 2012

Web References

- www.indianspices.com/pdf/major_spice_statewise_area_production_2013_new.pdf

Chapter 16

Transformation of A Jamun Farmer into an Agripreneur

INTRODUCTION

Jamun (*Syzygium cuminii*) is a tall evergreen tree of tropical and subtropical regions and had its origin probably in India and the East Indies. It is one of the hardiest fruits and best suited for wastelands and drylands. At the same time it also tolerates water stagnation and marshy lands, where other fruit crops cannot be grown to the desired level. The wood is used as a timber in building and railway sleepers. The fruits of the tree are loved by almost all age groups from children to the aged. The fruits of the tree are good to the sugar patients and the nut is used in siddha formulations and hence it is of high demand in home and in the industries. It is also used to make beverages, squash, jam, jelly and wine. Fruit extract is also used in curing intestinal disorders.

CLIMATIC REQUIREMENTS

Jamun is adapted to tropical and subtropical conditions requiring a dry climate during flowering and fruiting seasons. Early rains resulting in better soil moisture will help in growth and development and ripening of fruits. Well drained deep loamy soil would be ideal though it can grow on a wide range of soil conditions. In the initial stages of its establishment there should not be any drought and hence one should take utmost care in providing irrigation periodically for its establishment. When the tree has grown sufficiently it can tolerate any type of drought and flooding (Veerar-agavathatham et al.2004).

THE CULTIVARS

In northern parts of India, a cultivar known as 'Ra Jamun' with big sized fruits is being cultivated widely. One seedless type (with underdeveloped ovule), isolated at Horticultural College and Research Institute, Periyakulam is being grown in Agricultural Research Station, Paramakudi, Tamil Nadu which is small in size. Preferably, the farmer, in Dindigul district has the cultivar named 'Jambu Naval' which drawn from Rajahmundhry of Andhra Pradesh is also cultivated in Tamil Nadu. An improved variety Konkan Bahdoli has been recommended for cultivation in Maharashtra.

SILVICULTURAL PRACTICES

The propagation of true to the type quality planting material is an integral part of Jamun cultivation, since Jamun has a long gestation period. The utmost care is required in the selection of genuine planting material at the time of planting. Jamun is monoembryonic. It is heterozygous in nature hence it does not produce true to the type planting material. However, seedlings are being raised as rootstocks. Hence, knowledge of both methods of propagation is important.

SEED PROPAGATION

Seed propagation is the most common method of propagating Jamun. Jamun seeds have no dormancy; hence fresh seeds can be sown 4-5 cm deep in the nursery within 10-15 days. The seeds germinate 10-15 days after sowing. The seedlings become ready for transplanting in spring or next monsoon. If the seeds are sown too deep, seedling emergence is delayed and there may be some rotting due to poor aeration. Seeds may also be sown in polythene bags, as it facilitates easy handling of rootstocks and grafted plants. There is occurrence of polyembryony in Jamun to the extent of 20 to 50 per cent, hence nuclear seedlings may be utilized to produce true to the type plants (Singh and Thakur, 1977). The seeds of Jamun took 24 to 61 days for total germination under Bihar conditions, whereas in Tamil Nadu, it took only 15 to 20 days for germination. Studies revealed that seed extraction of Jamun after heaping the fruits for a single day was better for getting good quality seeds in comparison to the extraction of seeds immediately after collection (Srimathi et al. 2003). Sasthri et al. (2001) recorded that large sized seeds had higher germination percentage (98-99 per cent) than the small sized seeds (79-89 per cent).

The seeds of Jamun are sown in the poly ethylene bags filled with soil and FYM at the rate of 3:1 for raising the rootstocks. Seeds can also be sown in the raised nursery beds in open or polyhouses. Transplanting of young seedlings of 2-4 leaf stage in the polythene bags gives 80 per cent success. About 9-12 months old seedlings of uniform size having stem of pencil thickness are used as rootstock for budding and grafting. Plants raised in the polybags can easily be transported to distant places and give higher transplanting success (Singh et al. 2011).

SOFT WOOD GRAFTING

About 15-20 cm long mature shoots of 2-3 months old are defoliated 12-15 days prior to grafting operation. These shoots are detached from the mother plant with the help of secateurs or sharp grafting knife for grafting by cleft method. For this, soft portion of seedling rootstock is cut at 20-35 cm height and the top portion is removed. With the help of knife, 5 cm long vertical downward incision is made in the center of the rootstock. A sharp cut of 5 cm is made on both the sides on the base of the scion shoot to make wedge shape. Thereafter, prepared scion was carefully inserted in vertical slit of the rootstock and tightly secured with the help of 200 gauge thick and 2 cm wide polythene strips. The polythene strips should carefully be removed after the completion of the union. This method is proved to be better for *in-situ* grafting under rainfed conditions. Chovatia and Singh (2000) recorded 41.67 per cent success in soft wood grafting in June under Gujarat conditions.

HIGH DENSITY PLANTING

High density planting system ensures better utilization of land, labor and solar radiation and higher yield in the initial years of planting, because of accommodating higher number of plants per unit area. High density orcharding appears to be the most appropriate answer to the problem of low productivity and for early economic returns from the Jamun orchards. Goods soil conditions, planting geometry and manipulation in the spacing are important means in obtaining higher production and productivity.

WATER MANAGEMENT

Irrigation is not normally practiced in Jamun cultivation. But it promotes better growth during the period of establishment and the early stages of growth, especially during the dry seasons. In the early stage, plants require 8 to 10 irrigations per year while, bearing trees require 4-5 irrigations during the time of fruit development and ripening. In dry areas, the use of water harvesting techniques during the rainy season will be useful for ensuring proper irrigation to improve subsequent growth and yield of Jamun. Use of mulch is very much beneficial for successful cultivation of Jamun. Mulching can be done with polythene or any suitable organic material. Mulching with grasses, paddy straw and rice husk reduces the weed population and conserves the moisture in the soil. Microbial and earthworm population in the basin soil increases with the use of paddy straw and grasses as mulch. Leaf litter of Jamun under the canopy is effective to retain soil moisture during the summer.

CANOPY MANAGEMENT

The basic concept in canopy management of a perennial tree is to make the best use of the land and the climatic factors for increased productivity in a three dimensional approach. Tree vigor, light, temperature and humidity play a vital role in the production and quality of the fruits. Young plants should be trained with 3-5 well spaced branches to develop into the main scaffold structure of the tree. Framework of branches is allowed to develop above 60-100 cm from the ground level. Jamun trees may be trained through the following training systems.

Central Leader

The trees are trained on a main central leader with 6 to 10 strong lateral scaffold branches, spread in all directions. This system of training provides unnecessarily large tree which creates harvesting problem.

Open Centre

The central leader is removed about one meter above ground level and four to six well spaced tertiary branches are retained. Secondary scaffold branches are encouraged to develop the fruiting canes. Both primary and secondary scaffold branches produce fruit bearing laterals. Training by this system keeps the centre of the tree open, permitting entry of adequate sunlight throughout the tree, but due to development of weak framework, plants do not resist high velocity winds resulting in breaking of branches.

Modified Leader

In this system, the main central leader is allowed to grow for few years, until 8-10 scaffolds develop around the central leader. The central leader is then cut to develop the side laterals, which in due course, will grow as a modified leader. In this type of training, the tree develops well spaced limbs with strong crotches. The top being open, allows more sunlight to penetrate deep inside the tree canopy. This system is appropriate for training of Jamun plants (Singh et al. 2011).

Pruning

Pruning is a tool to regulate tree size and shape to achieve a desired architecture of the canopy and also to reduce the foliage density by removing the unproductive branches to make the tree open. Regular pruning in Jamun plant is not required. However, dry, weak and diseased branches should regularly be removed. To maintain the dwarf framework of the Jamun tree, toping of main stem (4-6 meters) is needed. It will facilitate easy harvesting of the fruits. It is also observed that pruning of 50% annual extension growth after harvesting was effective to reduce the plant canopy and to improve the fruit quality attributes.

PLANT PROTECTION

Leaf eating caterpillars can be controlled by spraying dimethoate/malathion. White fly damages all parts and even fruits get wormy. Affected, dropped fruits should be collected and burnt. At the time of flowering, if spraying has to be taken up, only a safe insecticide to honey bees like endosulfan may be sprayed. Leaf spot and fruit spot caused by *Glomerella* can be controlled by Indofil Z.78 (2 grams per litre) (Veeraragavathatham et al. 2004).

HARVESTING PRACTICES AND YIELD

The fruit ripens in June-July. The main characteristic of ripe fruit at full size is deep purple or black in color. The Jamun fruit should be picked immediately when it ripens. It cannot be retained on the tree at the ripe stage. The ripe fruits are picked singly by hand and in all cases care is taken to avoid all possible damage to fruits. For harvesting, the picker climbs on the tree with bag of cotton slung on the shoulder. When the bag is full, the picker comes down from the tree and empties bag in the basket or hangs the bag down to the ground with the help of rope and another person standing below the tree catches the bag and empties it in baskets. The fruits of Jamun are generally harvested daily and sent to market on the same day. The average yield of fruits from a full grown Jamun tree is about 80-100 kg and from a grafted one the yield will be 60-70 kg per tree per year (Singh et al. 2011).

PROCESSING OF JAMUN FOR VALUE ADDITION

Jamun exhibits heavy post-harvest losses. It is therefore, necessary to process the fruits into different value added products to utilize marketable fruits and add value to fresh fruits to avoid price crash during the peak season due to gluts. The fruits may be utilized for jam, jelly, beverages, wine, vinegar and pickles (Kadam, 2001).

MARKETING PRACTICES

Marketing practices are more in Jamun due to their high degree of perishability and short season of fruit availability. An analysis of the marketing cost had indicated that the commission agent and the transport charges account for 80 to 90 per cent of the total marketing cost (Krishnamurthy and Sudhakar Rao, 2001). About 75 per cent of the farmers sell their produce at the farm level to the village merchants, retailers, big producers or pre-harvest contractors. They cannot afford to transport their produce to distant markets on account of the non-availability of transport facilities, expensive transport and malpractices in the market. Information regarding demand, supply, price, market outlook, knowledge of the consumer's preference, marketing channels is important for marketing the produce like Jamun.

PROBLEM FOCUS

Above marketing constraints revealed that the studies are not much focused and conducted towards production and marketing of minor fruits like Jamun. Amidst these constraints, there were progressive farmers who entered into these ventures and are making inroads to promote the underutilized fruits. In Tamil Nadu, Jamun orchard established by one such progressive farmer in Dindigul District, was specially approached and the production and value addition practices were analyzed and documented here as a case study to highlight the experiences of the progressive farmer as a paradigm to other locale. With this in mind, the study focuses on the pattern of investment made by the progressive farmer in Jamun orchard and the value addition practices practiced by the farmer to earn higher dividend from the Jamun orchard.

CASE STUDY OF A SUCCESSFUL FARMER

Jamun is one of the most hardy fruit corps and can easily be grown in the neglected and marshy areas, arid and semiarid areas, resource poor areas and wastelands where other fruit trees cannot be grown successfully. Information regarding area and production of this fruit in India is not available because it is not grown on plantation scale but trees in patches are common in the villages in many states. In Tamil Nadu, orchards in Jamun are being practiced in the farm lands by the progressive and innovative farmers. One such farmer is Mr. Jeyakumar hailing from J. Mettur village of Dindigul District of Tamil Nadu who is cultivating the Jamun in one acre with the plant density of 86 trees per acre in an espacement of 20 X 20 feet and also having an idea of expanding the area under Jamun additionally to an area of 3.50 acres at an espacement of 25 X 25 feet with the intention of cultivating guava as intercrop. With regard to the marketing, the farmer is having the practice of selling the fruits directly to the retail outlets located in important cities and also involved in extracting Jamun fruit squash from the residual Jamun fruits and hence a case study approach is practiced to learn the aspects of investment made in the Jamun plantation and its value addition aspects practiced in the farm. For that purpose, a questionnaire was prepared and pre-tested with the farmer and the details of land use, costs involved in establishing the plantation, returns realized and the details of value addition, etc., were collected, tabulated and interpreted using conventional percentage analysis.

Details of the case study under suitable headings

India is emerging as a major stakeholder in the global horticulture scenario accounting for 10 per cent of the world production of fruits. This has been made possible due to concerted efforts of scientists and progressive farming community. Jamun fruit is a good source of nutrients and it is used for curing diabetes for its hypoglycemic properties. The case farmer has visited the plantation in Andhra Pradesh after the advice of an expert from Tata Energy Research Institute (TERI). TERI had conducted a training at Tanjore in which the case farmer was a trainee. After coming to know that the 'Jambu' variety of Jamun is capable of yielding higher output per unit area and capable of yielding from the fourth year onwards, he purchased and planted the variety in his establishment. The farmer is found to be an innovator because of adoption of newer technologies from different parts of the globe. Because of its demand in the open market, the farmer has started cultivating the Jamun in his farm lands and hence special effort has been taken to share the experiences of the farmer following a case study approach and the results are discussed in the following headings.

- Land Use and Cropping Pattern Prevalent among the Case Farm
- Cost and Return associated with Jamun Orchards
- Details of Output Produced in the Case Farm
- Details of Distribution of Output
- Constraints associated with Production and Distribution of Output

LAND USE AND CROPPING PATTERN PREVALENT AMONG THE CASE FARM

A case study approach normally deals with A to Z practices practiced in the farm lands. In this respect, first, the land use and cropping practices prevalent among the case farm is analyzed and the results are presented in Table 1.

Table 1 Land Use Pattern Prevalent in the Case Farm of Dindigul District

Sl. No	Particulars of Land Use	Spacing in Feet	Area in Acres	Age of the Plantation/ Infrastructure	Percentage to Total Area
01	Jamun (Naval)	20 X 20	01.20	11.00	09.16
02	Jamun (Naval) And Guava (L 46) (Intercrop)	25 X25	03.50	00.01	26.72
03	Amla	20 X 20	03.60	12.00	27.49
04	Coconut (in Rows)	20X 20	00.40	12.00	03.05

Sl. No	Particulars of Land Use	Spacing in Feet	Area in Acres	Age of the Plantation/ Infrastructure	Percentage to Total Area
05	Other Crops		01.30	00.00	09.92
05	Farm House		00.10	11.00	00.76
06	Fruit Processing Unit		01.00	08.00	07.63
07	Parking Area Shed and Farm Road		02.00	00.00	15.27
	Total		13.10 Acres or 05.20 Ha		100.00

Table 1 reveals that the size of holding available with the case farm is 13.10 acres (5.20 Ha). Among this, the Jamun plantations were established in 1.20 acres (0.48 Ha) and 3.50 acres (1.40 Ha) which are having the respective age group of 11 years and very young plantation of one-month-old respectively accounting for 9.16 per cent and 26.72 per cent to the total size of holding. The 11-year-old Jamun orchards and the young plantations of one-month-old were established in an espacement of 20 X 20 feet and 25 X 25 feet respectively. The very purpose of establishing Jamun at higher spacing is mainly for establishing guava (L 46) as an intercrop in the inter-row spaces of Jamun.

In addition to Jamun, the case farm has also established amla plantation in an espacement of 20 X 20 feet in an area of 3.60 acres (1.44 Ha) which accounted for 27.49 per cent to the total size of holding and the coconut was established in rows in an espacement of 20 feet along the periphery of the case farm which accounted for 3 per cent to the total size of holding. Other field crops which were also in the farm occupied 1.30 acres (0.52 Ha) accounting for 10 per cent to the total size of holding. Remaining areas are earmarked for processing unit, farm road, parking area and shed and residential building which in toto accounted for 24 per cent to the total size of holding.

To sum up, the cropping program occupied 76 per cent of the total size of holding and the remaining 24 per cent of the land was under farm roads and permanent infrastructure like buildings, parking area, shed and processing units which are essentially required for production and value addition of output produced in the case farm.

COSTS AND RETURNS ASSOCIATED WITH JAMUN ORCHARDS

Most of the farmers are unaware about the costs and returns associated with the establishment of perennial trees especially Jamun and hence the spread of Jamun orchards was poor in Tamil Nadu and other parts of India amidst increasing demand. Analyzing and presenting the details of costs associated with the establishment of Jamun plantation and the possible returns realized from Jamun orchards could motivate other progressive farmers also to take up such venture. In this respect, the details of cost and the returns are analyzed and are presented in the following subheads as a motivation tool.

- Establishment Cost
- Maintenance Cost Over the Years
- Revenue realized from Jamun Orchards

Establishment Cost involved in Jamun Orchards

The researchers normally classify the costs involved in perennial crops into the establishment, maintenance and harvesting costs. According to Sekhar et al. (1990), the costs are classified into three, viz., (i) establishment costs, (ii) maintenance and (iii) harvesting costs. Establishment cost involved in jamun plantations were ploughing of land, preparation and layout, manures, cost of seedling, pitting and planting costs, cost of irrigation, soil working and weeding. Maintenance costs included cost of irrigation, weeding and soil working, gap filling. Harvesting costs included collection, packing, forwarding and transportation costs.

Harswarup Singh (1988) classified costs into two, viz., (i) fixed costs (establishment cost) and (ii) recurring and maintenance cost. The fixed cost consisted of all the expenditure on goods and services (excluding the value of land) required for the establishment of trees. These included preparations of land and layout, digging and filling of pits, cost of plant material, transportation of seedlings including loading and unloading charges, planting cost, manures and fertilizers, plant protection and miscellaneous costs. Recurring and maintenance costs were those which are incurred every year. These expenses included manures and fertilizers, irrigation, plant protection, labour, interest on establishment and opportunity cost of land (rental value). Harvesting costs were not included as standing trees were sold. In this study, the establishment cost concept given by Sekhar et.al is used.

In establishing Jamun orchards, the establishment costs were presented as those costs involved prior to the bearing of fruits and the maintenance costs presented were those costs incurred towards maintenance of the orchards from the year of bearing of fruits onwards and hence the establishment costs were categorized item wise and the details are presented in Table 2.

Table 2 Establishment Cost Involved in Jamun Plantation Established in the Case Farm of Tamil Nadu

Sl. No	Particulars of Cost	Year			Total Cost	Percentage to Total Est. Cost
1.0	Establishment Cost	01	02	03		
1.1	Preparation of Land – Ploughing 5 Times @ Rs 1500 per Plough	7500	00	00	7500	04.96
1.2	Cost of 90 Pits (1X1X1 Feet) per Acre @ Rs 5 per Pit	0450	00	00	0450	00.30
1.3	Cost of 100 Jamun Seedlings @ Rs 100 per Seedlings	10000	00	00	10000	06.61
1.4	Cost of Fertilizers and its Application	1100	1100	1100	03300	02.19

Sl. No	Particulars of Cost	Year			Total Cost	Percentage to Total Est. Cost
1.5	Watering to Plants on Weekly Basis @ Rs 300 per Man Day per week. Assume that per Watering, the Man Day requirement will be half Man Day only	5200	4300	4300	13200	08.72
1.6	Pruning of Plants during the Eighth Month after planting @ Rs 300 per Man Day	0600	0900	0900	03000	01.98
	Sum of Working Capital	24850	6300	6300	37450	**24.76**
1.7	Interest on Working Capital	01491	0378	0378	02247	01.48
	Total Variable Cost	**26341**	**06678**	**06678**	**39697**	**26.24**
2.0	Fixed Expenses					
2.1	Annual Cost of Installation of Electrical Implements	5000	00	00	5000	03.30
2.2	Cost of Ladder Type of Furniture	2100	00	00	2100	01.39
2.3	Rental Value of Land used for Jamun Plantation	32000	32000	32000	96000	63.44
2.4	Insurance and other Cost associated with Fixed Items	1200	1200	1200	3600	02.38
2.5	Land Tax	1500	1500	1500	4500	02.97

Sl. No	Particulars of Cost	Year			Total Cost	Percentage to Total Est. Cost
2.6	Depreciation Charges	0142	0142	0142	0426	00.28
	Total Fixed Cost	41942	34842	34842	111626	73.76
3.0	Total Establishment Cost	68283	41520	41520	151323	100.00

Table 2 reveals that the total establishment cost involved in establishing the Jamun plantation is arrived at 1.51 lakhs. In that, the first year of establishment of Jamun consumed Rs 0.68 lakhs; second and third years of establishment consumed respectively of 0.41 and 0.41 lakhs to the total establishment cost. Among the total establishment cost, the rental value of land alone consumed Rs 96000 which accounted for 63 per cent followed by watering to the Jamun orchards accounting for 8.72 per cent and the cost of Jamun seedlings drawn from Rajahmundry of Andhra Pradesh state consumed Rs 10000/ for one hundred saplings accounting for 6.60 per cent to the total establishment cost.

In respect of fertilizer application at the time of planting, the farmer added 5 kgs of farmyard manure plus 5 ml Azospyrillum, Phosphobacteria and Pseudomonas per pit to enhance the easy establishment of saplings of Jamun. After pruning, 150 kgs of farmyard manure or 15 kgs of vermicompost is also applied for healthy establishment of trees. Associated costs of fertilizer are presented in Table 2 which accounted for only 2 per cent of the total establishment cost. The case farmer travels abroad periodically to Indonesia, Malaysia, Singapore, Thailand and Sri Lanka and the newer technologies were brought to India from abroad and adopted. One such technology is extraction of jamun squash and he is exporting the same to different parts of India.

Maintenance Cost

Under the head of maintenance cost, weeding and basin formation, pruning of branches, watering, fertilizer and its application to each trees, pesticide and its application and harvesting of Jamun fruits are the main items under the variable expenses. Under the head of fixed expenses, the rental value of land, insurance premium charges, depreciation and the land tax payable annually are incorporated and the details are analyzed and presented in Table 3.

The maintenance cost over the years included are classified as variable expenses and the fixed expenses. The variable expenses per annum during the 10th year of establishment is arrived at Rs 32224/- and the fixed expenses was arrived at Rs 34842/- which respectively accounted for 48 per cent and 52 per cent to the total maintenance cost involved during the tenth year.

Among the variable expenses under maintenance cost, harvesting cost alone consumed 18 per cent to the total maintenance cost followed by watering the Jamun orchards consuming 9.69 per cent of the total maintenance cost and the pruning and fertilizer applications found to consume respectively 6.26 per cent and 5.96 per cent of the total maintenance cost.

Among the fixed expenses included under the maintenance cost, the rental value of land alone consumed 48 per cent of the total maintenance cost followed by the insurance premium, depreciation and the land tax found to consume only around 4 per cent of the total maintenance cost required for the Jamun orchards.

To sum up, the maintenance cost involved in the Jamun plantation was found to be almost equal for the variable and fixed expenses included in the analysis. Among the fixed expenses, the rental value of land alone consumed around 48 per cent of the total maintenance cost incurred during the tenth year of establishment of Jamun orchards.

Revenue Realized from Jamun Orchards

Revenue realization is possible from Jamun orchards through sale of fruits, value addition of Jamun fruits and sale of fuel wood which is derived from periodical pruning of orchards. The details of price realized per unit of fruits and the gross income derived are analyzed and the details are presented in Table 4.

Table 4 reveals the details of price realized per kg of Jamun, gross income realized and the net income realized from Jamun plantation and the cost of production per kg of Jamun. The price of Jamun fruit being Rs 100 per kg during the fourth and fifth year, Rs 120 during the sixth and seventh year of maintenance, Rs 130 per kg during the eighth and ninth year of maintenance and Rs 140 per kg during the tenth year of maintenance have generated a gross income of Rs 35600, Rs 87600, Rs 156800, Rs 260400, Rs 415400, Rs 673600 and Rs 845800 during the period of 4th to 10th year respectively.

The net income generated from the Jamun orchard was found to be negative during the fourth year and the same was found to be on the positive side from fifth year to the tenth year. The net income generated during the tenth year of the plantation was found to be 8.45 lakhs per acre. The cost of production of Jamun per kg was arrived at Rs 137 during the fourth year due to less yield and the same was arrived at Rs 11.15 per kg during the tenth year of establishment of Jamun orchards due to abundant yield during the tenth year and the yield started to sustain after the tenth year and hence the profit from the plantation was found to be appreciable after sixth year of establishment of the Jamun plantation. The average cost of production per kg (4th to 10th Year) was arrived at Rs 42.66. The average sale price of Jamun fruit per kg was arrived at Rs 120 revealing that the farmer is able to reap 185 per cent increased return over the cost of production of Jamun per kg.

To sum up, the average cost of production of Jamun fruit was arrived at Rs 43 per kg. However, the average sale price was arrived at Rs 120 per kg which revealed that the case farm is able to reap 185 per cent increased return over the cost involved in producing the Jamun per kg and the farmer is able to sustain his annual income from the fifth year of establishment of Jamun orchards.

Table 3 Maintenance Cost Involved for Establishing the Jamun Plantation in the Case Farm

Sl.N0	Particulars of Variable Expenses	Fourth Year	Fifth Year	Sixth Year	Seventh Year	Eighth Year	Ninth Year	Tenth Year
1.1	Weeding and Basin Formation	1200	1200	1200	1200	1200	1200	1200 (01.79)
1.2	Pruning of the Branches	600	600	900	1900	2600	3200	4200 (06.26)
1.3	Watering to Plants	5200	5200	5200	6500	6500	6500	6500 (09.69)
1.4	Cost of Fertilizer and its Application	1450	1650	1850	2400	2800	3200	4000 (05.96)
1.5	Pesticide Charges and its Application	1400	1550	1700	1900	2050	2200	2300 (03.44)
1.6	Harvesting of Jamun Fruits	1800	2400	2900	6200	8800	10800	12200 (18.19)
	Sum of Working Capital	11650	12600	13750	20100	23950	27100	30400 (45.33)
1.7	Interest on Working Capital	0699	0756	0825	1206	1437	1626	1824 (02.72)
1.8	Total Variable Cost	12349	13356	14575	**21306**	**25387**	**28726**	**32224 (48.05)**
2.0	Fixed Cost							
2.1	Rental Value of Land	32000	32000	32000	32000	32000	32000	32000 (47.70)
2.2	Insurance Premium and other Associated Costs	1200	1200	1200	1200	1200	1200	1200 (01.79)
2.3	Depreciation	0142	0142	0142	0142	0142	0142	0142 (00.21)
2.4	Land Tax	1500	1500	1500	1500	1500	1500	1500 (02.25)

Sl.N0	Particulars of Variable Expenses	Fourth Year	Fifth Year	Sixth Year	Seventh Year	Eighth Year	Ninth Year	Tenth Year
2.5	Total Fixed Cost	34842	34842	34842	34842	34842	34842	34842 (51.95)
	Total Maintenance Cost (1.8 + 2.5)	47191	48198	49417	56148	60229	63568	67066 (100)

(Figures in Parentheses Indicate Percentage to Total Maintenance Cost)

Table 4 Revenue Realized from Jamun Orchards

Sl.No	Details of Revenue Realized	04	05	06	07	08	09	10
1.0	Price of Jamun (Rs/Kg)	100	100	120	120	130	130	140
2.0	Yield of Jamun (Kgs per Acre)	0344	0860	1290	2150	3440	5160	6020
3.1	Revenue from Sale of Fruits	34400	86000	154800	258000	412800	670800	842800
3.2	Revenue from Sale of Fuel Wood	1200	1600	2000	2400	2600	2800	3000
4.0	Gross Income in Rs	35600	87600	156800	260400	415400	673600	845800
5.0	Net Income in Rs	(-) 11591	39402	107383	204252	355171	610032	778734
6.0	Cost of Production (Rs/Kg)	137.18	56.05	38.31	26.12	17.51	12.32	11.15

The cost incurred in establishing the Jamun plantation and the returns realized from the Jamun orchards over the years were analyzed using the discount factor at 12 per cent to find out the present value of cost and benefit stream and the results are presented in Table 5.

Table 5 Discounted Cash Flows from Jamun Orchards

Year	Cost in Rs	Gross Income in Rs	Net Income in Rs	Discount Factor 12%	Discounted Costs	Discounted Benefits
0	41942	0	-41942	1.000	41942	0
1	26341	0	-26341	0.893	23519	0
2	41520	0	-41520	0.797	33099	0
3	41523	0	-41523	0.712	29555	0
4	47191	35600	-11591	0.636	29991	22624
5	48198	87600	39402	0.567	27349	49707
6	49417	156800	107383	0.507	25036	79440
7	56148	260400	204252	0.452	25399	117792
8	60229	415400	355171	0.404	24325	167773
9	63568	673600	610032	0.361	22923	242907
10	67066	845800	778734	0.322	21593	272325
Sum	543143	2475200			304732	952568
Net Present Value (NPV)				647836		
Benefit Cost Ratio (BCR)				3.13		
Internal Rate of Return (IRR) (%)				43.98		

Table 5 reveals the details of cost incurred and benefits accrued from the Jamun orchards over the years. The data on cost and benefits were discounted using the 12 per cent discount rate. The Net Present Value (NPV) realized from the Jamun orchards is arrived at Rs 6.48 lakhs. The Benefit Cost Ratio (BCR) derived from the cost and benefit stream derived from Jamun orchards is arrived at 3.13 revealing that for a rupee of investment, the Jamun orchards could generate a return of Rs 3.13. The Internal Rate of Return (IRR) realized from the plantation is arrived at around 44 per cent which is above the cut off rate indicating that the investment in Jamun orchard is highly profitable according to the parameters. The farmers who are intending to establish Jamun type of plantations can invest confidently which could generate profit on a sustainable way. The only aspect that every farmer should care is watering and fertilization and other periodical maintenance could enhance income further besides arranging the distribution mechanisms as and when harvest is over. If not the fruits can well be value added and the same may be converted into squash and then tie-up mechanisms may be made with Pasumai Angadies, Retail Bazaars and Ayurvedic and Siddha medicine formulating companies.

DETAILS OF OUTPUT PRODUCED IN THE CASE FARM

Besides establishment of plantation developmental activities by the case farmer, other inputs which are used in the plantations were also produced and distributed to other farmers in the district. He is the leader for 70 farms in and around Dindigul District and hence whatever the inputs and outputs, he could sell it in the open market very easily and hence these details are analyzed and the results are presented in Table 6.

Table 6 Details of Output Produced in the Case Farm

Sl. No	Details of Output Produced in the Case Farm	Quantity Produced	Cost of Production Per Unit	Sale Price Per Unit
01	Production of Bio Boost (in Litres)	60000	060.00	0200.00
02	Production of Bio Complex (in Tonnes)	0100	6000.00	9500.00
03	Production of Jamun Fruit (Kgs) per Annum From 10th Year	6020	042.66	140.00
04	Production of Jamun Juice (in Litres)	1500	NA	240.00

(NA: Not Available from the Case Farm)

Bio Boost

Table 6 reveals the details of output produced in the case farm. It reveals that the case farm is producing bio boost of around 60000 litres which is used as an input mainly for foliar application. The cost of production of bio boost per litre is arrived at Rs 60 and the same was sold to other farmers for around Rs 200 per litre. From that he is generating a net return of Rs 140 per litre. However, the secrets and its detailed cost involvements are not disclosed as he is preparing the liquid based on his own formulations after the abroad visits.

Bio Complex

Bio complex is another output produced by the case farmer which is used as fertilizer for the plantations produced to the tune of 100 tonnes per annum. The cost of production of bio complex fertilizer per ton is arrived at Rs 6000 and the same is priced and sold at Rs 9500 per tonne. Around 80 per cent of the quantity is consumed in the farm itself and only 20 per cent of the output is sold to the farms under his consultancy.

Jamun Fruits

Production of Jamun fruit is another output in the case farm. The total production of Jamun fruit per annum from tenth year onwards is arrived at around 6 tonnes. The average cost of production of Jamun fruit per kg is arrived at Rs 43 and the fruit is sold to the retail outlets like Big Bazaar, Pazhamudhir Nilayam, Pasumai Angadies and the Departmental Stores in Theni, Madurai, Coimbatore and Dindigul Districts at the rate of Rs 140 per kg.

Jamun Squash

Production of Jamun squash is another output produced in the processing unit established in the case farm itself.

Requirements for the Preparation of Jamun Squash

- Jamun pulp: 1 kg

- Sugar: 01.80 kg
- Citric acid: 38.00 g
- Sodium benzoate: 02.40 g
- Water: 1150 ml

Method of Preparation of Jamun Squash

- Select the sound ripe Jamun fruits which are free from infection and decay
- Wash the fruits thoroughly without any damage to the fruits
- Crush the fruits into pulp without causing any damage to the seeds
- Separate the seed and extract the juice
- Prepare the sugar syrup by mixing the sugar and water and boil
- The dirt is skimmed off. Filter the syrup and cool it
- Add citric acid to the cool Jamun syrup
- Jamun juice is blended with clean sugar syrup
- Add preservative in small portion of juice and later add to the whole part of the juice
- Fill the juice in the pet bottles of 700 ml size leaving the head space of 1.20 cm to 2.50 cm.
- Keep the pet bottles filled with juice in the appropriate storage temperature and equipment.

The processing unit available with the case farm is producing around 1500 litres during the season and a litre of Jamun juice is priced at Rs 240. However, the cost of production of Jamun juice is unable to be arrived for want of data as the farmer declined to reveal almost all the details needed in this regard and hence certain data are unable to be given in a detailed manner.

Health Tips to Humans Jamun fruit is very useful for curing diarrhea. It is stomachic, carminative and diuretic, apart from having cooling and digestive properties. Jamun is used to treat diabetes and dysentery. It markedly lowers blood pressure. The seed powder reduces sugar in urine very quickly and permanently. It is also used for curing ring worm infections in the stomach.

To sum up, the outputs produced in the case farm were found to be bio boost, bio complex which are on the input side and, the Jamun fruit and Jamun squash produced in the farm are on the output side. Both inputs and outputs were sold to the other farms under his consultancy.

DETAILS OF DISTRIBUTION MECHANISMS

Such a dynamic case farm sustaining its output means the distribution mechanisms should be sound enough and more supportive and hence the distribution details are discussed and the details are presented in Table 7.

Table 7 Details of Distribution Centers for the Jamun Juice Produced in the Case Farm

Sl. No	Details of Value Added Output Sold to the Dealers/ Stores	Number of Dealers	Quantity Sold in Litres	Sale Price Per Unit
01	Chennai	03	450	240.00
02	Karaikudi	02	150	240.00
03	Nagercoil	02	150	240.00
04	Coimbatore	03	250	240.00
05	Kerala	02	500	240.00
	Total	12	1500	240.00

Table 7 reveals that the Jamun juice is distributed to different dealers present in Tamil Nadu and Kerala. In Tamil Nadu, around 10 dealers on an average are receiving 1000 litres from the case farm. Chennai is able to receive 450 litres of Jamun juice which is supplied through 3 dealers who are regularly receiving the juice from the case farm. Karaikudi is another town capable of receiving 150 litres of Jamun Juice through two dealers. Similarly, the Nagercoil town and Coimbatore city are receiving 150 and 250 litres of Jamun juice respectively from the case farm. All these centres are receiving the juice at the rate of Rs 240 per litre. On sale of 1500 litres of Jamun juice, a gross income of Rs 3.60 lakhs is received which is a handsome income to the case farm. One would be interested to know why this Jamun Juice is important and is well received among the consumers of different locales.

NUTRIENTS AVAILABLE IN JAMUN FRUITS

Health awareness is more among the consumers of different city. They wanted to consume quality output which is nutritionally enriched, sweet to the taste buds of different consumers of different age and income groups. If these features are met out, whatever the produce, all will be acceptable to the consumers at a price prescribed in the produce. The nutrients available in the Jamun fruit are analyzed and presented by Veeraragavathatham et al. during 2004 in the form of a table shown below.

Table 8 Nutrients Available in One Hundred Grams of Jamun Fruits

Sl. No	Particulars of Nutrients in 100 grams of Jamun	Unit in Grams	Percentage of Nutrients to Total
01	Carbohydrates	19.70	87.83
02	Protein	00.70	03.13
03	Iron	01.00	04.46
04	Calcium	00.02	00.08
05	Phosphorous	00.01	00.04
06	Fat	00.10	00.45
07	Fibre	00.90	04.01
	Total Nutrients in Grams	22.43	100.00

(Source: Veeraragavathatham, 2004)

Table 8 reveals the details of nutrients available per one hundred gram of Jamun fruit. Among the different nutrients, carbohydrates form 19.70 grams accounting for 87.83 per cent to the total nutrients available in the fruit. Presence of iron and fibre respectively accounted for 4.46 per cent and 4.01 per cent. Protein available in the Jamun fruits accounted for 3.13 per cent to the total available nutrients in the fruit. Fat, calcium and phosphorous are also present in the fruit only to the very little amount revealing that the essential nutrients are present in the fruit and could improve the health of the consumers.

The only constraint the case farmer revealed is that the output produced could not be stored well for considerable point of time and hence a storage facility is required at rural settings. Hence, the Department of

Agricultural Marketing and Agribusiness can explore the possibilities of erecting a cold storage facility in the important production sites in the rural environs so as to help the farming community further after exploring the needs analysis.

SUMMARY AND CONCLUSION

The cropping program occupied 76 per cent of the total size of holding and the remaining 24 per cent of the land was under farm roads and permanent infrastructure like buildings, parking area, shed and processing units which are essentially required for production and value addition of output produced in the case farm. The plantations established were Jamun and Amla in the case farm. Since the study is focusing only on Jamun, the details of Amla are ignored in this study. Among the total establishment cost, the rental value of land alone accounted for 63 per cent followed by watering to the Jamun orchards accounting for 8.72 per cent and the cost of Jamun seedlings drawn from Rajahmundry of Andhra Pradesh state consumed Rs 10000/ for one hundred saplings accounting for 6.60 per cent to the total establishment cost. The average cost of production of Jamun fruit was arrived at Rs 43 per kg. However, the average sale price was arrived at Rs 120 per kg which revealed that the case farm is able to reap 185 per cent increased return over the cost involved in producing the Jamun per kg and the farmer is able to sustain his annual income from the fifth year of establishment of Jamun orchards.

The Net Present Value (NPV) realized from the Jamun orchards is arrived at Rs 6.48 lakhs. The Benefit Cost Ratio (BCR) derived from the cost and benefit stream from Jamun orchards is arrived at 3.13 revealing that for a rupee of investment, the Jamun orchards could generate a return of Rs 3.13. The Internal Rate of Return (IRR) realized from the plantation is arrived at around 44 per cent indicating that the investment in Jamun orchard is highly profitable according to the parameters. The outputs produced in the case farm were found to be bio boost, bio complex which are on the input side and, the Jamun fruit and Jamun Juice produced in the farm are on the output side. Both inputs and outputs were sold to the other farms under his consultancy.

The only constraint the case farmer revealed is that the output produced could not be stored well for considerable point of time and hence a storage facility is required at rural settings. Hence, the Department of

Agricultural Marketing and Agribusiness can explore the possibilities of erecting a cold storage facility in the important production sites in the rural environs so as to help the farming community further after exploring the needs analysis.

References

- Chovatia R.S. and Singh S.P. (2000). "Effect of time on budding and grafting success in Jamun" ***Indian Journal of Horticulture***, 57 (3): 255-258
- Harswarup Singh,"Comparative Economics of Trees and Crop Production in Haryana", ***Indian Journal of Agricultural Economics***, 43(3): July-September, 1988. p. 290.
- Kadam, S.S. (2001). "New Products from arid and semi arid fruits", ***Indian Journal of Horticulture***, 58 (1-2): 170-177
- Krishnamurthy S and Sudhakar Rao D.V. (2001). "Status of Post Harvest Management of Fruits", ***Indian Journal of Horticulture***, 58 (1-2): 152-163.
- Sasthri G, Srimathi, P and Malarkodi K (2001). "Effect of Seed Size on Seed Quality in Jamun", ***Madras Agricultural Journal***, 88(7-9): 524-526.
- **Sekhar C.**, 0. Anchanam Alagia Pillai, 0. Timothy Randhir and G.Kumara Velu (1990). "Economic Analysis of Kapok under Agroforestry Conditions of Tamil Nadu", ***Agricultural Situation in India***, November 1990, pp. 537-540.
- Singh R.K and Thakur S (1977). " Seed Germination and Seedling Growth of Jamun", ***Proceedings of Bihar Academy of Agricultural Sciences***, 25(1): 139-142.
- Singh S., Singh A.K., Singh H.P., Bagle B.C. and T.A. More (2011). "***Monograph on Jamun***" (Indian Council of Agricultural Research: New Delhi), 01-48.
- Srimathi P, Ramanadane T, Malarkodi K and Natarajan K. (2003). "Seed Extraction in Jamun", ***Progressive Horticulture***, 35(2): 221-223.
- Veeraragavathatham D, M. Jawaharlal, S.Jeeva, R. Rabindran and G. Umapathy (2004). ***Scientific Fruit Culture***,(Suri Associates: Coimbatore) 187-189.

Chapter 17

Production and Trading Potentials in Curd Chilli

INTRODUCTION

Chilli is one of the most valuable crops of India. It is grown almost throughout the country. Different varieties are grown for use as vegetables, spices, condiments, sauces and pickles. Chilli is also known as 'hot pepper' and capsicum as 'bell pepper.' The Portuguese brought capsicum from Brazil to India during the year 1584. Chilli is a fruit of the plants '*Capsicum annum*' and '*Capsicum fruitescens*' that came from the genus 'capsicum,' belonging to the family of Solanaceae, which also includes tomato and potato. These fruits are small in size and known for their sharp acidic flavor and color. Currently, chillies are used throughout the world as a spice and also in the making of beverages and medicines. If some varieties of chillies are famous for red color because of the pigment 'capsanthin,' others are known for biting pungency attributed to 'capsian.' India is the only country which is rich in many varieties with different quality parameters.

Chillies are said to have originated in the Latin American regions of the New Mexico and Guatemala as a wild crop around 7500 BC. The people native to these places domesticated this crop in 5000 BC, as per the remains of the pre-historic Peru. Chilli is said to be the first ever domesticated crop in America. At that time, chillies were cultivated by the farmers together with a primary crop to protect the primary crop from any damage that the birds could do. Chillies gained popularity in the American continent for flavoring and have been largely cultivated since then. When America was discovered, and the Spaniards and the Portuguese explored the South American continent, this pungent-flavored fruit gained much more recognition. Christopher Columbus, the founder of America, was one of the first Europeans who encountered and consumed chilli, and called it pepper due to the similarity in taste. Crushing the dried pods gave chilli powder, which was later identified as a substitute of 'peppercorn.' The chilli crop came to the Asian continent as late as the 16th century with the identification of new sea routes by the Portuguese and the Spanish explorers. It became popular in the whole of Asia rapidly and native Asians started cultivating this crop as well. The south Asian climate suited this vegetable crop, and since then a large percentage of Chilli production has shifted to Asia. Today, the most sharp and valued varieties of Chillies are grown in Asia only.

Table 1 Vernacular Name of Chilli in Different Languages of India

Sl. No.	Name of the Language	State	Common Name of Chilli
01	Bengali	West Bengal	Lanka, Lanka - Morich
02	Gujarati	Gujarat	Marcha
03	Hindi	Hindi Speaking States	Lal Mirch
04	Punjabi	Punjab	Lal Mirch
05	Urdu	Urdu Speaking States	Lal Mirch
06	Kannada	Karnataka	Menasina Kayi
07	Kashmiri	Jammu and Kashmir	Martsu, Waungum
08	Malayalam	Kerala	Mulaku
09	Marathi	Maharashtra	Mirchi
10	Oriya	Odhissa	Lanka
11	Tamil	Tamil Nadu	Milagai
12	Telugu	Andhra Pradesh	Mirapakayi

(Source: J.S. Pruthi, 1976)

PRODUCTION OF CHILLIES IN INDIA

As already mentioned, India is the largest producer of Chillies in the world. Though it is called in different vernacular name in different languages, in English it is pronounced as Chilli. The details are presented in Table 1. The production level of Chilli hovers around 1.1 million tonnes annually. India also has the maximum area dedicated to the production of this crop. Chilli is a universal spice of India. It is cultivated in all the states and union territories of the country. As per the latest statistics, India produced 800,100 tonnes of dry Chilli from an area of 930,000 hectares. Andhra Pradesh stands first in the list of Chilli-producing states in India and also has the maximum acreage under Chilli cultivation in the country. Andhra Pradesh alone commands 49 per cent of the Chilli production in India, with a production of around 2.7 lakh tonnes of chillies followed by Karnataka contributing 14 per cent of the country's production. The major Chilli producing states in India, namely, Andhra Pradesh, Karnataka, Maharashtra, Orissa, Rajasthan and Tamil Nadu, contribute around 86 per cent of the total area under chilli crop cultivation in the country and 90 per cent of the total Indian production (Subbiah, 2009).

INDIAN CHILLI MARKET

Currently, chillies are produced throughout India, making our country the most dominant player in the world market. The varieties of chillies produced by India are Sannam, LC 334, Byadgi, Wonder Hot and Jwala. The market for Chillies is affected by seasonal price fluctuations, overall production in the country, world demand, stocks available in cold storages and hedging among the various varieties of Chillies. The major trading centers of Chilli and Chilli powder in India are Guntur (Andhra Pradesh), Warangal (Andhra Pradesh), Khammam (Andhra Pradesh), Hindpur (Andhra Pradesh), Raichur (Karnataka), Bellary (Karnataka), Unjha (Gujarat), Chennai, Kolkata, Mumbai, Delhi, Ahmedabad and Nagpur. Guntur is the largest Chilli market in the world.

In Tamil Nadu, Dindigul District, Thirumalairayapuram is the village in which several cottage industries at household level entered into the process of making salted curd Chilli for the domestic and international markets. There are around 350 units functioning in the village itself and producing high quality varietal curd Chilli for different market segments and these are in different brands. Looking at the production potentials of salted curd Chilli in Tamil Nadu and its trade potentials in domestic and international markets, the study was initiated at micro level case study by identifying a potential cottage industry involved in such agribusiness with the following specific objectives

OBJECTIVES OF THE STUDY

1. To assess the sources of supply of raw materials for the cottage industries involved in production of curd Chilli in Tamil Nadu
2. To assess the methodology of curd Chilli making in vogue with the Cottage Industries of Tamil Nadu
3. To document the constraints faced by the cottage industries involved in production of curd Chilli and to suggest appropriate strategies and policy measures for the same.

DESIGN OF THE STUDY

The design of the study is an important component of research. In order to fulfill the objectives enshrined in the study, an appropriate methodology for conducting the study is inevitable. In this chapter, the methodology adopted for the present study including the selection of study area, selection of respondents, the methods of data collection and the different tools of analysis used are discussed.

Selection of Study Area

In Dindigul District, the Oddanchatram is a small town which is well known for its vegetable market across Tamil Nadu because of a variety of vegetables being supplied to different parts of Tamil Nadu at a cheaper rate.

Oddanchatram Vegetable Market

Oddanchatram, facilitated by the awesome weather conditions, has turned to be one of the largest areas of vegetation. With copious amount of vegetables being harvested, Oddanchatram vegetable market, with an excellent market value, occupies the red page in the diaries of vegetables and agro industries. Oddanchatram market has stood out as the second largest market in Tamil Nadu. Almost every vegetable harvested in India is available at the most economical rate in this market. People from the nearby villages and towns, rely on this market for vegetables and flowers. Thousands of people drop in at various agencies to buy the fresh vegetables that are procured right from the farmers (Web Reference 1). In the normal season about 3000 tons of vegetables are transported from the market to neighboring areas every day. Between June and December, around 5000 tons of vegetables are transported per day (Web Reference 2). Hence, Oddanchatram taluk was chosen in the first stage to collect the data relevant to this study.

In Oddanchatram Taluk, there are 57 revenue villages. Among the 57 revenue villages, Thirumalairayapuram and Boliammanur are the two potential revenue villages located in the foothills of Oddanchathiram where around 350 traders at household level are involved in processing of green Chilli into salted curd Chilli. In this respect, the two villages which had inhabited traders at household level were selected in the second stage.

In the third stage, among the 350 traders, only one cottage industry found to operate at higher level by employing several labor force for processing of green Chilli at Thirumalairayapuram village of Oddanchathiram taluk of Dindigul District was selected for a case study approach to document the production and trading practices prevalent in respect of curd Chilli production in Tamil Nadu for primary data.

From the website, the secondary data related to institutions involved in trading of curd Chilli, details of countries importing the curd Chilli from Tamil Nadu and the export of curd Chilli to different destinations and the number of shipments are downloaded and analyzed in this study using the conventional percentage analysis.

Period of Study

The data related to the agricultural year 2012–2013 were collected and the field enquiries were conducted during the months of October and November in the year 2013 (Plate – 1)

Methods of Collection of Data

In order to get an insight into physical and economic environments of the villages, a reconnaissance survey of the villages was undertaken. The primary data required for the study were collected through personal interview with the help of pre-tested comprehensive interview schedule (Plate – 2). The schedule covered aspects such as family size, educational status, asset position, cropping pattern, availability of land - both irrigated and rainfed, labour availability, machine and animal power, cost of cultivation, input–output relationships, net returns from the crops, marketing of maize, problem in production, etc. With respect to marketing, the details of operations of the merchants, expenses incurred on marketing, problems in marketing were collected from the traders. Besides, information about the cost incurred and profits realized by the different market functionaries was also gathered for working out the marketing cost and price spread. The data collected were tabulated, processed and subjected to statistical analysis.

Plate 1 Researcher and the Project Guide are Interviewing the Manager of R.S. Cottage Industry Thirumalairayapuram

Plate 2 Conducting Survey at R.S. Cottage Industry

To understand the basic characteristics of agricultural system in the study area, data on soil conditions, land utilization, cropping pattern, agro-climatic features and other available facilities were collected from published and unpublished records available in various government departments and are discussed in the description of the study area, the next chapter of the project work.

Methods of Analysis

The data collected in the schedules were tabulated for subsequent analysis. Keeping in view the objectives of the study, conventional percentage analysis was employed and the results are discussed as detailed below.

PRODUCTION PRACTICES OF CURD CHILLI

The project work on production and marketing of salted curd Chillies in Tamil Nadu with a special reference to Thirumalairayapuram village of Dindigul Taluk is mainly focused on analyzing the sources of supply of green Chillies for value addition, marketing and distribution practices prevalent in curd Chilli and the firms involved in facilitating the trade are analyzed and discussed under the following main headings. They are:

3.1 Sources of Supply of Raw Materials to the Cottage Industries

3.2 Methodology of Processing for Preparation of Salted Curd Chilli

3.3 Trading Practices Prevalent in Salted Curd Chilli

SOURCES OF SUPPLY OF RAW MATERIALS TO THE COTTAGE INDUSTRIES

Under the section of sources of supply of raw materials to the cottage industries, one would be interested in furnishing certain general features which are favoring the farming community to raise certain agricultural / horticultural crops which are causative one for reaping higher income so as to raise their economic well being. In this respect the results under this section are analyzed and the details are presented and discussed in the following sub sections.

Agro-Climatic Zones of Tamil Nadu

To procure the raw materials, cultivation of crops in different agro climatic region and learning on the potentials of agro-climatic zone will be a precursor to the supply of raw materials. Hence the details of agro-climatic zones available in Tamil Nadu are discussed and the results are presented in Table 2.

Table 2 Agro-climatic Zones of Tamil Nadu

Sl. No	Zone	Name of the Zone	Districts Covered	Number of Districts	Average Rainfall in mm
01	I	North Eastern Zone	Kancheepuram, Thiruvallur, Cuddalore, Villupuram, Vellore, Thiruvannamalai	06	1105
02	II	North Western Zone	Dharmapuri, Krishnagiri, Salem, Namakkal, Perambalur, Ariyalur	06	0875
03	III	Western Zone	Coimbatore, Erode, Thiruppur	05	0715
			Dindigul, Theni		
04	IV	Cauvery Delta Zone	Tiruchi, Thanjavur, Thiruvarur, Nagapattinam,	04	0984
05	V	Southern Zone	Madurai, Sivagangai, Pudukottai, Karur Ramanathapuram, Virudhunagar, Tirunelveli, Thoothukudi	08	0857
06	VI	High Rainfall Zone	Kanyakumari	01	1420
07	VII	Hilly and High Altitude Zone	The Nilgiris	01	2124
Total Number of Districts / Mean Rainfall of the State				31	1154

Source: Season and Crop Report of Tamil Nadu – 2010. P.6

Table 2 reveals that Tamil Nadu is blessed with seven agro-climatic zones and she is in possession of 32 districts which are very well administered by the district administration. Since the study area is coming under the Dindigul District of Tamil Nadu, the zonal classification had incorporated major part of Dindigul District with the Western Zone of Tamil Nadu and few portions under the Southern Zone of Tamil Nadu. However, in this study, the district is included in the western zone as per the records of Season and Crop Report of Tamil Nadu and the results are discussed in detail. The western zone is composed of 5 districts which are industrially advantageous and capable of generating higher employment.

This zone is receiving an annual rainfall of only 715 mm. However, the actual rainfall received by Dindigul District is around 881 mm. From that one could understand that the district is having comparative advantage within the zone to raise agricultural and horticultural crops particularly the Chilli which is an important raw material for making curd Chilli. Before discussing the results at micro level, the area production and productivity of Chilli at national and regional level becomes important and hence the following sub section discusses on the details of area, production and productivity of Chilli in India.

Area, Production and Productivity of Chillies in India

At the national level, among the spices, production of Chilli is found to be 23 per cent to the total production of spices (Government of India, 2011). Table 3 is enriched with the data on area, production and productivity of Chillies in India over years.

Table 3 Area Production and Productivity of Chillies in India

Sl. No	Year	Area in Thousand ha	Production in Thousand Tonnes	Productivity (Tonnes / Ha)
01	2000-01	836.50	983.70	01.18
02	2005-06	654.00	1014.60	01.55
03	2008-09	779.05	1269.85	01.60
04	2009-10	767.23	1202.94	01.60
05	2010-11	792.10	1223.40	01.50

Sl. No	Year	Area in Thousand ha	Production in Thousand Tonnes	Productivity (Tonnes / Ha)
	Percentage Change Between 2000–2010	(-) 05.31	24.37	27.12

Source: Agricultural Statistics at a Glance – 2012: p. 192

Table 3 reveals that the area under Chilli in India between the years 2000 and 2010 decreased to the level of 5.31 per cent. The production of Chilli is found to increase to the level of 24 per cent between the same period of 2000 and 2010. Though the area under Chilli is decreasing, the production is found to have increased due to the technological interventions made by the National Horticultural Mission. When we compare the productivity of Chilli between the period of 2000 and 2010, the productivity increased from the level of 1.18 tonnes during 2000 to 1.50 tonnes during the year 2010. The productivity of Chilli between the years 2008-09 and 2009-10, is found to be increased to 1.60 tonnes per ha. But only during 2010-11, the productivity is found to be reduced to the level of 1.50 tonnes per ha. It is understood that the increase in productivity of Chilli is very marginal out of higher efforts made towards expenditure on developing the technology and the infrastructure through the National Horticulture Mission across the states. As the National Horticulture Mission focuses on Area Expansion Programs and Technology Development Programs to enhance productivity, the decline in productivity revealed that the technology is not effectively percolated to the farm communities. In this context, the monitoring mechanisms need to be intensified and made accountable to achieve the desired targets in terms of production and productivity of Chillies in India.

Area Production and Productivity of Chilli in Tamil Nadu

The data on area, production and productivity of Chilli in Tamil Nadu is analyzed and the results are presented in Table 4. When we see the area under Chilli, a decrease in the area of 11 per cent is noticed in Tamil Nadu, whereas, the area reduction visible at national level, is only 5 per cent which conveys the meaning that the Tamil Nadu horticulture authorities need to take much efforts in augmenting the area under Chilli.

Table 4 Area Production and Productivity of Chillies in Tamil Nadu

Sl. No	Year	Area in ha	Production in Tonnes	Productivity (Kgs Per Ha)
01	2008-09	65412	32924	503
02	2009-10	58476	31230	532
Percentage Change during 2009-10 over 2008-09		(-) 10.60	(-) 05.15	05.77

Source: Season and Crop Report of Tamil Nadu 2010: p.31

In respect of production of Chilli in Tamil Nadu, a five per cent reduction is observed revealing that technology adoption is very poor and the package of practices should reach the farmers concerned at the earliest to reap the benefits of production of Chilli in Tamil Nadu. When one could compare the productivity at national and regional level particularly in Tamil Nadu, a 64 per cent reduction is observed. It is a wide technology gap. Authorities in horticulture should find the ways and means of bridging the gap to achieve at least 1.50 tonnes of Chilli per ha in Tamil Nadu. It should be made as a mission and the same should be accomplished.

Area Expansion Programs for Chilli in Tamil Nadu

Under National Horticultural Mission, the Government of Tamil Nadu has prepared an action plan with the support of a Non Governmental Organization. The action plan incorporates area expansion program as one of the main activity focused crop wise and district wise in Tamil Nadu. In respect of Chilli, the area expansion details are analyzed and the results are presented in Table 5.

Table 5 Area Expansion Program for Chilli in Tamil Nadu

Sl. No	Name of the District Targeted	Area in Ha (Target)	Percentage to Total
01	Coimbatore	025.000	06.25
02	Dindigul	025.000	06.25
03	Ramanathapuram	250.000	62.50

Sl. No	Name of the District Targeted	Area in Ha (Target)	Percentage to Total
04	Tiruchirappalli	050.000	12.50
05	Sivagangai	050.000	12.50
	Total	400.000	100.00

Table 5 reveals that the area expansion target fixed under National Horticultural Mission in respect of Tamil Nadu is arrived at 400 ha. Only five districts were targeted as potential areas for area expansion of Chilli in Tamil Nadu. Among the five districts chosen, the Ramanathapuram District was chosen as the potential district and hence 62.50 per cent of the area is fixed for the district which arrived at 250 ha. Since Ramanathapuram district has rich soil quality and mostly dependent on rain, the rainfed chilli is targeted in the district. Tiruchirappalli and Sivagangai districts had equally shared 12.50 per cent of the area expansion targets. However, the Coimbatore and Dindigul districts have commanded only meagre area under the area expansion program of Chilli which arrived at 25 ha for each district. From this, one could understand that the action plan envisages only five potential districts which have scope for area expansion under Chilli and hence only the 400 ha have been targeted for achievement.

Sources of Supply of Green Chillies

So far the discussion on Chilli rested with production of Chilli, trading and assembling centers for Chilli at national and regional levels. These are the sources that could facilitate supply of Chilli for the cottage industries. The details with regard to the sources are analyzed and the results are presented in Table 6.

Table 6 Sources of Supply of Green Chillies for Curd Chilli Preparation

Sl. No	Name of the Markets	Variety	Quantity Purchased per Annum (Tonnes)	Price per Ton of Green Chilli
01	Oddanchatram	Samba (Long Pungent Type)	23.000 (20.35)	10714.00
02	Dindigul	Samba	36.000 (31.86)	11430.00

Sl. No	Name of the Markets	Variety	Quantity Purchased per Annum (Tonnes)	Price per Ton of Green Chilli
03	Varushanadu	Samba	21.000 (18.58)	12143.00
04	Thiruppur	Gundu (Non-Pungent or Mild)	19.000 (16.82)	12500.00
05	Kangeyam	Gundu	14.000 (12.39)	12857.00
Total Quantity Purchased Per Annum			113.000 (100.00)	
Average Price Per Ton of Green Chilli in Rupees				11929.00

The cottage industries who are processing green Chilli for the preparation of salted curd Chilli are depending Oddanchathiram market and Dindigul Market in a big way. Both of these markets are capable of supplying around 50 per cent of the requirement of cottage industries in the village of Thirumalairayapuram of Oddanchathiram Taluk. The three markets as given in the Table 6 are capable of supplying the Samba variety of green Chilli. Thiruppur and Kangeyam areas are capable of supplying the Gundu Chilli which accounted for 29 per cent to the total requirement of the cottage industries chosen for the study.

Among the markets and their offerings, the Oddanchatram market is found to offer the Chilli at the lowest price to the consumers. All other markets are charging higher price per unit of the produce purchased. The average price per ton of green chilli purchased by the cottage industry is arrived at Rs 11929. In this respect, an effort has been taken to assess the market prices prevalent at a particular point of time in different markets like Oddanchatram and the Dindigul market and the prices are compared and the results are presented in Table 7.

Table 7 Price of Green Chilli in Different Markets of Dindigul District

Sl.No	Date	Price of Gundu Chilli per Kg in Oddanchathiram Market	Price of Gundu Chilli per Kg in Dindigul Market	Percentage Change in the Price
01	01.11.2013	17.00	30.00	76.47
02	02.11.2013	17.00	30.00	76.47
03	05.11.2013	17.00	32.00	88.24
04	06.11.2013	17.00	34.00	100.00
05	07.11.2013	17.00	32.00	88.24
06	08.11.2013	17.00	32.00	88.24
07	09.11.2013	17.00	32.00	88.24
08	10.11.2013	17.00	36.00	111.76
Average Price Per Kg		17.00	32.25	89.71

Table 7 reveals that the price of Gundu Chilli in Oddanchatram market was found to be stable at Rs 17 per kg in a week time. Almost all the vegetables sold in the market were found to be very cheap when compared to all other markets visited. In respect of Dindigul vegetable market, the price of Gundu Chilli is found to be on the higher side because the Chilli is being supplied from Oddanchathiram market in a big way and being the Municipal Corporation, the price per unit of the Chilli is found to be on the higher side. On an average, the percentage change in the price of green Chilli is found to be 89.71 per cent higher in Dindigul vegetable market when compared to the price of Oddanchathiram vegetable market price of green Chilli. On wholesale purchase, even the vegetable Chilli is reduced to the level of Rs. 10714 per ton of Chilli, according to the cottage industry. However, the price prevailing in other markets are not accessible due to the paucity of time to the researcher. Hence, the data available for Oddanchatram and Dindigul market alone were relied upon for the study.

To sum up, under the section of sources of supply of green Chilli, the area under Chilli both at national and regional level is decreasing. However, the production of Chilli is found to be increasing at national level. But the production of Chilli at regional level is found to be reduced to certain extent and hence the Department of Horticulture and Plantation Crops,

Government of Tamil Nadu should take appropriate efforts to enhance the area under Chilli and the production and productivity of Chilli by adopting suitable technologies at field and monitoring the same for realizing the fruitfulness of providing higher income to the farmers. The Government of Tamil Nadu is also poised to triple the income of the farming community through appropriate efforts, it is time to concentrate on the green Chilli too, one of the important commercial and useful crops and hence an area of 400 ha is earmarked under Chilli in the action plan prepared for implementation under National Horticultural Mission by mitigating and extending the support services to make the sector flourish.

METHODOLOGY OF PROCESSING FOR PREPARATION OF SALTED CURD CHILLI

In the preceding section, an elaborate discussion is availed regarding the agro-climatic zone wise potential to raise Chilli, the area, production and productivity of Chilli both at national and regional level, assembling markets for Chilli and the price prevalent in different markets and the sources of supply of green Chilli to the cottage industries who are processing the green Chilli for value addition. One should be interested in knowing the methodology of processing the green Chilli to prepare the salted curd Chilli which is in demand regionally, nationally and internationally and hence the methods adopted are inquired and the stages of processing is delineated in Table 8.

Table 8 Methods of Preparation of Curd Chilli in Tamil Nadu

Sl. No	Method of Preparing Curd Chilli
01	It is a widely preferred recipe among the Chettinadu consumers and it was developed by them and are preferred in many nations now-a-days
02	Buy green chillies that are slightly bigger. Preferably Samba Chilli or Gundu Chillies that are to be slit open by making a small hole on both the sides.
03	Do not wash the chillies in the water. If washed, see that there are no water particles on the Chilli. If possible wipe out the water particles with clean cloths

Sl. No	Method of Preparing Curd Chilli
04	Slit each Chilli using a needle in two or three places so that diffusion of salt and buttermilk or curd in diluted form is possible
05	Keep the stem intact and do not throw it out which adds taste to the output
06	Take a bucket or mud pot or bowl or cement tank (on large scale) and put the slit opened Chilli. Preferably mud pot is desirable (Plate – 3)
07	Add salt and sour buttermilk and see that the chillies are covered with buttermilk + salt mixture (Plate – 4)
08	Mix thoroughly till each Chilli is dipped into the salt and buttermilk mixture
09	You can test whether the mixture is with appropriate salt and buttermilk concentration
10	Do not over do the salt. If you feel that the salt is high, please balance it by adding some more quantity of buttermilk
11	Leave this mixture overnight for curing
12	Next day take the chillies out and spread the same under sunlight in a clean environment using clean sheets of tarpaulin if it is in large scale processing
13	Keep the curd or buttermilk in which the Chilli were soaked safely
14	Dry the Chilli in the hot sun for four to five hours and take them and dip the partially dried Chilli in the vessel in which the buttermilk and salt mixture is available (Plate – 5)
15	Repeat the process for few days till the salt and buttermilk mixture gets absorbed by the Chilli and got dried
16	This process will consume four to five days in a week and depends on the sunlight. Even it may extend to 15 days during the colder months
17	Once done, you can store the produce in the well cleaned containers
18	Heat a little oil in a frying pan, fry 10 Chillies (or more if you like) in oil any time you want a spicy, crunchy accompaniment with curd rice, sambar rice, etc.

Sl. No	Method of Preparing Curd Chilli
Important Points for Consideration while Preparing Curd Chilli	
19	Do not add water or any extra curd or buttermilk once you start the drying process. Use the same buttermilk or curd the chillies have been mixed in till it dries up.
20	Buttermilk is always desirable instead of curd because the fat content present in the curd will develop rancid odor on long storage and hence buttermilk is always desirable

Table 8 reveals that the recipe of making salted curd Chilli originated from Chettinadu cuisine. Chettinadu is the place known for quality and tasty foods which are preferred by almost all the consumers at regional, national and international level. Many of the restaurants, hotels and messes are having the chef originating from Chettinadu areas and hence the recipe of making curd Chilli also would have spread to the length and breadth of the nation and the value added curd Chilli is finding an important position in the diet of people of Tamil Nadu, the people of South India and the people in other states and are of good demand in Singapore and Malaysia where there is considerable Tamil population.

Plate 3 Checking the Cement Tank used for Processing of Green Chilli

Plate 4 Processing of Green Chilli is on using the Cement Tanks

Plate 5 Drying the Processed Chilli in the Drying Yard

Table 10 also clearly depicts the do's and don'ts in the process of making curd Chilli. There are two important points which are delineated under the don'ts category. They are

- Do not add water or any extra curd or buttermilk once you start the drying process. Use the same buttermilk or curd and salt mix for the chillies which have been mixed-in till it dries up and the contents of buttermilk or curd along with salt are not wasted and used up in the curd Chilli itself and
- The buttermilk is always desirable instead of curd because the fat content present in the curd will develop rancid odor on long storage and hence buttermilk is always desirable for making salted curd Chilli.

Once the processing is over, you can store the produce in the well cleaned containers. Whenever needed, heat a little oil in a frying pan, fry 10 Chillies (or more if you like) in oil any time you want a spicy, crunchy accompaniment with curd rice, sambar rice, etc. For its rich taste, certain consumers will be searching the hotels for the lunch where the fried curd chilli is one of the side dish along with other vegetables. Let us see the trading opportunities for curd Chilli in the ensuing sections.

TRADING PRACTICES PREVALENT IN SALTED CURD CHILLI

Before entering into the detailed marketing practices prevalent in curd Chilli, learning on the marketing concepts becomes very important for better understanding of the marketing and other related aspects.

Chillies are brought to the regulated markets in different parts of the state mostly by primary producers. Unlike the other perishable commodities, dry chillies are sold in the market by many producers in a phased manner. The producers try to get the best advantage of the prices. They store the produce as long as they can and bring to the market when prices are favorable to them. Only during market glut the producer-seller come to the market with entire produce and try to dispose off the lots as quickly as possible because storing and selling may result in further loss due to crash in prices. The other agencies selling the produce in the market yard are village merchants, wholesalers, commission agents and cooperatives. Unfortunately, the role of cooperatives in bringing the Chillies into the market is very less. Only in Tamil Nadu it is reported that one per cent of the total arrivals of chillies in

Thirunelvelli and Kovilpatti markets are brought by the local cooperative societies. The cooperatives were also active in Kozhikode and Sanghli markets of Kerala and Maharashtra States (Government of India 2009).

In Tamil Nadu, the Farmers' Market was created in almost all the town areas with one or more Farmers' Market to promote the practice of direct marketing whereby the producers get better advantage in price by providing free logistics. In these markets, along with fruits and vegetables other commodities are also marketed. Green Chilli is one of the important commodities sold in the farmers' market. However, the cottage industries are not much interested in procuring the green Chilli in the farmers' market owing to the small arrival of the produce and hence they mainly depend on the Oddanchathiram wholesale vegetable market which serves the customers at low price per unit of the produce.

Dried and salted curd chillies (Moar Milagai) are one of the classic condiments which is very popular in South Indian cuisine. It is also being consumed in countries like Malaysia, Singapore, Sri Lanka, etc. Moar milagai (curded green chilli) is prepared from either long dried chillies or with small chilly based on requirement. Product is prepared in a hygienic condition, having long shelf life, crispy and less hot. Moar Chillies (chilly vathal, moar milagai) are fried and consumed along with meals. This can also be called as milagai vathal. Product packing can be customized as per the requirement.

Details of Product Types

Product Type Single Herbs and Spices.

Product Style Cured and Dried.

Taste of the Product Salty and Spicy.

Product Types

- Small Dried and Salted Curded Chillies,
- Long Dried and Salted Curded Chillies (Plate – 7).
- Dried Bold and Gundu Type of Curd Chillies

Brand Name Many brands are available in curded chilli exported to different destinations

Place of Origin The Chettinadu, Tamil Nadu, India.

The marketing and trading practices including both export and import details prevalent for salted curd chilli in Tamil Nadu are discussed under the following heads.

5.3.1 Cost of Production of Salted Curd Chilli

5.3.2 Details of Realization of Curd Chilli and other Associated Products

5.3.3 Important Assembling Centers for Curd Chilli in Tamil Nadu

Cost of Production of Salted Curd Chilli in the Case Farm

The green Chilli procured from different places are processed by adding the curd and salt in the prescribed combination and are value added by doing alternate wetting in the curd plus salt solution and dry the same in the sunlight for a day. Similarly, this process is extended for 6 to 8 days till the curd plus salt solution gets exhausted and the produce is dried under the sunlight and is preserved in insulated gunny bags safely. The cost of conversion or production of curd Chilli is processed and the results are presented in Table 9.

Table 9 Annual Fixed Cost Involved in Production of Salted Curd Chilli in the Case Farm

Sl. No	Particulars of Fixed Cost	Cost (Rs. In Lakhs)	Expected Life (Years)	Annual Fixed Cost	Percentage to Total
01	Rental Value of Owned Land (0.25 Acre)	00.063	01	6250.00	04.11
02	Erection of Raised Cement Platform for Drying (2500 Sq. Feet) (Plate – 8)	15.00	40	37500.00	24.63
03	Erection of Processing Tanks for Chilli	00.05	05	1000.00	00.66

Sl. No	Particulars of Fixed Cost	Cost (Rs. In Lakhs)	Expected Life (Years)	Annual Fixed Cost	Percentage to Total
04	Construction of Store Cum Office Room (1000 Sq. Feet)	10.00	40	25000.00	16.42
05	Cost of a Pick-up Van for Material Transportation	06.50	10	65000.00	42.69
06	Electrical Installation Charges	00.25	05	5000.00	03.28
07	Miscellaneous Expenses Under Fixed Cost Items	00.25	02	12500.00	08.21
08	Total Annual Fixed Cost in Rs.			152250.00	100.00
09	Cost Per Production Period (Assuming that summer months attracts 23 -25 production periods and the winter months attracted 12 production periods)			4350.00	00.00

Plate 8 Drying Yard Erected By R.S. Cottage Industry

Table 9 highlights the details of fixed costs incurred in production of salted curd Chilli in the study area. It revealed that an amount of Rs

1.52 lakhs is needed for producing curd Chilli per annum to the level of 36 tonnes. Assuming that the summer and the winter in the study environ prevail for six months, production period is manipulated. In summer month, four production period is possible and hence a total of 24 production periods and during the winter months only two production period is possible and hence the winter months had attracted only 12 production periods. Put together, the production period per annum is arrived at 36 each commanding to the level of one ton of output. In this respect, for one production period, the annual fixed cost is arrived at Rs 4350 treating the average production of 35 tonnes.

Table 10 reveals the details of variable costs incurred in producing the curd Chilli per ton. The procurement price of green Chilli is arrived at Rs 11929 per ton which accounted for 53 per cent to the total cost of production of curd Chilli.

Table 10 Variable Cost Involved in Production of Salted Curd Chilli in the Case Farm

Sl. No	Particulars of Variable Cost	Units	Actual Cost in Rs.	Percentage to Total
01	Procurement Cost of Green Chilli	1000 Kgs	11929.00	52.83
02	Women Labor for Nailing the Green Chilli of One Ton	01.00 / Kg	1000.00	04.43
03	Men Labor for Processing Activities like Mixing, Dipping and Drying @ 3 Man Days	350 per Man Day	1050.00	04.65
04	Transportation Cost of Green Chilli	1000 Kgs	1500.00	06.64
05	Cost of Curd for 1000 Kgs	60 Litres	0450.00	01.99
06	Cost of Salt for 1000 Kgs	50 Kgs	0300.00	01.33
07	Miscellaneous Cost per Operations including Packing	One	2000.00	08.86

Sl. No	Particulars of Variable Cost	Units	Actual Cost in Rs.	Percentage to Total
08	Total Variable Cost (TVC)	Rs	18229.00	80.73
09	Annual Fixed Cost (AFC) per Production Period	Rs	4350.00	19.27
10	Total Cost (TVC + AFC)	Rs	22579.00	100.00
11	Cost of Production per Ton of Curd Chilli	Rs	22579.00	100.00
12	Cost of Production per Kg of Curd Chilli	Rs	22.58	37.63
13	Sale Price of Curd Chilli @Rs 50 per Kg	Rs	60.00	100.00
14	Net Return per Kg of Curd Chilli		37.42	62.37

The total variable cost incurred in producing one ton of curd Chilli is arrived at Rs 18229. The total cost of converting one ton of green Chilli into curd Chilli is arrived at Rs 22579. For producing one kg of curd Chilli, the producer has to invest Rs 22.50 or 23 per kg. From one kg of curd Chilli, the producer is able to generate a net return of Rs.37.42 which accounted for 62.37 per cent to the sale price per kg of curd Chilli. The Curd Chilli is of high demand at the production site itself. Sometimes, the producer has to take the produce during the peak season to store the produce in the cold storage facilities which are available at Madurai and Dindigul which commands additional expenditure of around Rs 2000 per ton. The curd Chilli produced is segregated as Grade-I and Grade-II and is separately packed in quality packaging for easing exports through Thoothukkudi port to the different import destinations. The Grade-II produce is supplied to the regional and national level. In important cities, the Grade-I type is also supplied at higher price per unit. The details of Grade I and Grade II are depicted in the Plates – 9 and 10.

Plate 9 Exportable Curd Chilli – Grade I

(Grade - I) **Plate 10** Grading of Curd Chilli (Grade – II)

Details of Realization of Curd Chilli and other Associated Products

In addition to the curd Chilli, the R.S. Cottage Industry is also generating other outputs. The details of those outputs are also analyzed and the results are presented in Table 11. Similar to the curd Chilli, the other products like Chunda Vathal, Mithukka Vathal, Mango Vathal, and Kothavarai Vathal are also produced by utilizing the available resources in the village.

Table 11 Details of Realization of Curd Chilli and other Products Produced by the Cottage Industry

S.No	Name of the Produce	Input (Qty. in Kg)	Out Turn (Qty. in Kg)	Producer Price/Kg (in Rs.)	Consumer Price/ Kg (in Rs.)
1	Mooar Milagai	100	33.000	60.00	140.00
2	Chunda Vatthal (*Solanum*)	100	33.000	120.00	230.00
3	Mango Vathal	100	17.000	38.00	130.00
4	Kothavarai Vathal (Cluster Beans)	100	10.000	35.00	120.00
5	Bhendi Vathal	100	07.000	45.00	210.00
6	Midhuka Vathal (Coccinea)	100	108.000	80.00	180.00

Among the above mentioned products, Table 11 outlines that the Moar Milagai (Curd Chilli) is the preferred one in the market. Followed by the Mooar Milagai, the Chunda Vathal belonging to the Solanum group is also processed and sold for meeting the demand of the consumers. The Mooar Milagai and Chunda Vathal output could be realized to the level of only 33 kgs for the raw material of 100 kgs which are priced respectively at Rs 140 and Rs. 230 per kg. Mithukka Vathal is another variety belonging to the Coccinea group which is also processed in curd plus salt mixture, similarly dried in the sunlight and traded across the nation. The producer of the output is able to get Rs 80 per kg.

The number of outputs added in the production process is mainly to meet the consumer requirements and to tap the benefit of utilizing the cheaply available resources into the value added form to reap higher price per unit. The lean season of one will permit the processing of other produce and hence the producer industry is always busy with realization of different output from the production process.

Important Assembling Centers for Curd Chilli in Tamil Nadu

The produced curd Chilli are being supplied to different assembly markets which are available in Tamil Nadu and the details are analyzed and the results are presented in Table 12.

Table 12 Important Assembling Centers for Curd Chilli in Tamil Nadu

Sl.No	Name of the District	Number of Assembling Centers	Percentage to Total
01	Dindigul	02	18.18
02	Madurai	01	09.09
03	Thoothukkudi	01	09.09
04	Coimbatore	02	18.18
05	Chennai	03	27.28
06	Tiruchirappalli	02	18.18
	Total	11	100.00

Table 12 reveals that, in Tamil Nadu, there are 11 assembling markets for curd Chilli which are distributed in Dindigul, Madurai, Thoothukkudi, Coimbatore, Chennai, Tiruchirappalli districts. Among the 11 markets, Chennai city alone shared 27 per cent of the total assembling markets followed by Coimbatore, Tiruchirappalli and Dindigul districts each of which shared 18 per cent of the assembling markets for curd Chilli. Other districts like Madurai, Thoothukkudi shared only 9 per cent each in respect of curd Chilli. From this one could understand that the southern districts had the assembling centers of 36 per cent followed by the central part of Tamil Nadu sharing 18 per cent and the northern and western part of Tamil Nadu sharing 27 and 18 per cent respectively revealing that the assembling markets were distributed across the length and breadth of the state of Tamil Nadu. Western part of Tamil Nadu takes care of the supply to the Kerala markets and the southern part of Tamil Nadu takes care of the market in the southern part of Kerala and southern districts of Tamil Nadu.

SUMMARY AND CONCLUSIONS

In Tamil Nadu, Dindigul District, Thriumalairayapuram is the village in which several cottage industries at household level entered into the process of making Curd Chilli for the domestic and international markets. There are about 350 units functioning in the village. However, a case study approach has been adopted by selecting the cottage industry purposely which is operating at large scale and distributing to different local and national markets. The study found that the area under Chilli is decreasing at national level. However, the production of Chilli is found to be increasing at national level but at regional level, the production is decreasing. A proven methodology of processing curd chilli is narrated for further follow up by the cottage industries. The production period per annum is arrived at 36 and the annual fixed cost per production period is found to be Rs 4350. The total cost of converting one ton of green chilli into curd chilli is arrived at Rs 22579. The curd chilli so produced is graded into Grade I and Grade II and the same are distributed to different local and national markets. The percentage of out-turn of curd chilli is arrived at 33 for a kg of green chilli and the same are traded through 11 trading centers prevalent in Tamil Nadu. The residual salt from the process of curd chilli making has created some environmental issues and the same may be eliminated by adopting Environmental Legislation Prescriptions, viz., Consent To Establish and Consent to Operate the industry by suitably erecting the treatment units at micro level by the panchayat administration and the cottage industries at household level may be protected from law.

References

- Government of Tamil Nadu, "Season and Crop Report of Tamil Nadu – 2010" (Directorate of Economics and Statistics: Chennai)
- Government of India, "Agricultural Statistics at a Glance – 2011-12" (Directorate of Economics and Statistics: New Delhi)
- Pruthi, J.S. (1976). "Spices and Condiments" (National Book Trust: New Delhi) P. 48
- Shivasankar, K. (2007). "Marketing of Dry Chillies in Karnataka – A Management Appraisal" (*An Unpublished Ph.D. Thesis submitted to the Department of Agricultural Marketing, Cooperation and Agribusiness Management*, University of Agricultural Sciences, Dharward.)
- Subbiah A., and S. Jeyakumar (2009). "***Market Survey of Chillies***" (Department of Commerce, Sri S.R.N.M. College: Sattur) March 2009. Pp.1-3

Web References

- http://oddanchatramvegetablemarket.com/about-market/
- http://oddanchatram.in/vegetable-market-large-production-of-vegetables.html

Chapter 18

Bixa Plantations in Tamil Nadu – A New Agripreneurial Initiative

INTRODUCTION

Indians have been considered as forerunners in the art of natural dyeing. Natural dyes find use in the colouring of textiles, drugs, cosmetics, etc. Owing to their nontoxic effects, they are also used for colouring various food products. In India, there are more than 450 plants that can yield dyes. In addition to their dye-yielding characteristics, some of these plants also possess medicinal values. Colour in one form/another, has been added to our foods for centuries. Bixin is a seed-specific pigment obtained from *Bixa orellana,* widely used in foods and cosmetics industry. In the last twenty years, however, consumers have become increasingly aware of the ingredients in their foods and as such they require foods to be as 'natural' as possible (Plotkin, 1993).

DISTRIBUTION

Bixa orellana L. (Family: Bixaceae) is commonly known as annatto or roucan or lipstick tree. It is native to tropical Central and South America but it has become naturalized in many countries of Africa and Asia. In India, it is naturalized and cultivated to a small extent in Mysore, Kerala, along the

Coromandel Coast and in certain districts of Andhra Pradesh, Maharashtra, West Bengal and Assam.

DESCRIPTION

Its single or multiple stems are light brown. The bark is more or less smooth with many warty lenticels, but may become fissured in old individuals. If given ample space, annatto generally branches several times near the ground and develops a dense, spreading crown. The plant roots firmly with a thick taproot and finer laterals. The alternate leaves have long petioles, thin ovate blades with long-pointed tips. Panicles at the branch tips have few to many pink or white flowers. It is a multipurpose small evergreen tree, having three types of flowers with shades varying from white to pink, *viz.*

1. White flowers bearing green pods,
2. Pink flowers bearing red pods, and
3. Pinkish white flowers bearing brown pods.

The flowers and fruits of annatto appear in clusters. Its leaves resemble betel vine leaf and its veins are red in colour. The fruits of these trees are soft and thorny like that of Aranoli or dhatura which have four petals. The fruits bear nearly 50 red seeds in colour on ripening. It flowers during August to December. The fruit splits longitudinally in length in two parts as a result of which the red seeds become conspicuous. The fruits are spiny capsules that dry and split open in two parts to expose red seeds on their inner surfaces (Howard 1989; Liogier 1995). The red colour compound is extracted from the seeds which are coated with thin pulpy, bright orange resinous coating commonly known as Annato dye or Bixin. The quantity of bixin in the seed varies from 0.70 to 3.50 per cent by weight and contains carotenoid of various types out of which *cisbixin* alone accounts for 82 per cent. Since bixin is the principal colouring matter, the chemistry and performance of annatto colour is essentially of the bixin, which is a highly unsaturated compound.

***Bixa orellana* Tree with Pods**

SILVICULTURE OF BIXA ORELLANA

Bixa orellana can adapt to a wide range of conditions in the tropics and in the frost-free sub-tropics. It thrives best at a temperature range of 20-36^0 C in areas with an annual rainfall of 1,250-2,000 mm, preferably well distributed but with a dry season for seed ripening. Tropical climate with humid atmosphere is suitable for the cultivation of annatto plant. Nutritional requirements are not high and the tree can be grown on a variety of soil types, provided that drainage is good. Annatto fits well into an agro-forestry system, provided that it is not shaded by other large trees; an open sunny position is necessary for its good performance.

NURSERY TECHNIQUES

The plant can be propagated by seeds or through other vegetative methods.

Propagation Through Seeds

Best time for raising the nursery is during April-May months. Seedlings are raised in polythene bags containing a mixture of soil, sand and manure. Seeds germinate in about 8-10 days. Seedlings are transplanted when they are 20 cm tall. Pits of size 30 cm^3 are dug in early June at a spacing of 3.0 m x 3.0 m. They are filled with a mixture of soil and compost before the onset of monsoon. Six months old healthy seedlings are planted in the main field at the onset of rains.

Pods and Seeds of *Bixa orellana*

Propagation Through Vegetative Methods

It can also be propagated through stem cuttings and cleft grafting methods. The optimum season for these methods is during November to December. The semi hardwood stem cuttings of size 15 cm (length) x 2.0 mm (diameter) treated with 4000 ppm of IA results in good rooting percent (85%). The rootstock in the diameter class of 10.10 - 11.50 mm, with the scion collected from superior mother tree and the cleft grafting during November-December recorded the maximum success per cent of 85 (Kala, 2009).

PLANTATION MANAGEMENT

The planting season is generally the monsoon period (June to September). However, it can be planted till October provided assured irrigation source exists for protective irrigation. The pits of 30 cm^3 size are dug at 3 x 3 m or 3 x 2.5 m spacing for accommodating 1100 to 1200 plants per ha. Though the plantation is required to be done during rains, in case of dry days during monsoons it is always desirable to provide 2-3 irrigations immediately after planting. Therefore, irrigation is desirable at twice a month taking into consideration the soil type, texture, water holding capacity and atmospheric temperature. Better results are obtained when moisture stress is absent from flowering to capsule maturity stage. The weeding, mulching and soil working around the pits are always desirable for better yield. Two weeding up to 3rd year is essential, i.e., one weeding and soil working before and one after monsoons. There is no specific pest attack reported in this species. But they sometime show die back symptoms under prolonged waterlogged conditions. It is advisable to spray the plants with a mixture of Bavistin and copper oxy-chloride during the initiation of symptoms of die back.

PHENOLOGY

Flowering generally starts in 2nd year of planting. However, plantation raised using vegetative propagules are reported to initiate flowering after 12 months. Blooming starts from July-August to end of October. The capsule (fruit) formation starts after 30 days of flowering. The fruiting therefore may be seen from September-October.

Variation in Flowers of *Bixa orellana*

YIELD AND BIXIN CONTENT

Harvesting of fruits may be initiated from October onwards. The right stage of harvesting is determined by stage of drying of capsules and development of cracks thereon. Capsules are harvested in bunch and spread on polythene sheets or clothes under semishaded condition for 6-7 days. The dried pods are beaten to remove the seeds. The separated seeds are winnowed and stored in gunny bags in cool and dry places. The yield of *Bixa orellana* increased with plant age from 528.6 kg/ha for 2-year-old plants to 2483.3 kg/ha for 3-year-old plants. For individual trees, the dried seed yield may vary between 0.5 and 4 kg per year. Seed contains 0.07 per cent essential oil and 1.008 per cent bixin (Kanjilal and Singh 1995). However, the bixin content of seeds present in the assembled 34 genotypes ranged between 1.13 per cent (KABi 5) and 3.13 per cent (TNBi 1) (Kala, 2009).

PRUNING

Pruning is important to get better yield from *Bixa* plant. It is recommended to prune the branches/twigs every year after harvest by sharp pruning saw or secateurs. Pruning should be done in early morning. A light spray of fungicide after pruning is also required to avoid the fungal attack. The pruned twigs sprout into 3-5 new shoots which becomes ready for blossoming in next season.

USES

The seed colouring matters known as bixin and orellin are isolated from *Bixa orellana*. Annatto seeds are the world's second most important natural colorant after caramel yielding yellow to red colour (Mercandante and Ptander, 1998). The dye which is obtained from the seeds is used for colouring food, dyeing the cloth and painting the skin. The pulp surrounding the seeds is widely used in herbal medicine to treat burns, dysentery, constipation and fever (Parrotta, 2001).The colour may reach up to 7 per cent of the seed's dry mass (Kutzer, 1999). Annatto is grown in garden as ornamental and hedge crop in social forestry/live fence. The crop is now fetching importance in food industry as the seeds are directly powdered and mixed with food products for their attractive colour. The branches with dry pods are used in dry floral arrangements and from the bark, ropes and twines are made. Restrictions on the use of many synthetic colorants and the relative instability of most other carotenoids, are leading to the increasing use of bixin, especially in the dairy industry. World production, estimated at about 3000 tonnes of annatto seed in 1983, is now thought to have risen rapidly to over 10000 tonnes about half of which comes from Brazil (Anand 1983). Hence it is cultivated throughout the tropics for food coloring and use as dye for cheese, soap, cloth, paint, lipstick, margarine, etc. A few seeds can be added to rice, soups, or to flavor chocolate and meats. Seeds are eaten and cooked in butter. Bark provides fiber. It is a good hedge and honey plant. The roots are used as flavouring for meats to which they impart the colour and flavour of saffron. They are also used for dyeing hairs and abound in floor polishes, shoe polishes, hair oils, pharmaceutical ointments, etc.

STEPS TO INCREASE BIXIN PRODUCTION

The demand for seeds of Bixa is increasing over a period of time. The supply source is shrinking which leads to the dependence on unnatural dye and food colorant adding health hazards to the human. If produced and supplied sufficiently to meet the industrial requirements by enhancing its area, production and productivity, the demand could be met and organic source of food colorant and other uses could be anticipated. Being a new crop, an awareness on economics of establishing the crop in farm fields is of paramount importance and hence special effort has been taken to document the details of area under Bixa, cost of cultivation and its associated economic benefits to aid the farming community and the entrepreneurs.

STUDY OF THE AREAS UNDER BIXA CULTIVATION

The multifarious uses of seeds of Bixa has urged the few farmers from Tamil Nadu to establish the Bixa plantations and the seeds are being demanded in dye making industries as natural dye. The farmers have not realized the importance of this plant and the risk taking farmers have come forward to establish the plantations. In South India, the Andhra Pradesh state had higher area. However, the details of area under Bixa plantations in different states are delineated in Table 1.

Table 1 Area under Bixa Plantations in South India

Sl. No	Name of the State	Number of Locations	Area Under Bixa in Acres	Percent-age to Total
01	Andhra Pradesh	29	320	48.93
02	Kerala	08	023	03.52
03	Karnataka	16	120	18.35
04	Pondicherry	06	015	02.29
05	Tamil Nadu	23	176	26.91
	Total	82	654	100.00

The area under Bixa plantations is very poor in Tamil Nadu and the demand for seeds of Bixa is also increasing for natural dye. Hence the Government of Tamil Nadu through the National Agricultural Development Project permitted the Tamil Nadu Agricultural University, Coimbatore to establish a research project on Bixa with a mission to enhance the area under Bixa and to enhance the industrial utility for deriving Bixin which is used in the processing industries of textile and food. The project entitled "Popularization of Annatto (*Bixa orellana*) for Natural Edible Dye in Tamil Nadu" is the sub-project sponsored by Government of Tamil Nadu under the NADP project. Dr. K. Kumaran, Professor in Forestry, Horticultural College and Research Institute, Periyakulam being the Principal Investigator of the said project is taking all efforts to increase the area expansion and to mitigate the industrial utility of Bixa plantations. In this regard, few industries have agreed to enter into Memorandum of Understanding with Tamil Nadu Agricultural University and with the farmers of Bixa plantations regarding buy back agreement. Such a venture to be brought under higher area means, the researcher has to convince the farmers based on its economic viability and the marketability of the output drawn from Bixa. Hence, a special effort has been taken by the Principal investigator and the Co-Principal Investigators of the project in documenting the economic incentives associated with the Bixa plantations in Tamil Nadu by paying personal visit to the progressive farmers who are cultivating Bixa in Tamil Nadu and in Pondicherry. In Tamil Nadu, around 176 acres of plantation is visible, out of which, the project had made an area expansion of 150 acres which accounted for 85.23 per cent to the total area under Bixa Plantations. Old Bixa plantations already established by the farmers of Tamil Nadu under different locations as delineated in Table 1 accounted for only 26 acres out of which the area available under bixa with the contact farmers were arrived at around 12 acres which is in Kallar, Sirumugai of Coimbatore District, Gudalur of The Nilgiris District, Cuddalore and Kancheepuram Districts.

The Bixa was established totally in strips, boundary plantations, bulk plantations and as ornamental trees in different parts of Tamil Nadu. The details of its existence are delineated in Table 2.

Table 2 Area under Bixa Plantations in Tamil Nadu

Sl. No	**Name of the District**	**Number of Locations**	**Area Under Bixa in Acres**	**Percentage to Total**
01	Coimbatore	08	12.20	46.92
02	Cuddalore	04	01.20	04.62
03	Kancheepuram	06	08.90	34.23
04	Pudukottai	02	01.80	06.92
05	The Nilgiris	03	01.90	07.31
	Total	23	26.00	100.00

From Table 2, the Bixa established in 23 locations were visited and the data on cultivation and its economic incentives were collected and documented for this chapter using a questionnaire designed for that purpose. The data were analyzed using the descriptive statistics like mean and percentage analysis. Besides, the economics of bixa plantation is communicated and awareness created to the farmers who have recently planted with one day training program on Bixa. A good response is visible from the one day training programs conducted at Dry Land Agricultural Research Station, Chettinadu and at Horticultural College and Research Institute, Tamil Nadu Agricultural University.

With respect to analyzing the economics of plant like Bixa, the methodology advocated by the following authors were reviewed and appropriate methods have been incorporated in the analysis.

Sekar et al. classified costs into three viz., (i) establishment costs, (ii) maintenance and (iii) harvesting costs. Establishment costs involved in bixa plantations were ploughing of land, preparation and layout, manures, cost of seedling, pitting and planting costs, cost of irrigation, soil working and weeding. Maintenance costs included cost of irrigation, weeding and soil working, gap filling. Harvesting costs included collection, packing, forwarding and transportation costs.

Harswarup Singh classified costs into two viz., (i) fixed costs (establishment cost) and (ii) recurring and maintenance cost. The fixed cost consisted of all the expenditure on goods and services (excluding the value of land) required for the establishment of trees. These included preparations of land and layout, digging and filling of pits, cost of plant material, transportation of seedlings including loading and unloading charges, planting cost, manures and fertilizers, plant protection and miscellaneous costs. Recurring and maintenance costs were those which are incurred every year. These expenses included manures and fertilizers, irrigation, plant protection, labour, interest on establishment and opportunity cost of land (rental value). Harvesting costs were not included as standing trees were sold. In this study, the concepts used by Sekar (1995) is used for analysing the costs and returns realized from Bixa.

STUDY OF THE DATA UNDER THE FOLLOWING HEADINGS

The main objective of the present chapter was to assess the economics of Bixa plantations in Tamil Nadu. In harmony with the objective enunciated, the data obtained from Bixa plantation farms were analysed and the results are discussed as under the following headings.

1. Details of Size of Holding Available with the Sample Farms
2. Factors Favouring Bixa Cultivation among the Sample Farms
3. Costs and Returns in Bixa Plantations
4. Constraints Faced by the Sample Farms
5. Strategies for Better Management of Bixa Plantations

SIZE OF HOLDING AVAILABLE WITH THE SAMPLE FARMS

The National Sample Survey Organization (NSSO) through "The Hindu" newspaper reported that nearly 90 per cent of India's farmers have less than two hectares of land (The Hindu 2014). But the tree cultivation practices are not widely observed in small and marginal farms. In this respect, the details of size of holding available with the sample farms were analyzed and presented in Table 3.

Table 3 Details of Size of Holding Available with the Sample Farms

Sl. No	Particulars of Holding	Average Size of Holding in Acres	Percentage to Total
01	Owned Land under Cultivation	19.00	76.77
02	Land under Non Agricultural Uses	00.75	03.03
03	Owned Land (1+2)	19.75	79.80
04	Leased-in Land for Agriculture	05.00	20.20
05	Total (3+4)	24.75	100.00
06	Area Under Other Crops	22.40	93.33
07	Area Under Bixa	01.60	06.67

Table 3 reveals that the average size of holding available with the farms was found to be 19 acres which accounted for 77 per cent to the total size of holding under cultivation. The remaining 20 per cent of land is under lease with the farms considered and the sample farms are coming under the category of medium farm which is within 10 ha classification as practiced by the Directorate of Economics and Statistics in its Statistical Hand Book (Government of Tamil Nadu, 2013). Because of availability of higher area, the farms have come forward to establish the Bixa plantations in their farm lands who can infuse all type of resources in protecting the plantations. However, the small farmers are not much interested in taking up such plantations because of poor size of holding available with them for allocation of land for such new crops because of risk averting nature. In India, it is reported that 58 per cent of the rural households are agricultural households and over half of the agricultural households are in debt due to poor income generation opportunities (The Hindu, 2014). In this respect, highlighting the economic incentives associated with the Bixa plantation means, some of them among the poor lot who are ready to take risks might come forward to take up such plantations. But guidance and counselling through the awareness creation training is imperative. Such effort is taken up by the Principal Investigator of the project on popularization of Bixa in Tamil Nadu.

FACTORS FAVORING BIXA CULTIVATION

Analysis on size of holding revealed that the medium and large sized farms alone prefer the Bixa plantation as they have higher land holding with them. It might be one of the reasons for preference of new technologies. The real reasons behind adoption of such new technology is analyzed and the details are presented in Table 4.

Table 4 Factors Favoring Bixa Cultivation

Sl. No	Factors in Favour of Bixa	Number of Farms Reported	Percentage to Total
01	Bixa Plant Adds Scenic Beauty	16	69.57
02	Attractive Colour of Pods	18	78.26
03	Larger Size of Holding with the Farms	20	86.95
04	Source of Natural Edible Dye	09	39.13
05	Influence of Private Companies	03	13.04
06	Scope of Higher Income Generation Possibilities	15	65.22
Total		23	100.00

Table 4 reveals that there were six principal reasons behind the adoption of cultivation of Bixa. The reasons are it adds scenic beauty, attractive colour of the pods, larger size of holding with the farms, source of natural edible dye, influence of private companies to plant Bixa and the scope of higher income generation. Among these factors, larger size of holding was found to be the principal one, decided based upon adoption of newer technologies, according to 20 farms who accounted for 87 per cent to the total locations in which the plants are cultivated followed by attractive colour of pods and scenic beauty which respectively accounted for 78 per cent and 70 per cent to the total locations. The assurance of higher income generation given by the private companies at the time of planting is also one of the motivating factors which accounted for 65 per cent to the total locations. However, Bixa as a source of natural dye is known only to 39 per cent of the farms who have adopted Bixa cultivation in their farm lands.

COSTS AND RETURNS IN BIXA PLANTATIONS IN TAMIL NADU

Sekhar (1995) suggested that the tree crops in general commands the fixed cost which are often referred to as establishment costs, annual maintenance costs and the harvesting costs. In this respect, the costs and returns realized from Bixa plantation is analyzed and discussed under the following sub heads

1. Annual Fixed Cost Incurred for Establishment of Bixa Plantations
2. Maintenance Cost involved in Bixa Plantations
3. Details of Harvesting Cost incurred in Bixa
4. Details of Revenue Realized from Bixa Plantations
5. Discounted Cash Flow in Bixa Plantations

Annual Fixed Cost Incurred for Establishment of Bixa Plantations

The details of annual fixed cost or establishment costs incurred in respect of Bixa plantations of different locations are analyzed and the results are presented in Table 5.

Table 5 Annual Fixed Cost Incurred for Establishment of Bixa Plantations

Sl. No	Particulars of Fixed Expenses	Units	Price Per Unit (Lakhs)	Annual Fixed Cost (Rs in Lakhs)	Percentage to Annual Fixed Cost
01	Land Revenue	02.50	00.40	01.00	62.50
02	Bore Well With Accessories	01.00	01.00	00.10	02.50
03	Barbed Wire Fencing	02.50	01.00	00.25	06.25

Sl. No	Particulars of Fixed Expenses	Units	Price Per Unit (Lakhs)	Annual Fixed Cost (Rs in Lakhs)	Percentage to Annual Fixed Cost
04	Erection of Drip Irrigation Structures	02.50	00.25	00.63	15.75
05	Annual Cost of Construction Value of one Store Room of 10 X 11Feet	1400.00	00.15	00.37	09.25
06	Small Implements for the Farm	10.00	00.005	00.05	01.25
07	Miscellaneous Items Including Cess	00.05	00.02	00.10	02.50
Total Annual Fixed Cost for Establishment of Bixa				02.50	100.00

Table 5 reveals that the annual fixed expenses included the land revenue, cost of erection of bore well, erection of barbed wire fencing, erection charges for drip irrigation structures, annual cost of construction of one store room, small implements like crow bar, pad and other soil working implements. Barring the land revenue, all other fixed expenses were included in the establishment cost which arrived at 1.50 lakhs accounting for 37.50 per cent to the total annual fixed cost. Because of the resource scarcity, the drip irrigation becomes part of the investment cost now-a-days in establishment of plantations and hence included which accounted for 15.75 per cent to the annual fixed cost. The annual cost of construction of store room to preserve the small implements and other assets is arrived at Rs 37000 per ha which arrived at 9.25 per cent to the total annual fixed cost.

Table 6 mainly addresses other establishment costs which are one-time expenses in plantation development activities like land preparation, lay out and planting design, pitting, filling with manure and planting of saplings, watering and basin formation. These costs are separately analyzed and are presented in Table 6.

Table 6 Other Establishment Costs Incurred in Bixa Plantations

Sl. No	Particulars of Fixed Expenses	Units	Price Per Unit in Rs	Establishment Cost (Rs)	Percentage to Total
01	Land Preparation and Cleaning of Bushes	02.50	4000.00	10000.00	26.58
02	Cost of Layout and Planting Design	02.50	1000.00	2500.00	06.65
03	Pitting Charges	1111.00	0005.00	5550.00	14.75
04	Seedlings Cost	1111.00	0010.00	11110.00	29.53
05	Transportation of Seedlings up to 40 Km Radius	01.00	1500.00	1500.00	03.99
06	Farm Yard Manure @ 01Kg per Pit	1111.00	0001.00	1111.00	02.95
07	Cost of Filling the Pit and Planting	1111.00	0005.00	5550.00	14.75
08	Cost of Irrigation @ one Man Day	01.00	300.00	300.00	00.80
Total Establishment Cost for Bixa per Ha				37621.00	100.00

Table 6 reveals that among the various establishment costs included in the analysis, the cost of seedlings is found to be the prime expense accounting for 29.53 per cent to the total establishment cost followed by the land preparation and bush cleaning expenses found to be the next important expenditure accounting for 26.58 per cent to the total establishment cost.

The next important item of expenditure under the head of establishment cost is pitting and planting expenses both of which equally shared 14.75 per cent to the total establishment cost. Put together, the total establishment cost is arrived at Rs 37621 per ha in respect of bixa plantations.

Maintenance Cost Involved in Bixa Plantations

Unlike other plantations, Bixa did not command higher maintenance and involves protected irrigation. Hence the cost of irrigation is not included in the estimation as it is free of cost for the farming community in Tamil Nadu. However, the cost of labor used for irrigating the crop is furnished. Actual maintenance costs included involved are weeding, soil working and basin formation, pruning of branches and replacement of casualties in Bixa plantations. These details are analyzed over the years and the details are presented in Table 7.

Table 7 Maintenance Cost Involved in Bixa Plantations

Head of Maintenance	Years									
	01	02	03	04	05	06	07	08	09	10
Weeding @ 5Women Days	1000	1000	000	1250	000	1500	000	1500	000	1500
Soil Working, Basin Formation and Watering	1450	1450	1200	1400	1400	1600	1600	1600	2000	2000
Pruning of Branches @ 4 Man Days	000	000	000	1600	000	000	1600	000	000	2000
Replacement of Casualties	1100	000	000	000	000	000	000	000	000	000
Total Maintenance Cost	3550	2450	1200	4250	1400	3100	3200	3100	2000	5500

Table 7 reveals that the average maintenance cost incurred per year is arrived at Rs 3000. However, it varies from year to year as the operations vary between the years. During the first year of maintenance, additional cost of replacement of casualties of Bixa seedlings is added because of mortality of seedlings during the first year of planting. For that around one thousand rupees is allocated on an average by the farmer. The soil working and basin formation are found to be the annual routine tasks in the farms to promote profuse growth of the plant by providing better aeration and tillage among the plants.

Pruning is one of the important activities which is carried out by the farmers during the third or fourth year of planting and it is repeated every three years to regulate the height of the trees which facilitates harvesting of Bixa pods. Similarly weeding is also carried out on an annual basis for better establishment of the crop. Such a maintenance will permit good growth and yield of pods in the plantations. In this respect, assessing the harvesting and related cost to get a good yield is important.

Details of Harvesting Cost involved in Bixa Plantations

Harvesting of Bixa pods is little difficult which consumes human labor to collect and the seeds or kernels should be extracted from the pods. Hence these details are separately analyzed and the results are presented in Table 8.

Table 8 Details of Harvesting Cost Incurred Per Ha of Bixa

Items of Cost	Years									
	01	02	03	04	05	06	07	08	09	10
Harvesting of Pods @ 10 Women Days	00	1000	2000	2500	2500	3000	3000	3500	3500	4000
Extraction of Seeds @ 5 Women Days	00	1000	1000	1500	1500	2000	2000	2500	2500	3000
Total	00	2000	3000	4000	4000	5000	5000	6000	6000	7000

Table 8 reveals that the yield of pods could be realized from the second year onwards considerably and hence one needs to pay attention on harvesting the pods before opening up of the pods. Careful monitoring of the plantations should be done to avoid kernel or seed loss. In a hectare of plantation, it is estimated that the farmers were using on an average of 10 women laborers to collect the pods from Bixa plantations. On an average, the harvesting cost is arrived at Rs 2500 per ha per year.

With respect to extraction of harvested pods for its kernels, five women labor force is needed. Annually, the sample farms have to spend Rs 2000 towards extraction of seeds or kernels form the Bixa pods. In toto, average harvesting cost spent for harvesting and extraction of kernels is arrived at Rs 4000 per ha per annum. When comparing the investment in other plantations, the maintenance and harvesting cost are almost minimum and hence one can invest in the establishment of the plantations.

Details of Revenue Realized from Bixa Plantations

Estimation of the revenue generated from the farmer fields become a cumbersome process. Even after putting several indirect questions before the farmers, sometimes detailed collection of revenue generation is difficult and hence the data have been estimated in such cases based on the experience shared by the progressive farmers. The revenue generation from Bixa plantations is from sale of fuel wood, pods and Bixa kernels or seeds. These details are separately analyzed and the results are presented in Table 9.

Table 9 Details of Revenue Generated from Bixa Plantations

Heads of Revenue	**Years**								
	02	**03**	**04**	**05**	**06**	**07**	**08**	**09**	**10**
Yield of Seeds (Ton)	00.50	01.00	01.00	01.30	01.50	01.50	01.60	01.70	01.80
Price of Seeds (Rs in Lakhs /Ton)	00.90	00.90	00.90	01.00	01.00	01.00	01.20	01.20	01.20
Revenue on sale of Seed (Rs in Lakhs)	00.45	00.90	00.90	01.30	01.50	01.50	01.92	02.04	02.16
Quantity of Shell (60% of Seed Weight)	00.30	00.60	00.60	00.78	00.90	00.90	00.96	01.02	01.08
Revenue Earned on Sale of Shell @Rs 1/Kg	30.00	60.00	60.00	78.00	90.00	90.00	96.00	102.00	108.00
Yield of Fuel Wood on Pruning @ Rs 1/Kg	00	00	5555	00	00	5555	00	00	5555
Gross Revenue in Rupees	45030	90060	95615	130078	150090	155645	192096	204102	221663

Table 9 reveals that the price of Bixa kernels or seeds is around Rs 100000 per ton. The revenue realized is found to be minimum during the initial years of establishment, but it increases with the age of the plantation. It is estimated that a five year old plantation is capable of generating Rs 100000 to Rs 1.30 lakhs as revenue from sale of seeds. It is increased to double during the eighth, ninth and tenth year of establishment.

The revenue earned through sale of shell is arrived at Rs 70 per annum which is very meagre and the revenue generation from sale of fuel wood arising from pruning of the plantations was arrived at Rs 1830 per annum from a ha of plantation of Bixa which is also a very meagre income. The fuel wood is consumed either at the farm or at the households who use them for heating and cooking purposes.

The gross revenue generated from the Bixa plantation is estimated to be around Rs One lakh which increased to 1.50 lakh during the fifth year of the bixa plantations and the same was capable of generating around Rs Two lakhs from the eighth, ninth and tenth year of establishment of Bixa plantations. From this one could conclude that the Bixa plantation is capable of generating higher income during and after the fifth year of establishment.

CASH FLOW IN BIXA PLANTATIONS

The income generated during the tenth year of establishment of Bixa plantation may be attractive when one could see in the present day environment. The tenth year revenue may be ideal for comparing between the economic viability of different plantations that urge the mind of the farmer. In this respect, the details of cash flows were discounted using the discounting techniques by employing the bank lending rate of loan for agricultural purposes at 12 per cent and the details are analyzed and the results are presented in Table 10.

Table 10 Discounted Cash Flow in Bixa Plantations

Year	Benefits From Bixa	Cost Incurred in Bixa	Discount Factor@ 12%	Discounted Benefits	Discounted Costs
00	00	124621	1.000	00	124621

Year	Benefits From Bixa	Cost Incurred in Bixa	Discount Factor@ 12%	Discounted Benefits	Discounted Costs
01	00	3550	0.893	00	3170
02	45030	4450	0.797	35889	3547
03	90060	4200	0.712	64123	2990
04	95615	8250	0.635	60716	5239
05	130078	5400	0.567	73754	3062
06	150090	8100	0.506	75946	4099
07	155645	8200	0.452	70352	3706
08	192096	9100	0.404	77607	3676
09	204102	8000	0.361	73681	2888
10	221663	12500	0.322	71375	4025
	1284379	196371		603443	161023
Net Present Value (Column 5 - 6)				442420	
Benefit Cost Ratio = (Column 5 / Column 6)				03.75	

Table 10 reveals that the sum of undiscounted costs incurred in the Bixa plantation is arrived at Rs 1.96 lakhs and the value of investment compared to present day situation is found to be reduced to Rs 1.61 lakhs if one could employ the discounting technique. A reduction of Rs 36000 is visible in investment costs in Bixa. Whereas, the sum of undiscounted benefits is estimated to be 12.84 lakhs over a 10 year period of establishment and the same could fetch Rs 6.03 lakhs on employing the discounting technique. The income generated at the end of tenth year is found to be only 6 lakhs instead of 12.84 lakhs. Now the farmer could compare which investment becomes worthy. For that, the Benefit Cost Ratio and Net Present Value realized from the Bixa plantation may be ideal.

The Benefit Cost Ratio realized from the Bixa plantation over a 10 year period is found to be 3.75. It revealed that for one rupee of investment the farmer makes in Bixa, he could realize Rs 3.75 as gross income from the plantation. A net return of Rs 2.75 could be realized for a rupee of investment. The Net Present Value is arrived at Rs 4.42 lakhs which reveals that the plantation could generate the income of Rs 4.42 lakhs alone over a period of ten years of its maintenance. If we maintain further by invigorating the plantation with manures and fertilizers and proper care, the return to be generated from Bixa could spiral manifold. What is needed is buy back arrangement for sale of seeds of Bixa. Before establishing the plantation, the farmer or the farmer associations/clubs can establish buy back arrangement with the company interested and should prosper. Otherwise, cheating may happen and sale of seeds become a problem to the farming community.

CONSTRAINTS FACED BY THE BIXA FARMS

Any production activity normally commands certain constraints and creates hardship in the smooth sailing. One has to identify those constraints which are affecting the production activities and the same should be finding appropriate strategies for alleviation of the constraints. The details of such constraints were analyzed and the results are documented in Table 11.

Table 11 Details of Constraints Faced by the Bixa Farms

Sl. No	Details of Constraints	Number of Farms Reported	Percentage to Total
01	Lack of Genotypes / Varieties with Higher Bixin Content and Yield	20	86.95
02	Absence of Technical Know-How	21	91.30
03	Market Accessibility to the Farmer is Very Poor	23	100.00
04	Lack of Mechanization	16	69.57
05	Lack of Supply and Value Chain for Seeds of Bixa	19	82.61
06	Absence of Dye Processing Units in India	06	26.09
Total		23	100.00

Table 11 reveals that there were six constraints involved in inhibiting the production activity of Bixa to some extent. Those constraints are market accessibility to the farmer is very poor, absence of technical know-how, lack of genotypes or varieties which could yield higher bixin content and yield of pods, lack of supply and value chain for seeds of bixa and the absence of dye processing units in India. Among these constraints, almost all the farmers have reported for absence of market for the produce of Bixa followed by absence of technical know-how and lack of genotypes for higher bixin content and yield respectively accounting for 91 per cent and 87 per cent to the total number of locations in which the Bixa is being cultivated in Tamil Nadu.

Lack of supply and value chain is another important constraint in taking the output without any interference for value addition. Lack of such value addition poses the traders not to infuse much demand for the Bixa seeds. Very few are accepting poor price for exporting the same to the developed nations who had good processing units both for dye and food colorant making. India has to concentrate on mechanization and enforcing the supply and value chain for such wonder crop so that the farming community could prosper by making buy back mechanisms.

STRATEGIES FOR BETTER MANAGEMENT OF BIXA FARMS

Above constraints quite naturally called for appropriate strategies to overcome the same so that the production and processing activity could move smoothly. In this regard, the following recommendations are found to be important.

- Superior genotypes of Bixa which could yield higher bixin content and higher productivity should be identified, developed at research institutions like Tamil Nadu Agricultural University (TNAU) and the same should be transferred to the user groups for better adoption. National Agricultural Development Project should fund the TNAU in the second term for screening such varieties and for creating a value chain in Bixa.
- Transfer of technology in Bixa is very poor. Hence, the findings emanating in Bixa from Research and Development Institutions should be popularized through demonstration, conduct of awareness creation workshops and trainings to the stakeholders of Bixa besides ensuring buy back mechanisms with the farmer and the user group firms.
- Market demand should be created for Bixin by popularizing the usage among the consumers and processors. The health benefits of Bixa should be taken into account while popularizing the

same, the value chain should be created and it should be seen that the chain of intermediaries should be protected somehow or other by infusing some subsidy and allowances component for takeoff.

- In India, the dye processing industries are almost absent and hence production of Bixa seeds quite naturally commands very low price owing to export demand which is also very poor from India. Bulk suppliers are there when comparing our quantity available for export. Hence processing industry should be established and the industry should come forward to establish tie up with the Bixa farmers on the lines of milk and Aavin. Such venture should be promoted for further takeoff in Bixa cultivation.

SUMMARY AND CONCLUSIONS

Bixa, a new crop is to be brought under higher area for cultivation. On examining the economic viability of the Bixa plantation, the plantations on maintaining could generate higher incentives beyond fifth and tenth years. The technology gap, awareness on cultivation and buy back mechanism related questions are left unanswered to the farmers and hence the farming community should be taught on awareness related to transfer of technology besides making buy back arrangements. An enterprise based approach for developing a value chain which includes cultivation techniques, harvesting and processing with precision and value addition is the need of the hour. This could be addressed by appropriate capacity building programs to the stakeholders.

References

- Anand, N. 1983. **The Market for Annatto and other Natural Colouring Materials with Special Reference to the UK.** (Report of the Trop. Dev. Res. Institute: UK). G 174.
- Harswarup Singh," (1988). "Comparative Economics of Trees and Crop Production in Haryana", ***Indian Journal of Agricultural Economics***, 43(3): July-September, 1988. p. 290.

- Howard, R.A. (1989). "Flora of the Lesser Antilles" ***Arnold Arboretum***, (Harvard University: Jamaica Plain MA) 6: P. 658
- Kala, S. 2009. **Screening, propagation and characterization of Annatto (Bixa orellana Linn.) using biometric traits.** Ph.D. Thesis, Submitted to Forest College and Research Institute, TNAU, Coimbatore, India
- Kanjilal, P.B. and R.S. Singh. (1995). "Agronomic Evaluation of Annatto (**Bixa orellana** L.)" ***Journal of Herbs, Spices, and Medicinal Plants*** 3(3): 13-17.
- Kutzer, G. (1999). *Annatto (**Bixa orellana** L.).* http://www-ang.kfunigraz.ac.at/~katzer /engl/Bixa_ore.html. p. 3
- Liogier, H.A. (1995). "Descriptive Flora of Puerto Rico and Adjacent Islands". Vol. 4. Editorial de la Universidad de Puerto Rico, San Juan, PR. P. 617
- Mercadante, A.Z. and H. Ptander. (1998). "Carotenoids from Annatto: a Review". **Recent Research Developments in Agriculture and Food Chemistry, 2(1)**: Pp. 79-91.
- Parrotta, J.A.(2001). ***Healing plants of Peninsular India***. (CAB international: Wallingford UK).
- Plotkin, M.J. (1993). ***Tales of a Shaman's Apprentice***, (Penguin Books: New York). P.233
- Sekhar, C., 0. Anchanam Alagia Pillai, 0. Timothy Randhir and G.Kumara Velu, (1990). "Economic Analysis of Kapok under Agroforestry Conditions of Tamil Nadu", ***Agricultural Situation in India***, November 1990, pp. 537-540.
- The Hindu (2014). "More Than Fifty Per Cent of Farm Households in Debt", The Hindu, December 21, 2014, p.1

Chapter 19

Training on Value Addition –The Case of Bio-Waste Management

INTRODUCTION

Madurai and Virudhunagar Districts of Tamil Nadu are blessed with vast resource potentials especially of soil and water which had motivated the farming community to adopt different seasonal and perennial crops like vegetables, cereals, pulses and fruits. After the harvest of farm produce, the residues are left unutilized and go waste and sometimes add to production problems like choking of canals and water ways. It is mainly due to the absence of manpower, take it easy approach with the farmers and lack of awareness on technology of effective usage of bio wastes.

Organic waste management through cultivation of milky mushroom will go a long way in harnessing the available agricultural waste materials, lesser use of plant protection chemicals and sustenance of subsidiary food fungi (mushroom). The technologies for bio waste management transferred to the farming community through capacity building programs will help in increasing awareness among the farming community on health, environment and ecological balance besides conserving and recycling of agricultural wastes. Utilization of locally available agricultural waste materials for making manure by utilizing spawn, production of milky mushroom and formulating bio-pesticides not only creates employment opportunities, but also builds confidence among the farmers. The efficiency of conversion of bio waste resources helps in income generation and eco-friendly management of crop diseases will sustain the healthy environment. If the benefit accrues, the target group or unemployed may find livelihood and become self employed thereby enhancing their economic status. Therefore the cycle of sustainability of improved technology and employment generation will create further social and economic development in this region.

Though the technologies were available to utilize the bio wastes, the transfer of technology to spread the same to the farming community for effective and efficient use is dismal. An institutional approach is needed to link the research lab and the farm front through appropriate capacity building programs so as to empower the rural poor which in turn enhance their entrepreneurial capabilities. Hence a study has been proposed in southern part of Tamil Nadu with the funding support from Department of Science and Technology, Government of India with the following specific objectives.

OBJECTIVES OF THE STUDY

- The unemployed SC/ST women will be exposed to demonstration and training on the technologies which are effective in bio waste management.
- To commercialize the product generated from the technology used for bio waste management
- To motivate and guide the trainees to establish micro level enterprises on bio waste management

METHODOLOGY

In southern Tamil Nadu, the district in which the Agricultural College and Research Institute is located in Madurai and the adjacent district, the Virudhunagar District, the scheduled caste and scheduled tribe women were very meager. They are put to lot of pressure both socially and economically and hence the rural Scheduled Caste and Scheduled Tribe women were the respondents aimed at for this project and hence the Madurai and Virudhunagar districts were purposively chosen to implement the project in the first stage.

The list of SC/ST women were obtained from the Department of Agriculture and Department of Economics and Statistics, Government of Tamil Nadu in Madurai and Virudhunagar districts. From the list, the villages in which higher number of Scheduled Caste and Scheduled Tribe women available were selected in the second stage.

The selected villages in the second stage were Pudutamaraipatty, Thirumalnatham of Madurai district and Sithumoondraippu and Chinnapparettiapatty of Virudhunagar districts. From each village, 25 participants were chosen as beneficiaries and thus, a total number of 100 beneficiaries were selected from the above villages in the third stage following a three-stage random sampling technique. The details of selected samples are delineated in Table 1.

Table 1 Details of Samples Selected in the Study Area

S.No	Block	Village	Total Number of Beneficiaries
1.	Madurai North	Puthuthamaaraipatty	25
2.	Cholavandan	Thirumalnatham	25
3.	Kariyapatti	Sithumoondaraippu	25
4.	Senkundrum	Chinnapparettiapatty	25

The respondents were invited for their opinion on the sustainability of the project by inviting responses for each questions delineated in the Table 2 and the number of persons revealing the word 'Yes' and 'No" were taken to stock and the same were analyzed using the percentage analysis. Besides, to know the socio-economic features of the respondents, a brief questionnaire is also designed to record their demographic profile, status of women employment, present earning strength per month and the extent of awareness on the selected technology to which the training is proposed. The details regarding both pre-training and post-training situations were analyzed and follow-up to assess the impact of training was also imparted in the project.

Table 2 Sustainability Aspects of the Project at Planning Level

Relevancy	**Yes**	**No**	**Not Sure**
Is the Project consistent with National Priority			
Is the Project consistent with Provincial / District Priority			
Is the Project consistent with the Sectoral Priority			
Acceptability			
Has the community been consulted regarding acceptability			
Has the project identified / defined clearly the targeted beneficiaries			
Does the community consider the project as beneficial			
Will the project disadvantage/discriminate any particular group			
Have the local elected representatives/community leaders been consulted and agreement obtained			
Have the local elected representatives / community leaders endorsed the project			
Have the implementation agencies been consulted and agreement obtained			
Have the capacity of the implementation agency / agencies been assessed			
Has the implementation agency / agencies got the necessary resources, logistics (trained manpower, vehicles etc) to implement the project.			
Economic / Financial Viability			
Would the post-implementation project operations be implemented without subsidy			
Are the project benefits greater than the project costs			

Relevancy	Yes	No	Not Sure
Will the project services be given free, continuously			
Is there a cost-recovery plan			
Will there be sufficient funds available to implement the project			
4. Environmental sustainability			
4.1.Is the project environmentally sustainable?			
5. Implementation and monitoring strategy			
Has the project developed an implementation plan with clearly defined goals and responsibilities			
Has the project developed monitoring and evaluation plan			
Has the project developed all the success indicators (Financial, Physical and its Impact)			
Has the project identified various levels of review mechanisms for monitoring the progress			
6.Operation and Maintenance (O & M)			
Has the project clearly delineated the responsibility of operation and maintenance			
Whether budgetary provisions have been made for Operations and Maintenance			

To assess the socio-economic profile of the selected respondents, income of the respondents is taken into account. The income classification is made on the basis of minimum income obtained by the individual respondent and the maximum income obtained, the same were compared and the four class interval was arrived using Rs 3000 as interval between the groups following the lower income category earnings as the base. The number of respondents in each interval is assessed and reported in the analysis. Similar methodology has been adopted by Sheela Margaret during the year 2013. However, the results were not discussed on the basis of income group but only on the basis of number of respondents reported in different characteristics because, the study is mainly aimed at imparting the training and its focus is to identify the impact created among the respondents.

TRAINING PROVIDED AND THE ANALYSIS OF RESULTS

Farm mechanization and modernization have resulted in accrual of bio wastes both at the production and processing sites. Bio waste is a potential resource if it is value-added properly. Most of the resource is left uncared and unutilized because of lack of awareness on the potentials of best utilization. If the farmers, particularly the downtrodden women, were offered with suitable training and capacity building programs, the bio waste resources could be best utilized to realize its fullest potential. On these lines the training was given to the respondents on the following themes.

- Production of milky mushrooms by utilizing the bio waste resources
- Production of bio pesticides
- Degradation of bio wastes and its management for value addition

These trainings are offered to the selected scheduled caste and scheduled tribe women respondents in Madurai and Virudhunagar districts and hence the results are analyzed and presented under the following headings.

- Socio-economic profile of selected respondents
- Details of trainings and technologies imparted to the trainees
- Sustainability aspects of the bio waste management project
- Perception of the sample respondents about the training
- Impact of the training programs offered

ANALYSIS OF RESULTS

Socio Economic Profile of the Selected Respondents

Before discussing the socio economic profile of the selected respondents, it is important to discuss the population and its related details as the respondents belonged to scheduled caste and scheduled tribe women. The details of these are analyzed and the results are presented in Table 3.

Table 3 Details of Scheduled Caste and Scheduled Tribe Population in the Study Area

S.No	Particulars of the District	Units	Madurai District	Virud-hunagar District	Total
1.	Male Population	Lakhs	13.00	08.70	21.70
2.	Female Population	Lakhs	12.80	08.81	21.61
	Total	Lakhs	25.80	17.51	43.31
3.	SC / ST Women Population	Lakhs	02.52	02.04	04.56
4.	SC / ST Women Population	Percent to Rural	19.70	23.10	21.40

Table 3 reveals that the population of Madurai district is 25.80 lakhs of which 13 lakhs are males and 12.80 lakhs are females. About 44 per cent of the population of the district lives in rural areas. The sex distribution of the population shows that 5.70 lakhs are males and 5.60 lakhs are females in rural areas. The proportion of SC and ST population is 12.70 per cent and 6.90 per cent of the urban and 19.70 per cent of the rural population belongs to SC/ST category.

The population of Virudhunagar is 17.51 lakhs with 8.70 lakh males and 8.81 lakh females. The total Virudhunagar district population living in rural areas is 9.63 lakhs of which males and females are 4.80 and 4.83 lakhs respectively. In this district, the people of rural areas belong to Backward Class (BC), Most Backward Class (MBC) and SC/ST. Majority of them are in MBC and BC categories. Only 23.10 per cent fall in SC/ST category. In the sample districts as a whole, the scheduled caste and scheduled tribe women population is accounted for 21.40 per cent to the total. Now let us analyze the socio economic profile of the selected trainees who are invited for the training program on bio waste management and their details are analyzed and the results are presented in Table 4.

Table 4 Socio Economic Profile of Selected Trainees for the Training Program

Sl.No	Variables	Particulars	Number of Respondents	Percentage to Total
01	Age	Below 20	10	10.00

Sl.No	Variables	Particulars	Number of Respondents	Percentage to Total
		20–30	46	46.00
		31–40	32	32.00
		41 and Above	12	12.00
	Total		**100**	**100.00**
02	Marital Status	Unmarried	07	07.00
		Married	72	72.00
		Widow	09	09.00
		Divorce and Separated	12	12.00
	Total		100	100.00
03	Place of Living	Rural	61	61.00
		Semi Urban	23	23.00
		Urban	16	16.00
	Total		**100**	**100.00**
04	Educational Status	Illiterate	24	24.00
		Primary	18	18.00
		Secondary	27	27.00
		Higher Secondary	23	23.00
		Collegiate	08	08.00
	Total		**100**	**100.00**
05	Occupa tional Status	Wage Labor	52	52.00
		Private Employee	21	21.00

Sl.No	Variables	Particulars	Number of Respondents	Percentage to Total
		Govt. Employee	02	02.00
		Business	14	14.00
		Others	11	11.00
	Total		**100**	**100.00**
06	Monthly Income	< Rs 3000.00	48	48.00
		3001–6000	23	23.00
		6001–9000	08	08.00
		9001–12000	12	12.00
		More Than 12000	09	09.00
	Total		**100**	**100.00**
07	Practice of Savings	Practiced	34	34.00
		Not Practiced	66	66.00
	Total		100	100.00

The word 'Empowerment' means giving power. According to International Encyclopedia (1999), power means having the capacity and the means to direct one's life towards the desired social, political and economic goals or status. Empowerment provides greater access to knowledge and resources, more autonomy in decision making, greater ability to plan lives, more control over the circumstances which influence lives, and freedom from customs, beliefs and practices (Uma Narang, 2012).

The training program on bio waste utilization has given way for empowerment to the SC/ST women or not is to be examined in this section. Under the head of socio economic profile, the variables like age, marital status, place of living, educational status, occupational status, monthly income earned and the practice of savings in vogue among the trainees during the pre-implementation of training programs were analyzed and reported in Table 4.

Among the socio economic variables included, the age classification is adopted based on the methodology practiced by Sheela Margaret during the year 2013. In that the author has given fivefold classification in age with an interval of 5 years. In this study, the classification is slightly modified because the unmarried classes were in the age group of less than 20 years and hence the above classification is presented in Table 4. It revealed that 46 per cent of the respondents were in the age group of 20 to 30 followed by 32 per cent of the respondents in the age group of 31 to 40 years. According to Nataraju (2013), 40 per cent of the women respondents were in the age group of 41-50 years. In toto, 78 per cent of the respondents invited for the training were in the age group of 21 to 40 years. Only 10 per cent of the respondents were in the age group of less than 20 years. It revealed that the experience of the respondents could motivate them to establish business units at household level after completing the program.

The second variable in the socio economic profile is the marital status of the trainees. It is also important in this study as the multiple commitments of the respondents towards their family might force them to attend the training so as to establish business unit at household level. Here, in the analysis, 72 per cent of the respondents were women involved with commitments and their married status could motivate them to establish business units.

The third variable included in the socio economic profile is place of living viz., rural, semi urban and urban. Among the regions, 61 per cent of the respondents were living in the rural areas followed by 23 per cent of the respondents hailing from semi urban and only 16 per cent of the respondents were from the urban environment. The urban respondents were having their residence mostly in the nearby villages and their business related dealings were in the urban zone.

The fourth variable included in the socio economic profile is educational status of the respondents. On assessing the educational status, 27 per cent of the respondents hold secondary education followed by 23 per cent of the respondents with higher secondary education. Illiteracy in the sample accounted for 24 per cent to the total respondents. The collegiate education was held by only 8 per cent of the respondents. From this one could understand that the respondents were having educational background that accounted for 76 per cent in toto having the scope of establishing the business units either in mushroom production and or in bio waste decomposition so as to develop the compost manure.

With regard to the occupational status of the respondents selected, wage labor was found to be dominant that accounted for 52 per cent followed by private employment and business both respectively accounting for 21 per cent and 14 per cent to the total. The respondents with other sources of income accounted only for 21 per cent.

In respect of the variable monthly income, the minimum income earned per month is arrived at Rs 3000 which accounted for 48 per cent of the respondents. The income range of Rs 3001 to Rs 6000 was earned by 23 per cent of the respondents followed by 12 per cent of the respondents holding the monthly income in the range of Rs 9001 to 12000. More than Rs 12000 income is prevalent only among the 12 per cent of the respondents. In general, 71 per cent of the respondents were earning only less than 6000 as their monthly income from wage labor and private employment. Absence of income generating opportunities to the common public resulted in poor wages and poor opportunities for employment to the labor force in the study environment. Due to the above mentioned factors, only 34 per cent of the respondents have inculcated the saving attitude from their monthly income and the 66 per cent of the respondents were reluctant to save due to subsistence income generated from different sources and they live in the level of poverty. Hence imparting the training to the respondents would have a motivational function among them to earn further.

DETAILS OF TRAINING AND TECHNOLOGIES IMPARTED TO THE TRAINEES

Technologies Offered to the Trainees

Three technologies viz., cultivation of milky mushroom, mass production of *Trichoderma viride* and *Pseudomonas fluorescens* and agricultural waste degradation already developed by Tamil Nadu Agricultural University, Coimbatore were imparted to the trainees selected from the Madurai and Virudhunagar districts. The details of trainees and period of training conducted are delineated in Table 5. Before discussing the details delineated in the Table 5, it is important to highlight the nature and type of technologies transferred to the trainees.

Technology 1 - Mass Production of Trichoderma viride

Molasses yeast medium (Molasses 30g + Yeast 5g + Water 1000ml) is prepared in conical flasks and sterilized at 1.1 kg /cm^2 for 20 minutes. *T.viride* culture is inoculated by taking a fungal disc from 10 day old culture and incubated for 10 days. This serves as mother culture. Molasses yeast medium is prepared in a fermentor and sterilized. Then, the mother culture is added to the fermentor @ 1.5 litre/50 litres of medium and incubated at room temperature for 10 days. The fungal biomass and broth are mixed with talc powder at 1:2 ratios. The mixture is air dried and mixed with carboxyl methyl cellulose (CMC) @ 5g / kg of the product. It is packed in polythene bag and used within 4 months. It is one of the components of training (Das et. al. 1997).

Technology 2 - Mass Production of P.fluorescens

P.fluorescens is multiplied in sterilized Kings 'B' broth for 48 hours. The pH of the substrate (peat soil or talc powder) is adjusted to 7 by adding calcium carbonate @150 g / kg. The substrate is then sterilized at 1.1 kg/cm^2 pressure for 30 minutes for two successive days. Four hundred ml of *P.fluorescens* suspension is added to 1 kg of substrate containing 5 g of carboxyl methyl cellulose and mixed well. The formulation is packed in polythene bag and can be stored for one month. It is another component under the Technology- 1 offered to the trainees (Dutly et.al. 1999).

Technology 3 – Bio Waste Degradation

A bulk quantity of agricultural waste viz., millets crop residue, cotton waste, coir waste, approximately one ton in quantity, is arranged in ten layers, i.e., sandwich technique in which one layer of waste of 100 kg is sprinkled with 250 g of *Pleurotus* spawn, followed by 100 kg of agricultural waste and sprinkled with 1 kg of urea. This procedure is repeated for 10 layers of waste so that, 5 layers have spawn and another 5 layers have urea. The degradation is done in aerobic condition. Watering is be done regularly twice a day to maintain 50–60 per cent moisture level. After 30 days, turning is to be given to maintain uniform degradation. After 45 days of composting, agricultural waste is ready as organic manure. It is also one of the component of training offered to the trainees (Theradi Mani et.al. 1992).

Technology 4 – Production of Milky Mushrooms

Milky mushroom can be cultivated on a wide range of cellulosic substrates namely, paddy straw, maize stalks, sorghum stalks, pearl millet stalks, Palmarosa grass, Vetiver grass, sugarcane bagasse, soybean hay, groundnut haulms, etc. Mushroom preparation is done in polythene bags system of 60 X 30 cm size by sandwich technique, i.e., rice straw followed by spawn. The beds are then incubated for spawn run under semi dark condition in a clean room. Spawn run will be completed in 10-12 days. After completion of spawn run, casing is done by cutting bed horizontally and beds after casing are kept in cultivation chambers. After casing, the beds are to be incubated over racks in partially sunken chamber lined with blue colored HDP sheet as roofing material. Inside the chamber the temperature should be around 30-35°C and RH > 85%. Light intensity of about 1600-3200 lux is essential in cropping room. The water should be sprayed regularly to maintain the 50-60 per cent moisture content in the casing medium. Pin heads appear in 8-10 days after casing and the first harvest can be made in 6-8 days after pin head formation. Mushrooms should be harvested before they start shedding the spores. It is also one of the components of training offered to the trainees selected from different locale (Theradi Mani et.al. 2001).

Table 5 Details of Training and Demonstrations Conducted in the Study Environment

S.No	Block	Village	Total Number of Trainees	Period of Training	Number of Days
1.	Madurai North	Puthuthamaaraipatty	25	23.01.12 to 28.01.12	06
2.	Cholavandan	Thirumalnatham	25	06.02.12 to 09.02.12	04
3.	Kariyapatti	Sithumoondaraippu	25	20.02.12 to 24.02.12	05

S.No	Block	Village	Total Number of Trainees	Period of Training	Number of Days
4.	Senkundrum	Chinnapparettiapatty	25	05.03.12 to 09.03.12	05
Total			100		20

Table 5 reveals that in the Puthuthamaraipatty village, the training was conducted for a period of 6 days during the month of January 2012 and in the Thirumalnatham village the training conducted was only for four days. In toto, the training conducted in the Madurai district villages accounted for 10 days incorporating all the technical components delineated in Technology – 1, 2 and 3.

In Virudhunagar District, training was conducted in the Sithumoonradaippu and Chinnapparettiapatty villages during the months of February and March 2012. The number of days taken to conduct the training component in each village accounted for 5 days. Training the respondents alone is not going to pay dividend, but the impact of training is equally important. The impact of the training could better be visualized through the demonstration of each technology package. In this respect, the details of technologies demonstrated are delineated in Table 6.

Table 6 Details of Field Day cum Demonstration Conducted for the Trainees

Sl. No.	Date	Place in which training was conducted	Name and Type of Program
1.	19.03.2012	Bio-control Lab, Department of Plant Pathology, AC &RI, Madurai	Field day on Mass Production of Bio-Pesticides
2.	20.03.2012	Central Farm, AC &RI, Madurai	Demonstration on Seed Treatment with Bio-Pesticides
3.	21.03.2012	Mushroom Unit, Department of plant Pathology, AC &RI, Madurai	Demonstration on Production of Milky Mushroom using Bio Wastes

Sl. No.	Date	Place in which training was conducted	Name and Type of Program
4.	22.03.2012	Wetland, AC &RI, Madurai	Field Day on application of Bio-Pesticides
5.	23.03.2012	Compost Yard Unit, Department of Plant Pathology, AC &RI, Madurai	Demonstration on Decomposing of Coirpith and other Bio Wastes

Table 6 reveals that the demonstrations were conducted during the month of March 2012. The field day on 'Mass Production of Bio Pesticides' was conducted on 19th March 2012 at the Bio Control Lab of Agricultural College and Research Institute, Madurai. The produced bio pesticides and its application for seed treatment was demonstrated on 20th March 2012 at the Central Farm of Agricultural College, Madurai.

The demonstration on 'Production of Milky Mushrooms' by utilizing the Bio Wastes' was conducted on 21st March 2012 at the Mushroom Production Unit and the demonstration of 'Decomposition of Coir pith and other Bio Wastes' was conducted at Compost Yard of Agricultural College and Research Institute, Madurai. The very purpose of conducting the demonstration is to infuse the ideas and to train them in the practice of technology to develop the hands-on experience to the trainees.

SUSTAINABILITY ASPECTS OF THE PROJECT AT THE IMPLEMENTATION STAGE

The project sustainability at the implementation stage was assessed by incorporating the following dimensions viz., economic dimensions, community participation, equity distribution, operation and maintenance and environmental dimensions. The project has set the targets in each dimension and hence these details are analyzed and documented in the Table 7.

Table 7 Sustainability of Project at the Implementation Stage

Dimensions	Indicators	Sustainability Status*	
		Target	**Achievement**
Economic	Success Rate of the Project	2 to 3	02.38
Community	Proportion of Community Participation	5 to 10 Per Cent	78 Per Cent
	Local Leader Support	Cordial	Cordial
Equity	Pattern of distribution of project target benefits, by income	Not mentioned	Not followed
	Pattern of distribution of project target benefits , by gender	Yes mentioned	Female alone were Selected
	Pattern of distribution of project target benefits by groupings	Small to Medium farmers	Spread of Different Category is visible
	Pattern of distribution of project target benefits by caste	Not Mentioned	Only SC/ST were Selected and Made to be benefitted
	Condition of Facilities Provided	Good	Moderately Good
Operation and Maintenance	Operating budget	Sufficient	Sufficient
	Cost recovery (% targeted for recovery)	Nil	Does Not Arise
Environmental	Positive / Negative Impacts	Less soil degradation, Use of Organic Biocides and Mushroom as Protein supplement	Soil fertility status improved by using degraded waste and disease can be managed by biocide application and yield increased. Protein supplement by Taking Mushroom

(* Indicates that the results are based on the Opinion Survey of the Respondents)

Table 7 reveals that the rate of return under the economic dimension is fixed as 2 to 3. But the project has realized the return to the tune of 2.38 revealing that the return realizable is 2 times more than the investment. With regard to the community participation, the target fixed is 10 per cent. But, the project has realized the participation to the tune of 78 per cent. According to Asha Sharma (2011), increase of outreach to the poor and marginalized state of the society had high work participation for marginalized groups like SC/ST (54 per cent) and women 48 per cent. Nataraju et al. (2013) reported that the women participation in the training programs accounted for 55 per cent. In terms of equity, different size class of farm households participated and enriched the program.

The environmental dimension was soil fertility status improved by using the degraded waste and disease can be managed with bio pesticides application and yield increase is visualized. Another aspect is utilizing the bio waste resources, the production technology of mushrooms and bio pesticides were demonstrated and the value addition of waste is visualized. In this respect, analyzing the opinion of the respondents about the training programs is imperative.

PERCEPTION OF THE SAMPLE RESPONDENTS ABOUT THE TRAINING PROGRAM

The training program was conducted to the respondents on three major technology components which are described in the preceding pages. However, imparting the opinion of the participants is equally important with respect to the parameters enunciated on sustainability aspects of the project at planning stage to know the success rate of the project and hence the opinion are analyzed and reported in Table 8.

Table 8 Opinion of the Sample Respondents about the Training Program

Sl. No	Parameters	Number of Respondents Reported		
		Yes	No	Not Sure
01	Project is Relevant to National and Regional Priority	100	00	00
02	Acceptability of the Technical Training Components	60	40	00
03	Economic Viability of the Project	90	00	10
04	Environmental Sustainability	100	00	00
05	Implementation and Monitoring Mechanism	70	20	10
06	Operation and Maintenance	80	10	10
07	Follow up Mechanism and Feed Back	45	20	35
Average		78	13	09

Table 8 reveals that cent per cent of the respondents have indicated that the project is relevant to national and regional priorities. With respect to the acceptability of the technical training components, only 60 per cent of the respondents have revealed that the components are acceptable to them and 40 per cent reported that the components are not acceptable.

With regard to the economic viability of the project 90 per cent of the respondents have revealed that the project is a viable one as it could generate considerable income increase on adoption of those practices and 10 per cent of the respondents have highlighted that they were not sure if the project is economically a viable one. In respect of environmental sustainability, cent per cent of the respondents have uttered that the project could meet all the environmental requirements because it decomposes the waste into value added manure and at the same time the waste were value added with TNAU technologies.

In respect of implementation and monitoring mechanism, 70 per cent of the respondents revealed that the monitoring and implementation mechanism is acceptable to them, 20 per cent of the respondents revealed that the monitoring system is not much convincing to them and 10 per cent of the respondents revealed that they are not sure about the monitoring mechanism which was practiced by the project team.

With regard to the operation and maintenance aspects of the project, 80 per cent of the respondents have highlighted that the operation aspects was quite acceptable to them, the remaining 10 per cent of the respondents did not wish to comment with respect to operation and maintenance aspects and another 10 per cent of the respondents revealed that they are not sure about the operational aspects of the project.

On an average 78 per cent of the respondents have revealed that the parameters included in the project is acceptable to them, 13 per cent of the respondents revealed that the parameters delineated are not acceptable to them and only 9 per cent of the respondents highlighted that they are not sure enough to comment on the overall aspects of the project implementation. In a nutshell, the project implemented by the Project Team is found to be a successful one.

IMPACT OF THE TRAINING PROGRAMS OFFERED

Training conducted to the respondents would have motivated few to take up business venture on small level or at household level. The impacts so created were analyzed and the results are presented in Table 9. Table 9 reveals that the monitoring mechanism practiced in the project motivated 76 per cent of the respondents to enhance their awareness related participation with the technical persons of Tamil Nadu Agricultural University and to clear their doubts regarding the bio waste utilization for value addition.

Table 9 Details of Socio Economic Changes Made during Post Training Program

Sl. No	Socio Economic Changes	Number of Respondents	Percentage to Total
01	Mushroom Production Unit Established	56	56.00
02	Degradation of Bio Waste and Compost Unit Established	34	34.00

Sl. No	Socio Economic Changes	Number of Respondents	Percentage to Total
03	Improvements Made in the Existing House	14	14.00
04	Purchase of New Consumer Durables	29	29.00
05	Enhancement in the Awareness Related Participation with the Technical Persons of TNAU	76	76.00
06	Increased Investments Made in the Production of Mushroom Unit	42	42.00
Average		42	41.83

The number of mushroom production units established by the respondents in the Madurai and Virudhunagar Districts arrived at 56 per cent to the total respondents. Among the 56 Mushroom Production Units established, 42 units have expanded their operation because of good demand for the output produced and higher income generation becomes possible. Due to this the improvements made in the existing house was carried out by 14 per cent of the respondents. At the same time, purchase of new consumer durables like two wheelers, etc., was also visible among the 29 per cent of the sample respondents. All this is possible due to the enhancement in the income because of the training program conducted through the project. The compost manufacturing unit established by the respondents by utilizing the bio waste accounted for only 34 per cent to the total due to poor interest in the degradation of bio waste activity among the respondents. However, the project would have motivated the respondents further to take up the degradation activity by highlighting the economic incentives associated with the project and the need for bio fertilizers and compost which are dire need of the hour to promote organic agriculture. Mathur (2008) reported that the scheme entitled "Mahatma Gandhi National Rural Employment Guarantee Program" acts as a great agent of socio economic upliftment and provides livelihood security to the poorest of the poor by providing income earning opportunity.

As part of the impact analysis, analyzing the perception of the respondents on the components of the training program conducted is also immensely important and hence the perception of the respondents were analyzed and the results are presented in Table 10.

Table 10 Perception of the Respondents on the Component of the Training Program

Sl. No	Particulars	Perception of the Trainees (Per Cent)			
		Excellent	Very Good	Good	Satisfactory / Irrelevant
01	Training on Mushroom Production	81.00	14.00	05.00	00.00
02	Demonstration on Mushroom Production	74.00	16.00	10.00	00.00
03	Training on Production of Bio-Pesticides Unit	62.00	27.00	08.00	03.00
04	Demonstration on Bio-Pesticides	58.00	30.00	09.00	03.00
05	Degradation of Bio Waste from Agriculture	46.00	32.00	14.00	08.00
06	Demonstration on Bio Waste Degradation	39.00	34.00	21.00	06.00
Average		60.00	26.00	11.00	03.00

Table 10 reveals the details of perception of respondents who had attended the training program. The perception of the respondents was classified as Excellent, Very Good, Good and Satisfactory technology component. The training on Milky Mushroom Production is ranked as excellent technical component as perceived by 81 per cent of the respondents followed by Demonstration on Mushroom production technology that scored 74 per cent and is ranked excellent.

The training component on Production of Bio Pesticides was ranked as excellent by only 62 per cent of the respondents. At the same time, 27 per cent of the trainees have reported that the component was very good and only 8 per cent have revealed that the component is good. Three per cent of the respondents highlighted that the component is satisfactory because it was not much important to those trainees.

The Demonstration of Bio Pesticides scored 58 per cent and was graded as excellent and 30 per cent of the respondents revealed that the demonstration is very good. Among the components, the Degradation of Bio Waste from Agriculture and Demonstration on Bio Waste Degradation got poor rank. The reason might be that the importance and usefulness of the technology would not have delineated to the trainees at right time. In this respect, the role of economists is much important to give motivational exercise to the trainees for adoption and getting the feedback from the trainees.

In a nutshell, 60 per cent of the respondents have revealed that the technical components imparted through the training program was excellent while 26 per cent of them said that the components were very good and 11 per cent of them exercised their option as good and three per cent of the respondents expressed that the components were irrelevant and not much important to them due to their poor understanding. Hence, such type of projects should come forward which would promote the benefits realizable from them in different stages by coining suitable promotion packages. On doing so, the people who are less interested could be motivated to adopt the technology component with full spirit on highlighting the potential benefits that accrue from the project. Much take off is not visible because of some constraints available to the trainees and hence analyzing the details of constraints that de-motivate the trainees are spelt out and are delineated below. They are:

- Lack of money and assets for investment decisions among the respondents
- Lack of marketing skills in the trainees
- Poor price offered for their produce at doorsteps.

These constraints are to be appropriately handled to give a boost to the household enterprises like mushroom production and bio pesticides production activities.

SUMMARY AND CONCLUSION

A five-day training and field day cum demonstration programs on agricultural waste degradation, bio pesticide production and cultivation of milky mushroom was given to 100 women SC/ST beneficiaries of the following villages viz., Puthuthamarai patty, Thirumalnatham, Sithumoondraippu and Chinnapparettiapatty in Madurai and Virudhunagar districts. The training given to the respondents has given better results. The success rate of the project is higher than the prescribed one in the project. The impact of the project revealed that the socio economic changes were visible and the project components were rated as excellent by almost 60 per cent of the respondents.

RECOMMENDATIONS

- To redress the constraints, the project authorities could organize all the trainees into three producer commodity groups or associations and the same could be registered for tapping the fullest benefits of the Government. On doing so, the collective bargaining power to realize better price for their produce and availing the finance from financial institutions by sending appropriate proposal to establish small business venture at household level or by the commodity group becomes possible.
- The training program if it incorporates production of output, marketing practices and promotional measures and certification related issue means, the respondents could develop confidence building which further promotes enhanced production, certification and trade of outputs generated from bio waste resources.

References

- Asha Sharma, (2011). "***Impact of MGNREGA on the Socio Economic Upliftment of the Beneficiaries – A Case Study of Pali District of Rajasthan***" (Unpublished Ph.D. Thesis Submitted to the Department of Management and Commerce, Jayoti Vidyapeeth Women's University, Rajasthan).
- Das, B.C., S.K.Roy and L.C. Boro. (1997). "Mass Multiplication of Trichoderma Species on Different Media", ***Indian Journal of Agricultural Sciences***, 10(1): 95-100.
- Dutly,B.K., O.G. Defag (1999). "Environmental Factors Modulating Antibiotic and Siderophore Bio Synthesis by P. fluresence Bio Control Strains", ***Applied and Environmental Microbiology***, 65: 2429-2438.
- Jeffrey Ashe and Lisa Parrott (2001). "***Impact Evaluation of Women's Empowerment Program in Nepal***", (Institute for Sustainable Development: Brandeis University) pp. 43-56.
- Mathur, L (2008). "Employment Guarantee: Progress Made So Far" ***Economic and Political Weekly***, 52(29); December 2007 to January 2008, pp. 17-20.
- Nataraju, M.S., S.M. Raghavendra and R. Shasidhar (2013). "Women Empowerment through National Afforestation Program in Karnataka", ***Journal of Extension Education***, 25(3): pp. 5069-5077
- Sheela Margaret and N. Kala (2013). "Study on Impact of NGO Interventions on the Empowerment of Women", ***Journal of Business Management and Social Sciences Research*** 2(3): March 2013, pp.01-06.
- Theradi Mani, M and T. Marimuthu (1992). "Utilization of Plurotus Spp for Decomposing Coconut Coir Pith", ***Mushroom Research***, 1: 49-51.
- Theradi Mani, M., B. Meena and A.S. Krishnamoorthy (2001). " Innovative Techniques for the Improvement of Sporophore Size and Yield of Milky Mushrooom", ***Mushroom Research***, 10: 23-26.
- Uma Narang (2012). "Self Help Group: An Effective Approach to Women Empowerment in India" ***International Journal of Social Sciences and Interdisciplinary Research***, 1(8): pp.8-16.

Chapter 20

Match Splints Industry – A Gateway to Agripreneurship Development

INTRODUCTION

Bengal is the birth place of match industry in India. The first match wood industry was established in West Bengal in 1920s and second in Tamil Nadu during 1923 at Sivakasi producing lucifer matches. The small scale handmade match industries in Sivakasi, Kovilpatti, Sattur, Kalugumalai and Sankarankoil towns of Tamil Nadu state show a highly organized network (Saresh et.al. 2011). Wooden match production in India is split into three sectoral categories: the mechanized large scale sector; the handmade small scale sector; and the cottage sector. Around 82 per cent of the total match production is in the hand made, small scale accounted for 67 per cent and the cottage sector accounted for only 15 per cent where technology has remained relatively simple. These two non-mechanized sectors of the match industry are distinguished primarily by output size. The industry as a whole directly employs an estimated 2.50 lakh people with only 6000 of these in the mechanized sector. The cottage sector which involves totally manual operations and produces less than 75 million match sticks per year and is often household based, accounts for about 50000 workers. Thus, small scale, factory based match production units employ by far the largest number of people involved in the match sector (Tandon, 1990).

The major raw materials used in the production of safety matches are soft woods used to make the match sticks which is also known as "Splints" and boxes and chemicals for the match heads and the friction surface of the boxes. With the exception of the sulphur, all the basic raw materials are produced within India. A full appreciation of the employment potential of the match industry should also consider the workers involved in the production of all of these raw materials.

Historically, the Indian match industry depended on imported wood including aspen (*Populus tremula*) from Sweden, Canada, America and Russia; cotton wood (*Populus deltoids*) from Canada; balsam poplar (*Populus balsamifera*) from Manchuria; and linden (*Tilia japonica*) from Japan. But the government quickly moved to encourage the use of indigenous woods by restricting the import of foreign poplars. One result of the early use of poplar wood has been that the consumer continues to associate good quality matches with light colored wood, placing further limitations on the selection of indigenous species.

A large number of Indian tree species have been found suitable for use in the match industry. Among the most important Indian match woods are *Bombax ceiba* which is good for boxes as well as splints; Indian aspen, *Evodia roxburghiana* and white mutty, *Ailanthus excels* both suitable for high quality splints. Among many species tested, only a few are acceptable for making high quality boxes and match splints. One such species for match splints is *Ailanthus excelsa* which is drawn from farm lands, social forestry and agroforestry practices and from the forest fringes. Most of the match industries are having their own agents to supply the raw materials to the match splints industry operated in Tamil Nadu. The match industries are facing several constraints in getting the raw materials and hence special effort has been taken to assess the sources of supply of raw materials and the performance of match industries in the context of meeting the demand for the splints in the western part of Tamil Nadu as part of the students exposure visit to the Match Splints Industry and to infuse the confidence to the students in establishing such potential demand driven industry in Tamil Nadu.

SELECTION OF A SUITABLE MATCH SPLINT INDUSTRY FOR CASE STUDY

The western part of Tamil Nadu is blessed with the textile related business and it has the name of Cotton Zone of Tamil Nadu. Amidst this textile business, one industry in small scale is engaged in receiving the *Ailanthus excelsa* wood as raw material for making match splints and the finished output is being sent to different match making industries in the southern belt of Tamil Nadu which is an unrelated business in this zone.

The Horticultural College and Research Institute of Tamil Nadu Agricultural University, Periyakulam runs an academic program for four years duration in which Agribusiness Management course is taught regarding capacity building and confidence building exercise on how to establish an agribusiness enterprise and how efficiently one can run the enterprise with the available resources. In this circumstance, the second year students of Horticultural College and Research Institute were taken to different agribusiness enterprises to have an exposure visit to infuse confidence among them. The Ideal Match Splints Limited located in western part of Tamil Nadu is one such small scale industry aimed at for the students exposure visit.

The Match Splints Industry employs 15 casual wage laborers, one permanent supervisor and they were managed by the Managing Director of the company. All the respondents in the industry were discussed in person and their experiences were incorporated in this chapter following a case study approach. The data gathered were subjected to the descriptive statistics like mean and percentage analysis only. The survey was conducted, the results were analyzed and documented for guidance and follow up by the agripreneurs. The data related to the year 2014. Since the data presented are based on case study approach, one could understand the strength and the limitations associated with the findings.

CASE STUDY UNDER SUITABLE HEADINGS

The main objective of writing this chapter is to highlight the performance of Match Splints Industry in western Tamil Nadu and to highlight the strengths associated with the industry to guide and motivate the young agripreneurs in that field on the following heads.

- Sources of Raw Materials supplied to the Match Splints Industry
- Out-turn of Match Splints
- Methodology of Processing of Match Wood
- Details of Human Resources Employed
- Cost of Production of Match Splints
- Price of Output and Distribution of Match Splints and
- Constraints Faced by the Match Splints Industry

SOURCES OF RAW MATERIALS SUPPLIED TO THE MATCH SPLINTS INDUSTRY

The Government is the most dominant player in the forestry sector and directly own and manage about 97 per cent of all the forests. Whereas, more than 90 per cent of India's wood based products are manufactured in private sector alone (Dhinman, 2011). Wood based industries cannot survive without wood which is the basic raw material for their operations. The gap between its demand and supply is widening and the scenario is fast changing with increased wood requirement and reduced supply from the existing sources. FAO reports revealed the status on consumption of wood and wood related products, indicating that in a decade of time (1999 to 2009), the consumption increased constantly. National Forestry Action Program indicates a negative wood balance of 599.6 M cubic meter by 2015 (FAO 2009). Similar estimates are given by Rai and Chakrabarti (1996) while the ICFRE estimates were comparatively high which is arrived at 1873 M cubic meter (ICFRE, 2003). However, the major raw materials used in the production of safety matches are soft woods used to make the match splints and boxes. Among several species tried, *Ailanthus excelsa* is found to be the best fit and hence the details of sources of supply of raw materials are analyzed and the details are presented in Table 1.

Table 1 Details of Sources of Raw Materials Supplied to Match Splints Industry

Sl. No	Details of Sources	Name of the Tree Species	Quantity Supplied in Tonnes	Percentage to Total
01	Coimbatore District	*Ailanthus excelsa*	260	19.12
02	Thiruppur District	*Ailanthus excelsa*	310	22.79
03	Erode District	*Ailanthus excelsa*	190	13.97
04	Salem District	*Ailanthus excelsa*	220	16.18
05	Other Districts	*Ailanthus excelsa*	380	27.94
Total			1360	100.00

Table 1 reveals that the match wood species supplied is *Ailanthus excelsa* from different sources. The sources of supply were from the districts of Tiruppur, Coimbatore, Erode, Salem and other districts. Among these districts, Tiruppur is found to supply the raw materials to the match splints industry to the tune of 23 per cent followed by Coimbatore district capable of supplying the raw materials to the tune of 19 per cent of the total supply. Higher quantity of raw material supply is possible due to the establishment of Ailanthus plantations in the farm fields and under schemes of Social Forestry and Agroforestry which is practiced by the western zone considerably. Salem district is able to supply around 220 tonnes of Ailanthus followed by Erode district which respectively accounted for 16 and 14 per cent of the total supply. Other districts were able to provide around 380 tonnes of raw materials to the match splints industry. Put together the Match Splints Industry could receive the raw materials to the tune of 1360 tonnes per annum. In that only 1240 tonnes were consumed in a working day of 310 days per annum and the other stocks were kept under inventory to meet the demand during the peak seasons of match splints demand. The marketing channels present in the match wood supply are Producers (Farmers) → Industrial Consumers. The industry is purchasing the match wood at farm gate using the agents of the match wood firm (Web Reference – 1).

OUT TURN OF MATCH SPLINTS IN MATCH WOOD INDUSTRY

The quantity and quality of match wood supplied are the determinants of getting the quality output of match splints. Hence quality of raw material supply becomes important. The match wood which is received in the industry is fixed with the minimum girth of 18 inches and the same should be free from infection and deterioration. Such portions should be carefully examined and detected for deleting the same from the quality wood lot. After sorting the quality wood, one has to assess the out-turn of match splints per ton of match wood supplied and hence the details are analyzed and the results are presented in Table 2.

Table 2 Out Turn of Match Splints in Match Industry

Sl. No	Details of Output	Out Turn per Ton of Match Wood (Kgs)	Percentage to Total
01	Match Splints	340.000	34.00
02	Fuel Wood	410.000	41.00
03	Bark	180.000	18.00
04	Wood Residues	070.000	07.00
	Total	1000.000	100.00

Table 2 reveals the details of out turn of match splints per ton of match wood. There were four types of output that could be realized from one ton of match wood. They were match splints, fuel wood, bark and the wood residues. Among these outputs, the fuel wood accounted for 41 per cent with the actual quantity found to be 410 kgs followed by match splints. The quantity of match splints realized per ton of match wood arrived at 340 kgs accounting for 34 per cent. Ailanthus bark accounted for 18 per cent. The quantity of bark realized per ton of match wood is arrived at 180 kgs. The wood residue accounted for around 70 kgs which formed about 7 per cent of the total one ton of raw materials utilized in the industry. The extent of quality output realization is based on the expertise associated with the skilled manpower in the industry.

METHODOLOGY OF PROCESSING THE MATCH WOOD

In a mechanized industry, the entire process of match splints making takes place under one roof from log to match splints. In the hand made sector, the veneer for match boxes and splints are produced separately. Stage by stage processing of logs into the match splints is assessed and the details are presented in Table 3.

Table 3 Methodology of Processing of Match Wood

Steps	Methods of Processing of Match Wood for Match Splints	Remarks
01	Logs of *Ailanthus excelsa* (Minimum Girth of 18 inches)	Utilize the Wood within 24 Hours of Harvest
02	Billeting - Billets of 15 Inches Length	Machine Saw
03	Debarking	Manual
04	Peeling of Sheets from the Billets	Using Specific Machineries
05	Trimming the Sheets	Manual
06	Cutting the Sheets into Match Splints	Length of the Splints are 38, 39, 40 and 42 MM. For Crackers, different size of match splints viz., 50,60,80,120,180 MM is also supplied on special indents
07	Smoking the Splints with Sulphur to get Uniform Colour of the Splints	3 Kg of Sulphur is used per day
08	Drying the Match Splints	Sun Drying
09	Packing of Match Splints	Gunny or Plastic Bags are used
10	Dispatch the Match Splints to Different User Agencies	Match Industries

Table 3 reveals the details of different stages involved in processing of logs into match splints. In the first stage, the logs with more than 18 inches were selected and sent for billeting. In the second stage, the billets of Ailanthus logs were made with the length of 15 inches uniformly using the machine saw. In the third stage, debarking is carried out by the labor force who demand the bark for fuel wood which is supplied to them at free of cost. They can extract the bark from the billets and they can make use of the same for their fuel requirements.

In the fourth stage, the debarked billets were subjected to the peeling machine, the sheets were peeled from the billets to the extent possible and the sheets were trimmed manually for getting the uniformity for extracting the match splints in the fifth stage. In the sixth stage, the peeled sheets were cut into the match splints with the lengths of 38, 39, 40 and 42 MM based on the demand with the user industries. On a special occasion, the splints were also extracted with the lengths of 50, 60, 80,120 and 180 MM for preparing crackers by obtaining special indents from the match industries. After extracting the match splints, the same were smoked using the sulphur to get uniform color in the output of match splints.

After processing the match splints, the same were dried under the hot sun in the eighth stage and then the same were packed using the gunny or plastic bags of 60 kgs each and then the same are dispatched to the match industries located in Sivakasi and Kovilpatti. In each stage, timely processing is required to avoid damages.

DETAILS OF HUMAN RESOURCES EMPLOYED IN THE MATCH SPLINTS INDUSTRY

Delivering the quality match splints as output depends on the skilled manpower available with the match splints industry. In this respect, the number of persons employed in the industry, their roles and other details are analyzed and the results are presented in Table 4.

Table 4 Details of Human Resources Employed in the Match Splints Industry

Sl. No	Details of Human Resources Employed	Role Played by the HR	Number of HR Employed	Wages Paid per Month per Labor (Rs)	Percentage to Total
01	Supervisor	Overall Supervision	01	10000.00	06.67
02	Machine Operator – I	Peeling and Trimming the Sheets	01	8750.00	06.67
03	Machine Operator – II	Out Turn of Splints as per the Indents	01	8750.00	06.67
03	Wage Labor from Bihar	Removal of Barks	03	7500.00	20.00
04	Wage Labor from Odisha	Assembling, Drying, Packing and Forwarding	09	7500.00	60.00
	Total		15		100.00

Table 4 reveals that match splints industry is employing a total of 15 labor force for carrying out the different operations performed in the industry as revealed in the Table 3. Among the 15 labor force, only one employee is permanent and he is entrusted with the task of overall supervision of the activities of Match Splints Industry. Among the labor force category, two types of wages are prevalent. There are two Machine Operators taking care of the peeling of sheets from the log and trimming the sheets who are paid Rs 8750 per month at the rate of Rs 350 per day. The Machine Operator – II is able to deliver the quality match splints according to the indents made by the Match Industry.

The labor force drawn from Bihar and Odisha is entrusted with the task of removal of barks, assembling, drying and packing and forwarding of match splints. The labor force was provided with accommodation facility so as to have their own food and basic requirements. However, the labor force feel that the working hours are too long and the wages paid are poor for their executed tasks. The labor force from Bihar and Odisha is receiving Rs 300 per day and the monthly wages arrived at Rs 7500. In a nutshell, the match splints industry is employing 15 laborers for converting four tons of logs per day to get the quality match splints and meeting the requirements of laborers by providing them accommodation and other basic amenities.

COST OF PRODUCTION OF MATCH SPLINTS

Cost of production is a guiding force for the young agripreneurs who wanted to establish a business unit. In the preceeding chapters, we have discussed the potential businesses and the economic incentives associated with them. One can compare and establish the agribusiness venture in the desirable locale. In this unit, the cost of production of match splints is discussed in detail and the results are presented in Table 5.

Table 5 Cost of Production of Match Splints

Sl. No	Particulars of Cost	Number of Units	Item wise Cost in Rs	Percentage to Total Cost
A.	**FIXED COST**			
1	Rental Value of Land and Buildings	One Acre	38000	00.51
2	Depreciation Cost on Machineries	7 Per Cent	140000	01.87
3	Annual Electrical Installation Charges	10 Years	2500	00.04
4	Interest on Fixed Capital	5 Per Cent	75000	01.00
5	L and Tax and Cess		850	00.01
6	Salary on Permanent Staff	10000/ Month	120000	01.60
7	Share of Establishment		75000	01.00

Sl. No	Particulars of Cost	Number of Units	Item wise Cost in Rs	Percentage to Total Cost
8	Total Annual Fixed Cost		451350	**06.03**
B.	**VARIABLE EXPENSES**			
1	Cost of Raw Materials (Match Wood)	Rs 4000 / Ton	4960000	66.18
2	Cost of Sulphur for Fumigation of Splints	Rs 100 / Kg	93000	01.24
3	Labor Charges	Rs 117500/ Month	1290000	17.21
4	Electricity & Fuel Charges		107500	01.43
5	Cost of Marketing		96000	01.28
4	Miscellaneous Expenses	One Year	60000	00.80
5	Interest on variable cost		436590	05.83
6	Total Variable Expenses		7043090	**93.97**
7	Total Cost of production (A+B)		7494440	**100.00**
C.	**REVENUE REALIZED PER ANNUM**			
1	Sale of Match Splints @ Rs 41 per Kg	421.600 Tons	17285600	91.47
2	Sale of Fuel Wood @ Rs 3 Per Kg	508.400 Tons	1525200	08.07
3	Sale of wood residues @ Rs.1/kg	70 Kgs /ton	86800	00.46
4	Gross Income Realized	Rupees	18897600	**100.00**

Sl. No	Particulars of Cost	Number of Units	Item wise Cost in Rs	Percentage to Total Cost
5	Net Income Realized per Annum	Rupees	11403160	60.34
D.	**COST OF PRODUCTION OF MATCH SPLINTS**			
1	Cost of Production per Ton	Rs / Ton	17776.00	
2	Sale Price of Match Splints	Rs/Ton	41000.00	
3	Net Income Realized	Rs / Annum	23224.00	56.64
4	Output – Input Ratio	Ratio	02.52	

Table 5 reveals that the cost of production details are discussed under four heads viz., fixed cost involved in the production of match splints; variable expenses involved in the processing of match splints; revenue realized from sale of output of one ton of raw materials; and the cost of production of match splints per ton.

The fixed expenses involved in the production process of match splints are rental value of land and buildings, depreciation cost on machineries, annual electrical installation charges, interest on fixed expenses, land tax including cess, salary on permanent staff and the share of establishment cost. All these fixed expenses were arrived at Rs 4.51 lakhs which accounted for only 6 per cent to the total cost of production of match splints in the match wood industry. Among the fixed expenses, major investment is found to be on the machineries. The annual depreciation cost on machineries arrived at Rs 1.40 lakhs which accounted for only around 2 per cent to the total cost of production of match splints.

The variable expenses involved in the production process of match splints are cost of raw materials (match wood), labor charges, electricity and fuel expenses, cost of marketing which included sales promotion, packing and forwarding expenses, interest on variable cost and the cost of sulphur which is used for fumigation of match splints to get a uniform color to the match splints. The total variable expenses incurred were arrived at Rs 70.43 lakhs which accounted for 94 per cent to the total cost of production of match splints in the match wood industry. Among the variable expenses, the major investment goes to the purchase of raw materials from different zones of origin which arrived at Rs 49.60 lakhs per annum accounting for 66 per cent to the total cost of production of match splints. Another important component in the variable expenses was the labor charges or wages which arrived at Rs 12.90 lakhs accounting for 17 per cent to the total cost of production of match splints.

The total cost of production of match splints in the match wood industry included the fixed cost and the variable expenses both of which accounted for Rs 74.94 lakhs. The production cost of match splints per ton arrived at Rs 17776 and the per kg production cost arrived at Rs 18/-. The sale price of match splints on an average accounted for Rs 41 per kg. The efficiency of a firm in an industry could be evaluated or assessed by computing Output–Input Ratio (Web Reference -2). It revealed that the production of match splints is highly profitable according to the Output–Input Ratio of 2.52. It revealed that for one rupee of investment the company could realize a gross income of Rs 2.52 and the net income of Rs 1.52.

With respect to actual income generated from the production process, it is composed of three sources of income viz., sale of match splints, sale of fuel wood and the sale of wood residues. All these are priced at the market price prevalent during November 2014 and the same are taken into account for arriving at the gross income in the production of match splints. The sale of match splints alone is capable of generating 172.85 lakhs as gross income, the sale of fuel wood is able to generate a revenue of Rs 15 lakhs per annum and the wood residues are capable of generating only 0.87 lakhs per annum. Put together, all the three output could generate a total income of Rs 188.98 lakhs and the net income after deducting the total cost of production is arrived at Rs 114 lakhs per annum which is a considerable amount of income to the match wood industry and hence one could enter into the business with confidence by arranging the contract agreement with the farmer by promoting farm forestry and agroforestry schemes in the farm front.

BREAK-EVEN ANALYSIS FOR MATCH SPLINTS INDUSTRY

In an industry, the analysis of break-even quantity becomes very much important to know how much output is to be produced in the industry to break-even the no profit and no loss situation. In this respect, the details of break-even quantity are analyzed using the formula for that purpose, the results are analyzed and the details are presented in Table 6.

Table 6 Break-Even Analysis for Match Splints Industry

Sl. No	Particulars	Units	Amount (Rs in Lakhs)
01	Quantity of Match Splints Produced and Sold	Tonnes	421.600
02	Revenue from Sale of Match Splints	Rupees	17285600.00
03	Variable Costs involved in Production of Match Splints	Rupees	7043090.00
04	Contribution (2 Minus 3)	Rupees	10242510.00
05	Fixed Cost per Annum	Rupees	0451350.00
06	Price per Ton of Match Splints	Rupees	41000.00
07	Variable Cost Involved in Production of One Ton of Match Splints	Rupees	16705.00
08	Break-Even Point (TFC Per Annum/ (Price per Ton of Splints Minus Variable Cost / Ton	Quantity in Tonnes	18.578

Table 6 reveals that the total quantity of match splints produced by the industry is arrived at 421.60 tonnes per annum. On sale of match splints at the rate of Rs 41000 per ton, the total revenue realizable is arrived at Rs 172.85 lakhs. To produce the match splints of one ton, the company has to spend Rs 16705. Using the data presented in Table 6, the break even quantity is arrived at 18.578 tonnes. It revealed that the no profit and no loss situation could be realized by the match splints industry on production of 18.578 tonnes per annum. Whereas, the industry is capable of producing 421.60 tonnes by which the safety margin to the company is arrived at 403 tonnes. From this one could infer that the company is running profitably by producing 2160 per cent higher output over the break-even quantity and keeps running profitably and hence young agripreneurs could enter into this venture very boldly. However, experience makes the man perfect and capable of learning the in and outs in the business. Hence the agripreneur who wants to establish such a business venture can gain expertise for one or two years and then one can enter into this venture so as to develop the capabilities to address the emerging issues in the business.

PRICE OF OUTPUT AND DISTRIBUTION OF MATCH SPLINTS

The outputs realized from the production process are match splints, fuel wood, bark and wood residues. All the four products are distributed to the user agencies on time based on the indents made by the firms. The details are analyzed and the results are presented in Table 7.

Table 7 reveals that the match splints is one of the main output of high demand in Kovilpatti and Sivakasi of southern Tamil Nadu where 90 per cent of match industries of India is located (Saresh et.al. 2011) and the cracker industry that have mushroomed in these places could attract large volume of match splints for their use in their production. Though the match splints are priced from Rs 36 to 46 in a year, the average price of Rs 41 was taken into account in arriving at the gross income. When one could compare the cost of production per kg of match splints, the sale price is comparatively higher and the firm should be in a position to produce more during the demanding season.

Table 7 Price of Output and Distribution of Match Splints Industry

Sl. No	Details of Output	Places of Distribution	Average Price per Kg in Rs
01	Match Splints	Kovilpatti and Sivakasi	41.00
02	Fuel Wood	Coimbatore and Thiruppur	03.00
03	Bark	Annur, Coimbatore District	01.00
04	Wood Residues	Annur, Coimbatore District	01.00

Fuel wood is another important output consumed in brick kilns, hotels and restaurants and tea shops priced at Rs 3 per kg and which is of high demand in Coimbatore and Tiruppur districts. The bark is one of the outputs demanded in the households for heating the water during the cold months and the same is also used as fuel wood for cooking among the poor households. The households used to visit the industry, extract the bark while they are fresh and take the goods at free of cost. The dried bark is sometimes sold at the rate of Re 1 per kg of bark locally and the wood residue is another byproduct which is also used by the rural households and the same is received at the price of Re 1 per kg. The remaining is planned to be consumed at the industry itself for making briquettes. The industry is getting ready to install the briquetting unit to earn another output which is also demanded in the restaurants and hotels and bakery units heavily.

To sum up, the outputs generated from the match wood industry are consumed by different user agencies at appropriate prices agreeable to both the producer and the consumers. The outputs realized are capable of generating considerable income to the industry and motivate the firm to function by overcoming the shortcomings in this sector. The shortcomings are identified and delineated in the following section.

CONSTRAINTS FACED BY THE MATCH SPLINTS INDUSTRY

The constraints faced by the Match Splints Industry are classified as institutional constraints and individual constraints. The institutional constraints identified are non-availability of sufficient raw materials, higher wages to labor and non-availability of skilled labor and the frequent power failure. These details are analyzed and the results are presented in Table 8.

Table 8 Details of Constraints Faced by the Match Splints Industry

Sl. No	Details of Constraints	Number of Persons Reporting	Percentage
Institutional Constraints			
01	Non Availability of Sufficient Raw Materials	01	06.67
02	Higher Wages to Labor and Non Availability of Skilled Labor	01	06.67
03	Frequent Power Cut	01	06.67

Individual Constraints			
04	Heavy Work Load to Labor Force	14	93.33
05	Poor Wages Paid per Day	14	93.33
06	Poor Accommodation Facility Extended	12	80.00
	Total Respondents	15	100.00

The non-availability of sufficient raw materials scored only 6.6 per cent of the total constraints reported which could be overcome by establishing the Ailanthus plantations in the farm lands by entering into the contractual agreement by the match wood firm following the practice adopted by the WIMCO seedlings so that the raw material demand could be met.

The second constraint is higher wages to labor and non-availability of skilled labor. On the individual constraints side if we examine, the laborers were feeling that the length of working hours extends even up to 9 hours or 10 hours in a day. But these are not appropriately compensated and hence they demand higher wages. A compromise is needed between the labor force and the institution managers by weighing the individual difficulties in executing the tasks and realizing the quality output. Skilled labor could be easily absorbed in the open market if the company is able to pay appropriate wages to the employees.

Frequent power failure is another constraint, the institution is facing. This can be overcome by establishing the generator facility or by establishing the co-generation unit by utilizing the wood residues available with them.

Among the individual constraints, the poor accommodation facility could be overcome by providing good accommodation facility to them during the winter and summer season. The health of the labor force is a backbone for the success of the company and hence good rooms can be constructed and provided for smooth running of the business.

SUMMARY AND CONCLUSIONS

The Match Splints Industry located in the western zone of Tamil Nadu is capable of meeting the raw material requirements and able to generate various outputs through which it is generating higher income per annum. On examining the various issues of production to revenue generation by an enterprising candidate, he could easily convince on the aspects of establishing an agribusiness successfully.

RECOMMENDATIONS

The Match Splints Industry is able to get wood residues considerably. Instead of selling the same in the open market, the industry itself can enter into the venture of establishing a briquette making unit in the available space, some more employment generation could be met and at the same time additional revenue is also possible.

To meet the raw material requirements without any interference, the industry can enter into the contractual agreement with the producer farmers in the nearby districts following the pattern of WIMCO industries and Tamil Nadu News Print and Papers Limited, Karur.

Instead of sending the match splints to the Sivakasi and Kovilpatti for production of matches, the match wood industry can explore the possibilities of establishing additional units for production of match boxes and a mini match unit may be established with the technical support of experts in that field which in turn may promote SSI clusters in western part of Tamil Nadu.

References

- **Dhinman, R.C., (2011).** "Private Sector in Short Rotation Forestry: Opportunities and Challenges", ***Indian Journal of Ecology***, 38 (Special Issue): 27-34
- **ICFRE (2003).** "Forestry Statistics 2003", (Indian Council of Forestry Research and Education, Dehra Dun: Uttrakhand).
- **Rai, S.N., and S.K. Chakrabarti** (1996). "Demand and Supply of Fuelwood, Timber and Fodder in India", (Forest Survey of India: Dehra Dun)
- **Saresh, N.V., M. Sankanur, Rahul Singh, Archana Verma and D. Nuthan (2011).** "Comparative Ecological – Economic Study of Short Rotation Tree Cultivation and Its Impact on Livelihood of Local Stakeholders in Tamil Nadu", Indian Journal of Ecology, 38 (Special Issue): 199-201

Web References

- www. Academia.edu/6472448/markets and marketing of agroforestry products in India
- www.farmdoc.illinois.edu/irwin/archive/books/markets-prices/markets-prices_chapter21.pdf.

SECTION – III

MARKETING AND TRADE FOR AGRIBUSINESS

Chapter 21

Marketing of Processed Curd Chilli

INTRODUCTION

Chillies are brought to the regulated markets in different parts of the state mostly by primary producers. Unlike the other perishable commodities, dry chillies are sold in the market by many producers in a phased manner. The producers try to get the best advantage of the prices. They store the produce as long as they can and bring to the market when prices are favorable to them. Only during market glut the producer-seller come to the market with entire produce and try to dispose off the lots as quickly as possible because storing and selling may result in further loss due to crash in prices. The other agencies selling the produce in the market yard are village merchants, wholesalers, commission agents and cooperatives. Unfortunately, the role of cooperatives in bringing the chillies into the market is very less. Only in Tamil Nadu it is reported that one per cent of the total arrivals of chillies in Thirunelvelli and Kovilpatti markets are brought by the local cooperative societies. The cooperatives were also active in Kozhikode and Sanghli markets of Kerala and Maharashtra states (Government of India 2010). Role of government in managing the markets is on the decline worldwide. It is not easy to bring major changes in the traditional marketing system. The only way to modernize marketing is to promote alternative marketing system and that may operate parallel to and in addition to present marketing system. The purpose of the proposed alternative marketing is to promote modern trade practices, which in turn will pave way for transparency and efficiency in marketing.

Various forms of alternate marketing like (a) direct marketing, (b) marketing through Farmers Interest Groups (FIGs), (c) setting up of terminal markets, (d) forward and future market, (e) e-commerce, (f) setting up of mega markets, (g) negotiable warehouse receipt system, etc., have been suggested by expert committee on agricultural marketing. In respect of curd chilli marketing, what type of marketing practices is prevalent has not been documented so far. Farmers are practicing direct marketing to some extent. In this circumstance, the author has documented on direct marketing.

DIRECT MARKETING

Direct marketing is an innovative concept, which involves marketing of produce by the farmer/producer directly to the consumers/processors without any middlemen. Direct marketing enables producers and processors and other bulk buyers to economize on transportation cost and improve price realization. It also provides incentive to large scale marketing companies, i.e., processors and exporters to purchase directly from producing areas. Direct marketing by farmers to the consumers has been experimented in the country through *Apni Mandis* in Punjab and Haryana. The concept with certain improvements has been popularized in Andhra Pradesh through *Rythu Bazars*. Similarly, in Tamil Nadu, the Farmers' Market were created in almost all the town areas with one or more farmers' market to promote the practice of direct marketing whereby the producers get better advantage in price by providing free logistics. In these markets, along with fruits and vegetables other commodities are also marketed. Green Chilli is one of the important commodities sold in the farmers' market. However, the cottage industries are not much interested in procuring the green Chilli in the farmers' market owing to the small arrival of the produce and hence they mainly depend on the Oddanchatram Wholesale Vegetable Market which serves the customers at low price per unit of the produce.

Dried salted curd chillies (Moar Milagai) are one of the classic condiments which are very popular in South Indian cuisine. It is also being consumed in countries like Malaysia, Singapore, Sri Lanka, United States of America, etc. Moar milagai (curded green chilli) is prepared from either long dried chillies or with small gundu chilli based on the requirement. Product is prepared in the hygienic condition, having long shelf life, crispy and less hot. Mooar Milagai (chilly vathal) are fried and consumed along with meals.

The practice of making curd chilli though it is good old and confined to the region of Chettinadu in Tamil Nadu, for its quality taste and preferences, it is preferred widely and processed in different pockets of Tamil Nadu at household level and it is distributed to different markets both in domestic and international markets. The marketing practices prevalent in respect of green chilli and dried chilli are addressed in many institutions and the distribution channels are also documented well. But none of the institutions have addressed the marketing practices prevalent in curd chilli and its distribution channels and their brands. In this context, a special effort has been taken by the research team to assess the marketing practices prevalent in curd chilli and its distribution mechanism in Tamil Nadu which will be an incentive to the young agripreneurs to take up this as a new agribusiness venture in Tamil Nadu and elsewhere.

IDENTIFICATION OF A POTENTIAL COTTAGE INDUSTRY AND ITS CASE STUDY

In Tamil Nadu, Dindigul District, Thirumalairayapuram is the village in which several cottage industries at household level entered into the process of making salted curd Chilli for the domestic and international markets. There are around 350 units functioning in the village itself and producing high quality varietal curd Chilli for different market segments and these are in different brands. Looking at the production potentials of salted curd Chilli in Tamil Nadu and its trade potentials in domestic and international markets, the study was initiated at micro level case study by identifying a potential cottage industry involved in such agribusiness with the specific objective of identifying the marketing practices prevalent in salted curd chilli and to assess the price spread in marketing of the same. For that purpose, a well defined and pre-tested questionnaire is framed and administered with the case firm, R.S. Cottage Industry operating in a large scale in the village of Thirumalairayapuram, Oddanchatram Taluk of Dindigul District on a case study basis.

The case firm is having rich experience in marketing and distribution of curd chilli to different national and regional markets, his experiences in marketing of curd chilli is an experiential learning to the student researchers. This chapter is an outcome of the student researchers who have entered into identifying and documenting the results.

PRICE SPREAD ANALYSIS

Price spread in general, is referred to as difference between price paid by the consumer and that received by the producers per unit of the commodity.

Price spread analysis would estimate their share of different market functionaries in the consumer rupee and this would often facilitate their understanding of the relative efficiencies otherwise of alternative channels of marketing. The method of measurement of marketing margin is the most important aspect of price spread. It could be worked out in two ways.

1. Lagged Margins
2. Concurrent Margins

Lagged margins are the differences between the consumer's price obtained on a particular date and the producer's price obtained during the earlier date. The length of the time between the two dates, being the average period for which the marketing agency holds the produce. The concurrent margin refers to the differences between prices prevailing at successive stages of marketing on the same date. For the present study, lagged margin is used to analyze the price spread.

To analyze the price spread in the distribution of curd chilli, the following formula is used.

$$Ps = (Pp / Cp) X 100$$

Where,

Ps = Producer share in consumer money expressed in per cent

Cp = Consumer's price per unit of curd chilli in rupees

Pp = Producer's price per unit of curd chilli in rupees. Similar approach was used by Sekhar et.al. (1998).

The price spread analysis involved computation of different marketing cost and marketing margin at each stage and their expression as a percentage to the consumer's rupee. The various costs incurred in the marketing process were considered for each of the identified channels and separate price spread was worked out. The profit margin for each market functionaries in the different marketing channels was computed by subtracting the price paid and the marketing cost incurred by them from the price received by him on the sale of the commodity. Thus the marketing costs and marketing margins were distinguished.

Estimation of Marketing Efficiency

Marketing efficiency is the degree of market performance. The movement of goods from producers to the ultimate consumers at the lowest possible cost consistent with the provision of service desired by the consumer is termed as efficient marketing.

i. **Acharya and Agarwal's formula**

They compared relative efficiency of different markets by using the following formula:

$$E = (O/I)$$

Where,

E = Marketing efficiency expressed in percentage

O = Value added to output in marketing system

I = Input used in the marketing process

The lower the value, higher would be the efficiency.

ii. **Calkin's Index**

The Calkin's index of marketing efficiency was estimated using the following formula:

$$ME = 1 + \left[\frac{\text{Sum of profit or margin}}{\text{Sum of marketing cost}} \right]$$

The lower the value of the index, higher would be the efficiency.

CASE STUDY UNDER SUITABLE HEADINGS

Marketing is a process encompassing all the steps involved in the flow of goods and services from the producer to the consumer (Sharma 1980). It is a system operating within a network of channels, where middlemen and institutions not only influence but also play significant roles in the production, distribution and exchange of products. This research work is a case study of horticulture produce marketing especially the curd chilli. By understanding the existing marketing practices, it is hoped that the marketing efficiency and the income of the producers will increase.

The marketing practices and the price spread prevalent in different marketing channels dealing with salted curd chilli in Tamil Nadu are discussed under the following heads.

- Details of realization of curd chilli and other associated products
- Important assembling centers for curd chilli in Tamil Nadu
- Marketing channels prevalent in distribution of curd chilli
- Price spread in curd chilli marketing
- Marketing efficiency in curd chilli

DETAILS OF REALIZATION OF CURD CHILLI AND OTHER ASSOCIATED PRODUCTS

In addition to the Curd Chilli, the R.S. Cottage Industry is also generating other outputs. The details of those outputs are also analyzed and the results are presented in Table 1. Similar to the Curd Chilli, the other products like Chunda Vathal, Mithukka Vathal, Mango Vathal, Kothavarai Vathal are also produced by utilizing the available resources in the village.

Table 1 Details of Realization of Curd Chilli and other Products Produced by the Cottage Industry

Sl.No.	Name of the Produce	Input (Qty. in Kg)	Out Turn (Qty. in Kg)	Producer Price/Kg (in Rs.)	Consumer Price/Kg (in Rs.)
01	Mooar Milagai	100	33.000	60.00	140.00
02	Chunda Vatthal (*Solanum*)	100	33.000	120.00	230.00
03	Mango Vathal	100	17.000	38.00	130.00
04	Kothavarai Vathal (Cluster Beans)	100	10.000	35.00	120.00
05	Bhendi Vathal	100	07.000	45.00	210.00
06	Midhuka Vathal (Coccinea)	100	108.000	80.00	180.00

Among the above mentioned products, Table 1 shows that the Mooar Milagai (Curd Chilli) is the preferred one in the market. Followed by the Mooar Milagai, the Chunda Vathal belonging to the Solanum group is also processed and sold for meeting the demand of the consumers. The Mooar Milagai and Chunda Vathal output could be realized to the level of only 33 kgs for the raw material of 100 kgs which are priced respectively at Rs 140 and Rs. 230 per kg. Mithukka Vathal is another variety belonging to the Coccinea group which is also processed in curd plus salt mix, similarly dried in the sun light and traded across the nation. The producer of the output is able to get Rs 180 per kg.

The number of outputs added in the production process is mainly to meet the consumer requirements and to tap the benefit of utilizing the cheaply available resources into the value added form to reap higher price per unit. The lean season of one will permit the processing of other produce and hence the producer industry is always busy with realization of different output from the production process. In the process of making curd chilli and other associated products by the producers, the deposition of residual salt let into the soil has created some ground water problems and there exists some dispute among the local farmers and the producers of curd chilli. Hence the same may be alleviated by erecting common processing and drying yard and the waste water treatment mechanism need to be installed by the panchayat of the village in which the processing takes place. It needs to be looked into on priority basis.

IMPORTANT ASSEMBLING CENTERS FOR CURD CHILLI IN TAMIL NADU

The produced curd Chilli are being supplied to different assembly markets which are available in Tamil Nadu and the details are analyzed and the results are presented in Table 2

Table 2 Important Assembling Centers for Curd Chilli in Tamil Nadu

Sl.No	Name of the District	Number of Assembling Centers	Percentage to Total
01	Dindigul	02	18.18
02	Madurai	01	09.09
03	Thoothukkudi	01	09.09
04	Coimbatore	02	18.18
05	Chennai	03	27.28
06	Tiruchirappalli	02	18.18
	Total	11	100.00

Table 2 reveals that, in Tamil Nadu, there are 11 assembling markets for curd Chilli which are distributed in Dindigul, Madurai, Thoothukkudi, Coimbatore, Chennai, Tiruchirappalli districts. Among the 11 markets, Chennai city alone shared 27 per cent of the total assembling markets followed by Coimbatore, Tiruchirappalli and Dindigul districts each of which shared 18 per cent of the assembling markets for curd Chilli. Other districts like Madurai, Thoothukkudi shared only 9 per cent each in respect of curd Chilli. From this one could understand that the southern districts had the assembling centers of 36 per cent followed by the central part of Tamil Nadu sharing 18 per cent and the northern and western part of Tamil Nadu sharing 27 and 18 per cent respectively revealing that the assembling markets were distributed across the length and breadth of the state of Tamil Nadu. Western part of Tamil Nadu takes care of the supply to the Kerala markets and the southern part of Tamil Nadu takes care of the markets in the southern part of Kerala and southern districts of Tamil Nadu.

MARKETING CHANNELS PREVALENT IN DISTRIBUTION OF CURD CHILLI IN TAMIL NADU

Marketing channel is viewed as a set of interdependent organization involved in the process of making a product or service available for use or consumption. The interdependent organization included is the intermediaries who facilitate the flow of the produce for ultimate consumption from one level to another. In respect of curd chilli, there are six marketing channels identified and the same are presented in Table 3.

Table 3 Marketing Channels Prevalent in Distribution of Curd Chilli

Channel Number	Details of the Marketing Channel
I	Producers → Primary Wholesalers (Cochin) → Secondary Wholesaler (Thiruvananthapuram) → Wholesaler cum Retailers (Kerala Markets) → Retailers → Consumers
II	Producers → Primary Wholesalers (Chennai) → Wholesalers cum Retailers → Retailers → Consumers
III	Producers → Wholesalers (Tiruchirappalli and Thanjavur) → Wholesalers cum Retailers → Retailers → Consumers
IV	Producers → Primary Wholesalers (Madurai and Thoothukkudi) → Wholesaler cum Retailers → Retailers → Consumers
V	Producers → Wholesalers (Madurai) → Wholesaler cum Exporters → Retailers → Consumers
VI	Producers → Primary Wholesalers (Coimbatore) → Wholesalers cum Exporters → Wholesaler cum Retailers → Retailers → Consumers

Table 3 reveals that among the six marketing channels, the Channel IV and Channel VI are found to be traceable and hence these two channels were alone chosen for working out the price spread in respect of curd Chilli. The Channel IV takes the produce from the producer by the primary wholesalers positioned in Madurai and Thoothukkudi, the same was transferred to the wholesaler cum retailers in both the cities in required quantity, the same was transferred to the retailers positioned in these cities and the retailers are distributing the same to the consumers after incurring certain expenses as marketing cost to perform certain functions. These are documented in Table 4. Similarly, the Channel VI takes the produce from producer and the same was transferred to the primary wholesalers in Coimbatore city. The wholesalers of Coimbatore city then transferred the produce to the wholesalers cum exporters. The exporters do grading and standardization aspects in the produce to meet the consumer requirements, also do branding of the produce and then the same are transferred to the retailers and consumers. These details are analyzed and discussed in Table 5.

Apart from Channel IV and VI, the marketing channel – I feed to the state of Kerala markets like Cochin and Thiruvananthapuram through the primary wholesalers in Cochin and then the same is transferred to the secondary wholesalers at Thiruvananthapuram. From the secondary wholesalers, the wholesalers cum retailers take the produce, do some functions and then the produce gets transferred to the retailers and then to the ultimate consumers. The consumers' price is varying between Rs 100 to Rs 250 per kg of Curd Chilli and the details of prices in different markets were unable to be documented.

PRICE SPREAD IN MARKETING OF CURD CHILLI

Price spread is the broad spectrum which disclosed the properties of various components of marketing cost of the produce and thus explained the variance between the price paid by the consumer and price received by the producer. In short, it speaks about the price difference that prevailed between the consumers and the producers. Though there were six marketing channels identified, only two channels were traceable to work out the price spread prevailing in marketing of curd chilli and hence the results are analyzed and the details are presented in Table 4 and Table 5.

Table 4 Price Spread in Distribution of Curd Chilli through Fourth Channel IV. Producers → Primary Wholesalers (Madurai and Thothukkudi) → Wholesaler cum Retailers → Retailers →Consumers

Sl. No	Particulars of Cost	Amount (Rs/Kg)	Per Cent
01	Producer		
01	Producers' Price	50.00	50.00
1.1	Marketing Cost		
1.1.1	Grading and Packing Charges	00.30	
1.1.2	Loading Charges	00.30	
1.1.3	Transportation Cost	01.50	
1.1.4	Miscellaneous Expenses including Cold Storage Facilities	02.40	
	Marketing Cost at Producers' Level	04.50	04.50
1.2	Marketing Margin	05.50	05.50
02	Primary Wholesaler		
2.1	Procurement Price by the Primary Wholesaler	60.00	60.00
2.2	Marketing Cost		
2.2.1	Unloading, Storage and Reloading Charges	02.10	
2.2.2	Transportation Cost	03.40	
2.2.3	Weighing Charges	00.30	
2.2.4	Watch and Ward Charges	00.40	
2.2.5	Miscellaneous Charges	01.30	
2.2	Marketing Cost at Primary Wholesalers' Level	07.50	07.50
2.3	Marketing Margin to the Primary Wholesaler	07.50	07.50
03	Wholesaler cum Retailer		
3.1	Purchase Price Incurred by the Wholesaler cum Retailer	75.00	75.00

Sl. No	Particulars of Cost	Amount (Rs/Kg)	Per Cent
3.2	Marketing Cost		
3.2.1	Loading and Unloading Charges	01.10	
3.2.2	Transportation Charges	01.30	
3.2.3	Watch and Ward Charges	00.60	
3.2.4	Miscellaneous Cost	02.00	
3.2	Marketing Cost to the Wholesaler cum Retailer	05.00	05.00
3.3	Marketing Margin to the Wholesaler cum Retailer	07.00	07.00
3.4	Sale Price of Wholesaler cum Retailer	87.00	87.00
04	Retailer		
4.1	Purchase Price by the Retailer	87.00	87.00
4.2	Marketing Cost incurred by the Retailer		
4.2.1	Transportation Charges	01.60	
4.2.2	Packing Charges in Consumer Packs	01.50	
4.2.3	Miscellaneous Expenses incurred by the Traders	01.90	
4.2	Marketing Cost to the Retailer	05.00	05.00
4.3	Marketing Margin to the Retailer	08.00	08.00
05	Consumer		
5.1	Procurement Price by the Consumer	100.00	100.00

Table 4 reveals the details of price spread in Channel – IV. The marketing channel IV reveals that the producer of curd chilli is able to get Rs 50 per kg of the produce while the consumer price is at Rs 100 per kg. The price spread in this channel is found to be only 50 per cent to the producer of Curd Chilli. Since the producer is also making certain functions in distributing the produce to the primary wholesaler, he has incurred the marketing cost to the tune of 4.50 per cent and the marketing margin gained by the producer is arrived at 5.50 per cent. While taking into account the marketing cost incurred and the marketing margin gained by the producer, the producer is able to get 60 per cent share for his output. For this, he is incurring the marketing costs towards grading, packing in plastic cartons (Plate – 01), loading and transportation functions which has consumed an amount of Rs 4.50 per kg of curd chilli.

Plate 01 Process of Packing of Curd Chilli

The procurement price by the primary wholesaler is arrived at Rs 60 per kg of Curd Chilli and the primary wholesaler is also performing certain marketing functions like loading and unloading charges, transportation cost, weighing, watch and ward charges amounting to Rs 7.50 and he gains Rs 7.50 as marketing margin for the services performed and then he transfers the produce to the wholesaler cum retailer at the price of Rs 75 per kg of Curd Chilli.

The wholesaler cum retailer is also performing certain marketing functions and incurring marketing costs to the tune of Rs 5 per kg and gains the marketing margin of Rs 7 per kg and then the produce is transferred to the retailers on demand and the retailer procures the produce at the price of Rs 87 per kg which is accounted for 87 per cent to the consumers' rupee per kg of the produce and then the produce is sold to the ultimate consumers. The retailers are also performing certain marketing functions like transportation cost, loading and unloading, packing the produce in consumer packs for easy disposal and thereby incurring the marketing cost to the tune of Rs 5 per kg and the retailer is gaining a marketing margin of Rs 8 per kg of Curd Chilli and sold the produce to the consumers at the price of Rs 100 per kg.

The price spread in the channel is found to be Rs 50 which accounted for 50 per cent of the consumers' rupee in the Channel - IV. However, the consumer price varies in other channel where export of the produce is taking place which is discussed in the Channel – VI and the results are documented in Table 5.

Table 5 Price Spread in Distribution of Curd Chilli through Sixth Channel VI. Producers → Primary Wholesalers (Coimbatore) →Wholesalers cum Exporters → Wholesaler cum Retailers → Retailers →Consumers

Sl. No	Particulars of Cost	Amount (Rs/Kg)	Per Cent
01	Producer		
01	Producers' Price	60.00	42.86
1.1	Marketing Cost		
1.1.1	Grading and Packing Charges	00.60	
1.1.2	Loading Charges	00.50	
1.1.3	Transportation Cost	02.80	
1.1.4	Miscellaneous Expenses including Cold Storage Facilities	02.90	
1.2	Marketing Cost at Producers' Level	06.80	04.86
1.3	Marketing Margin	08.20	05.86
02	Primary Wholesaler		
2.1	Procurement Price by the Primary Wholesaler	75.00	53.57

Sl. No	Particulars of Cost	Amount (Rs/Kg)	Per Cent
2.2	Marketing Cost		
2.2.1	Unloading, Storage and Reloading Charges	02.60	
2.2.2	Transportation Cost	04.00	
2.2.3	Weighing Charges	00.50	
2.2.4	Watch and Ward Charges	00.60	
2.2.5	Miscellaneous Charges	01.80	
2.2	Marketing Cost at Primary Wholesalers' Level	09.50	06.79
Sl. No	**Particulars of Cost**	**Amount (Rs/Kg)**	**Per Cent**
2.3	Marketing Margin to the Primary Wholesaler	15.50	11.07
03	Wholesaler cum Exporter		
3.1	Purchase Price Incurred by the Wholesaler cum Exporter	100.00	71.43
3.2	Marketing Cost		
3.2.1	Loading and Unloading Charges	06.20	
3.2.2	Transportation Charges	08.40	
3.2.3	Grading and Branding related Charges	10.40	
3.2.4	Watch and Ward Charges	01.60	
3.2.5	Miscellaneous Cost including Licensing	08.40	
3.2	Marketing Cost to the Wholesaler cum Exporter	35.00	25.00
3.3	Marketing Margin to the Wholesaler cum Retailer	25.00	17.86
3.4	Sale Price of Wholesaler cum Exporter	160.00	
04	Wholesaler cum Retailer		

Sl. No	Particulars of Cost	Amount (Rs/Kg)	Per Cent
4.1	Purchase Price by the Wholesaler cum Retailer	100.00	71.43
4.2	Marketing Cost incurred by the Retailer		
4.2.1	Transportation Charges	01.90	
4.2.2	Loading and Unloading Charges	02.30	
4.2.3	Watch and Ward Charges	00.60	
4.2.3	Miscellaneous Expenses	02.80	
4.2	Marketing Cost to the Wholesaler cum Retailer	07.60	05.43
4.3	Marketing Margin to the Wholesaler cum Retailer	12.40	08.86
05	Retailer		
5.1	Procurement Price by the Retailer	120.00	85.71
5.2	Marketing Cost incurred by the Retailer		
5.2.1	Transportation Charges	02.35	
5.2.2	Packing Charges in Consumer Packs	01.95	
5.2.3	Miscellaneous Expenses	02.70	
5.2	Marketing Cost Incurred by the Retailer	07.00	05.00
5.3	Marketing Margin to the Retailer	13.00	09.29
06	Consumer		
6.1	Purchase Price by the Consumer	140.00	100.00

Table 5 reveals the details of price spread in Channel – VI. The marketing channel VI reveals that the producer of Curd Chilli is able to get Rs 60 per kg of the produce while the exporters' price is at Rs 160 per Kg. The price spread in this channel is found to be Rs 100 between the producer and consumers of Curd Chilli. Since the producer is also making certain functions in distributing the produce to the primary wholesaler, he has incurred the marketing cost to the tune of Rs 6.80 which accounted for 4.86 per cent and the marketing margin gained by the producer is arrived at Rs 8.20 which accounted for 5.86 per cent of the consumers' rupee. While taking into account the marketing cost incurred and the marketing margin gained by the producer, the producer is able to get Rs 75 per kg which accounted for 53.57 per cent share for his output from the consumers' rupee. For this, he is incurring the marketing costs towards grading, packing in plastic cartons, loading and transportation functions (Plate – 02) and stocking the produce in cold storage facilities which has consumed an amount of Rs 6.80 per kg of curd Chilli. By retaining the produce in the cold storage facilities, he is creating a demand in the market when the price goes below Rs 60 per kg of the produce.

The primary wholesaler has received the produce at the price of Rs 75 per kg and performed certain marketing functions like loading and unloading, transportation cost, weighing and other miscellaneous charges and thereby the marketing cost incurred by him was arrived at Rs 9.50 which accounted for 6.79 per cent to the consumers' rupee. The marketing margin gained by the primary wholesaler arrived at Rs 15.50 per kg which accounted for 11.07 per cent of the consumers' price and then the produce is received by the wholesaler cum exporter and the wholesaler cum retailer at the price of Rs 100 per kg of Curd Chilli.

The procurement price by the wholesaler cum exporter arrived at Rs 100 per kg of Curd Chilli and the wholesaler cum exporter is also performing certain marketing functions like loading and unloading charges, transportation cost, grading, branding of the produce and its related charges amounting to Rs 35 which accounted for 25 per cent of the consumers' price. The wholesaler cum exporter gains Rs 25 as marketing margin for the services performed and then he exported the produce to the wholesalers of importing countries at the price of Rs 160 ($ 2.62) per kg of Curd Chilli.

Plate 02 Loading of Curd Chilli for Storage

The wholesaler cum retailer is another stage in the marketing channel – VI and he is receiving the produce from the primary wholesaler at the price of Rs 100 per kg of Curd Chilli. He is also performing certain marketing functions and incurring marketing costs to the tune of Rs 7.60 per kg and gains the marketing margin of Rs 12.40 per kg which respectively accounted for 5.43 per cent and 8.86 per cent of the consumers' price and then the produce is transferred to the retailers on demand. The retailer procures the produce at the price of Rs 120 per kg accounting for 85.71 per cent to the consumers' rupee per kg of the produce and then the produce is sold to the ultimate consumers. The retailers are also performing certain marketing functions like transportation cost, loading and unloading, packing the produce in consumer packs for easy disposal and thereby incurring the marketing cost to the tune of Rs 7 per kg, gaining a marketing margin of Rs 13 per kg of Curd Chilli and sold the produce to the consumers at the price of Rs 140 per kg.

The price spread in the channel is found to be Rs 80 accounting for 57.14 per cent of the consumers' rupee in the Channel - VI. However, it is noticed that the exporters' price is high per kg which arrived at Rs 160 and the consumers' price at regional and national level is Rs 140 per kg of Curd Chilli. The consumers' price is also varying between Rs 100 to Rs 250 per kg when it is marketed to distant places and too many intermediaries between wholesaler cum retailer to the consumer is prevailing and many of the households buy at the price of Rs 100 to 140 per kg and the same is packed in 10 grams and 20 grams which are priced at Rs 2 per 10 gram (Plate – 03) and Rs 5 for 20 grams in rural and cities respectively and thereby creating a lot of employment opportunities to the people at low level. However, the quantum of employment generated in the sector is not addressed in detail which commands a separate study in this region at state level to assess the employment potential and prospects of agribusiness venture.

MARKETING EFFICIENCY

Marketing Efficiency is defined as the ratio of output (Consumers' Price) to the input (Marketing Cost). In this study, the consumers' price is varied in Channel – IV and Channel VI and hence the same are worked out using the Acharya and Agarwal's Method and the Calkin's Index. The details are analyzed and the results are presented in Table 6.

Table 6 Marketing Efficiency of Curd Chilli in Different Marketing Channels

Marketing Channel	Value of Output (Consumers' Price)	Input (Marketing Cost)	Marketing Efficiency Ratio (O/I)
Acharya and Agarwal			
IV	100.00	22.00	04.55
VI	140.00	65.90	02.12
Calkin's Index			
Marketing Channel	**Marketing Margin**	**Marketing Cost**	**Marketing Efficiency**
IV	28.00	22.00	01.27
VI	74.20	65.90	01.13

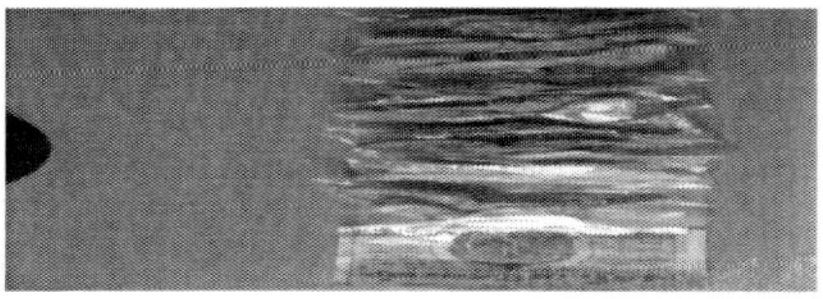

Palte 03 Consumer Packs of Curd Chilli – Each One Weighs 10 Grams

An increase in this ratio would represent improved efficiency and vice-versa. A reduction in the cost for the same level of satisfaction or an increase in satisfaction at a given cost would result in an improvement of efficiency. When comparing the data delineated in the Table 6, one could understand that Channel – IV is found to be the efficient one because at the lowest marketing cost, it fulfills several functionaries in the channel and keeps them satisfied.

CONCLUSIONS AND POLICY IMPLICATIONS

In the process of marketing of curd chilli, there were six channels that could be identified. Among the six marketing channels, Channel IV and Channel VI are found to be traceable and hence these two channels were alone chosen for working out the price spread in respect of curd Chilli. In Channel IV, the produce is taken from the producer by the primary wholesalers positioned in Madurai and Thoothukkudi, the same was transferred to the wholesaler cum retailers in both the cities on required quantity, the same were transferred to the retailers positioned in these cities and the retailers are distributing the same to the consumers after incurring certain expenses as marketing cost to perform certain functions. Similarly, in Channel VI, the produce is taken from producer and transferred to the primary wholesalers in Coimbatore city. The wholesalers of Coimbatore city then transfer the produce to the wholesalers cum exporters. The exporters do grading and standardization aspects in the produce to meet the consumer requirements, do branding of the produce and then the same are transferred to the retailers and consumers. Among the two channels traced, Channel – IV is found to be the efficient one because at the lowest marketing cost, it fulfills several functionaries in the channel and keeps them satisfied.

RECOMMENDATIONS

- The potential market and the price details are not known to the producer cum traders of value added Curd Chilli which often forces them to sell at low price per unit of the produce. In this regard, the farmers' can establish the Commodity Interest Group themselves (CIG). The CIG can procure entire commodity produced in the village and then the same may be sold to the demanding markets by grading the produce for export and domestic markets.
- On establishment of CIG, the stiff competition between the producer cum traders can be checked and force them to sell the product on higher price per unit. The benefits accrued from the process of trading through CIG can be distributed to all the stakeholder members.
- The farmers' agitation against the producer cum traders of value added Curd Chilli is taking place very vehemently. This problem can be solved by establishing a common effluent treatment plant by arranging for the drainage facilities and by making common drying yard for the traders; the waste water can be collected and treated and the treated water can be permitted for reuse in the same units after checking for the quality.
- Absence of cold storage facility is another serious problem to the producer cum traders of value added Curd Chilli. The traders are taking the produce to Madurai for storage. If storage facility is erected in the village of Thirumalairayapuram, the 350 traders will be benefitted

and the produce mobility can be planned from the production site itself which avoids unnecessary logistic expenditures.

References

- Acharya, S.S and N.L. Agarwal, "Agriculture Marketing in India", (New Delhi: Oxford and IBH Publishing Co., 1987), pp. 302-312.
- Government of Tamil Nadu, "Season and Crop Report of Tamil Nadu – 2010" (Directorate of Economics and Statistics: Chennai)
- Sekhar C., R.C. Panda, R.S. Vinaya Rai and P. Devaraj (1998). "Marketing of Forest Produce from Farm Land – A Case Study in Palani, Tamil Nadu, India" Journal of Tropical Forest Products, 4(1): pp.68-73
- Sharma L.C., (1980). "Forest Economics Planning and Management" (B.S.Mahendra Pal Singh: Dehra Dun, pp.121

Chapter 22

Garlic Trading – A Potential Agribusiness Venture

INTRODUCTION

Garlic belongs to the family Alliaceae; genus *Allium* and species *sativum*. It is an herbaceous annual for the bulb production. The edible underground stem is the composite bulb made of numerous smaller bulbs or bulblets called cloves (5-30/bulb). Leaves have solid thin blades. Some varieties produce flower but these flower do not set seed. There are some varieties which produce vegetative bulbs on the top called bulbils. They are also used for propagation in some cases.

Garlic has long been recognized all over the world as a valuable condiment for foods, and a popular remedy or medicine for various ailments and physiological disorders (Pruthi, 1998). The scientists have found in experimental farms that Garlic too can be produced in Khariff. Planting of Garlic by the last week of August gave the highest yield and the biggest size of bulbs. For better growth, bulb characteristics and marketable pulp yield in Garlic, integrated use of inorganic and organic sources of nutrients are preferred (Sevak et.al. 2012). For sustainable production of Garlic, the recommended dose of NPK supplemented with 20 tonnes of farmyard manure per ha of crop coupled with inoculation of phosphate solubilizing bacteria and Trichoderma can enhance the yield and profitability of Garlic (Nainwal et.al. 2015).

The crop is ready for harvesting when the tops turn yellowish or brownish and show signs of drying up, usually about a month or so after the emergence of seed stalks. The bulbs are lifted, freed from earth and the leaves tied at the top. The bulbs are cured for three to four days in the shade before storing them in a well ventilated room. Tops are removed before marketing the produce.

India is ranked second in area and third in production of Garlic in the world. The productivity of this crop is quite low which arrived at 5 tonnes per ha. The demand for Garlic in India is estimated to be 7.63 lakh tonnes whereas for the South India alone the demand for Garlic during the year 2016-17 is arrived at 1.38 lakh tonnes as the South Indians prefer spicy food (Srivastava, 2013).

Garlic is consumed in every home and use in processing industry has a fairly constant market demand. Bulbs are easily stored for 5-6 months after harvest which favors long marketing season. During the time, the physiological loss in weight (PLW) is also observed. To avoid this, the pre-harvest sprays of growth regulators and chemicals were found to be significantly superior in reducing the PLW and enhancing the recovery of healthy cloves (41.04 per cent) was recorded in Cycocel (1000 ppm) plus Dithane M-45 (1000 ppm) when compared to the control over a period of six months (Kumara et.al. 2015). Garlic is transported to the assembling centers in open bullock cart or tractor trolleys in open. It is also carried in jute bags in trucks. While loading the bags in trucks care should be taken in loading up to 5-6 bags high as overloading will damage the garlic bulbs in lower bags by pressure. Use of hooks should be avoided as it injures the bulbs /cloves and such bulbs when stored loose more weight and also rot easily. Garlic bulbs after reaching the assembling centers are generally sold to the local traders for distribution to the consumer markets. These transactions usually take place with the help of commission agents. Exporters either buy their requirements directly from assembling centers through local commission agents or from wholesalers in the important assembling or regulated markets. Garlic is sold by open auction in regulated markets. In other markets, sale often takes place by mutual negotiations which may either be direct between the buyer and seller or through commission agents or brokers. In certain markets undercover system is also present. It is observed that in the markets adequate facilities for handling the produce like sheds/stores, etc., are not present which ultimately results in increased loss. There is thus dire need to strengthen facilities at the assembling centers so as to handle the produce safely. There is considerable fluctuation in garlic price from year to year due to increased or decreased production. The prices of garlic are generally low during March to May due to peak period of arrivals which vary from Rs.30-40 per kg. There after prices of garlic increase and generally remain higher during the period of October to January. The rate prevailing during this period is normally more than Rs.100 per kg. Market support by government is necessary to encourage farmers in continuing production of quality garlic every year.

DEHYDRATION OF GARLIC

Some portion of garlic produce (surplus) is subjected to processing especially dehydration, i.e., drying under controlled conditions of temperature and humidity. The dehydration is important because of the concentrated form, low cost convenience and easy transportability and dehydrated garlic in the form of powder, etc., has become highly popular among the civilians, military forces, besides export. Garlic export makes excellent use of dehydrated products and its use as dry form has great scope particularly in foreign countries where there is scarcity.

EXPORT OF GARLIC

India has been exporting for many years to the major importing countries like Qatar, Saudi Arabia, Zambia, UAF, Bahrain, Mauritius, Kuwait, Bangladesh and Sri Lanka. The export has been ranging between 2-3 per cent of the total domestic production. The quantum of export of garlic has been fluctuating due to sudden change in the policy of garlic importing countries. China, France, Spain and Egypt are the major exporting countries. The country being at higher latitudes produces bigger cloved garlic though not on a big scale. It is picking up in northern hills and Nilgiris and Kodaikonal hills in view of increased demand in the export market. It is hoped that in the near feature increase in area under those varieties having bigger cloves and that have demand in the international market will be produced to the extent of 10-15 times easily. Hence, assessing the current trend in export of garlic becomes important and hence effort has been taken to present the export scenario in detail with the following specific objectives

OBJECTIVES OF THE STUDY

- To assess the Garlic marketing practices prevalent in the study area
- To document the status of export of garlic to different countries from India

MARKETING PRACTISES OF GARLIC IN INDIA AND TAMIL NADU

The objective of this chapter is to document the marketing practices and export of Garlic in India and to document the supporting infrastructure available in the state of Tamil Nadu to promote marketing. To document the same, the data were collected from the secondary sources of both Central and State Government reports. The reports taken into account for collecting the details on export of Garlic are furnished as follows.

- Agricultural Statistics at a Glance 2012 of Directorate of Economics and Statistics, Ministry of Agriculture, Government of India, New Delhi
- Food and Agriculture Organization Web Site
- Indian Horticulture Data Base 2011 of National Horticulture Board, Government of India, New Delhi
- Hort-Stat – 2008 of Government of Tamil Nadu, Chennai

The data gathered for different periods with respect to export of garlic, assembling and trading centers for garlic and the importing institutions and states involved in garlic export and import were gathered from the above reports and the data were analyzed using descriptive statistics like mean. Also the current change in the data in respect of export of Garlic over the base period was also documented to assess the technological and extension impact due to the government intervention through its planned programs in horticulture. The results of this analysis will be better utilized by the policy makers and the planners in horticulture and agriculture for further development of these sectors.

With respect to marketing of garlic, the traders are hailing from Vadugapatti Garlic Merchants Association. Vadugapatti which is one of the leading wholesale Garlic market functions twice in a week, particularly on Thursdays and Sundays throughout the year. The Vadugapatti Garlic Merchants Association has around 180 members to its credit who are involved in trading of garlic. Among the 180 traders, the active members involved in large scale transactions of garlic are numbering around 50. Among the 50 traders, only five traders have been interviewed as a case study approach to identify the assembling centers in Tamil Nadu, marketing channels in which the produce is moved and the price spread in garlic marketing. These are measured using the following formulae.

Price Spread Analysis

Price spread in general, is referred to as the difference between the price paid by the consumer and the price received by the producers per unit quantity of the commodity. Price spread analysis would estimate the share of different market intermediaries in the consumer rupee and this would often facilitate the understanding of the relative efficiencies otherwise of alternative channels in marketing.

To analyze the price spread in the distribution of Garlic, the following formula is used.

$$Ps = (Pp / Cp) X 100$$

Where,

Ps = Producer share in the consumer money expressed in per cent

Cp = Consumer's price per unit of Garlic in rupees

Pp = Producer's price per unit of Garlic in rupees. Similar approach was used by Sekhar et.al. (1998).

The price spread analysis involved computation of different marketing cost and marketing margin at each stage and their expression as a percentage to the consumer's rupee. The various costs incurred in the marketing process were considered for each of the identified channels and separate price spread was worked out. The profit margin for each market functionaries in the different marketing channels was computed by subtracting the price paid and the marketing cost incurred by them from the price received by him on the sale of the commodity. Thus the marketing costs and marketing margins were distinguished.

Estimation of Marketing Efficiency

Marketing efficiency is the degree of market performance. The movement of goods from producer to the ultimate consumers at the lowest possible cost consistent with the provision of services desired by the consumer is termed as marketing efficiency or efficient marketing. To analyze the marketing efficiency, the formula adopted by Shepherd is used.

$$ME = (V/I) - 1$$

Where,

ME = Marketing Efficiency Index

V = Value added to the output in the marketing system. Here it is r eferred to as the consumers price per kg of Garlic

I = Input used in the marketing process. Here it is referred to as the marketing costs incurred per kg of Garlic. Similar methodology was used by Sekhar et.al. during 1998.

The lower the value of marketing cost, higher would be the efficiency

Calkin's Index

Calkin's index of marketing efficiency was estimated using the following formula:

ME=1 + [Sum of profit or margin/Sum of marketing cost]

The lower the value of the index, higher would be the efficiency.

STUDY OF MARKETING PRACTISES UNDER SUITABLE HEADINGS

Garlic Markets and Marketing in Tamil Nadu

Garlic is used in various food preparations like chutneys, pickles, curry powders, curried vegetables, meat and meat product preparations, tomato ketchup, etc. The raw garlic is also used in second generation products like garlic powder, garlic salt, garlic vinegar, garlic cheese croutons, garlic potato chips, garlic bread, etc. It has also been extensively used as a popular remedy for various ailments and psychological disorders since vedic period.

Cost of seed cloves and labor for planting and harvest makes the initial investment for garlic production high in comparison to some other vegetable crops. Garlic returns are highly dependent on how the produce of crop is marketed. Management and markets will determine the profitability of garlic for the producer (Bachmann and Hinman, 2008). Properly cured or dried garlic can be stored for a period up to three months in a standard warehouse so as to make marketing decisions with respect to time and place utility.

Directorate of Marketing and Inspection (Ministry of Agriculture and Co-operation, Government of India) documented more than 15 major wholesale agricultural produce assembling markets involved in garlic marketing in different states of India (Table 1).

Table 1 Major Garlic Wholesale Markets in India

Sl. No	Name of the State	Name of the Markets	Number of Markets	Percentage to Total
01	Andhra Pradesh	Hyderabad and Nizamabad	02	09.52
02	Gujarat	Kondal	01	04.76
03	Himachal Pradesh	Bilaspur	02	09.52
02	Madhya Pradesh	Mandsaur, Neemuch, Ratlam,	07	33.33
03	Maharashtra	Mumbai	01	04.76
04	Tamil Nadu	Chennai, Mettupalayam and Vadugapatti	03	14.29
05	Uttar Pradesh	Eath, Farukhabad, Bhoagaon, Kuraoli	04	19.06

Sl. No	Name of the State	Name of the Markets	Number of Markets	Percentage to Total
06	West Bengal	Bara Bazaar	01	04.76
	Total Number of Wholesale Markets		21	100.00

Table 1 reveals that India is having 21 important organized assembling wholesale markets for garlic. Among these organized wholesale markets, large number of private and unorganized markets is functioning in all the major garlic producing states. These markets handle a significant quantum of garlic bulbs and acts as a wholesale market in those regions. One such market functioning over a long period of time in Tamil Nadu is the Vadugapatti of Theni District and the other one is Mettupalayam market wherein the Nilgiris Cooperative Marketing Society, Mettupalayam is one of the institution that facilitates garlic marketing and supplies to Vadugapatti also. Madhya Pradesh state is enriched with seven wholesale markets for garlic (33 per cent) followed by Utter Pradesh having four wholesale markets for garlic. All these markets contribute to the Vadugapatti garlic market.

WHOLESALE GARLIC AUCTION MARKET AT VADUGAPATTI

Vadugapatti is a small panchayat town located 5 km away from the Horticultural College and Research Institute, Periyakulam and it is 3 km away from Periyakulam town. Lack of essential market characteristics did not hinder this market to emerge as a hub for garlic marketing in South India due to its historical links and consistency in market functionaries.

The History of Vadugapatti Garlic Market

While tracing the historical developments of this market, it is a more than 150 years old market and started with the sale of small quantity of garlic. A group of people who cultivated garlic in and around Kodaikanal hills initially brought small quantity of garlic to Vadugapatti while they attended functions/visited relatives at Vadugapatti. At the same time, the labor force that migrated from this village to Kodaikanal hills to be engaged in garlic (Malaippoondu) cultivation used to bring the garlic in gunny bags. The market intermediaries were able to trace back the possible quantity of sale a century ago, which would be around 50 to 70 tonnes a week. In recent times, this market has grown enormously and handling as high as 350 tonnes a week and about 50 wholesalers are involved in this market. Initially a small group of wholesalers were involved in this marketing as they had links with the garlic producers who were relatives to them. However, this situation has been improving as one could find a mix of people engaged in the garlic wholesale marketing. This private market functions twice in a week, that is, on Thursdays and Sundays.

Marketing Channels Involved in the Distribution of Garlic

Marketing channel is defined as the path in which the produce travels from producer to the ultimate consumer through the involvement of intermediaries like wholesalers, commission agents and the traders. These details are analyzed in respect of garlic distribution and the same are presented in Table 2.

Table 2 Marketing Channels Prevalent in Distribution of Garlic

Sl.No	Channel Number	Marketing Channel Prevalent in Garlic Marketing
01	I	Producers → Consumers
02	II	Producers → Commission Agent (Vadugapatti) → Wholesaler Cum Retailers → Retailers → Consumers
03	III	Producers → Commission Agent (Mettupalayam) → Wholesaler Cum Retailers →Retailers → Consumers
04	IV	Producers → Primary Wholesalers (Other States) → Secondary Wholesalers (Vadugapatti) → Wholesalers cum Retailers → Retailers → Consumers
05	V	Producer (China) → Wholesaler cum Importers →Commission Agents (Vadugapatti) →Wholesaler cum Retailers → Retailers →Consumers
06	VI	Producers → Primary Wholesalers → Commission Agents (Vadugapatti) → Wholesaler cum Retailers → Retailers →Consumers

Table 2 reveals that the garlic marketing has commanded its distribution through six important channels. Among the six channels, the channel – I commands direct marketing in the local town area of Kodaikanal. During the peak tourist arrival, the garlic is distributed directly to the consumers. The producers of Garlic from Kodaikanal send the garlic through the regular service so as to reach the same to the Vadugapatti market. The commission agents at Vadugapatti receive the produce and conduct open auction on Thursdays and Sundays every week. The wholesalers cum retailers used to attend the open auction and take the title of goods to their place and the same was distributed to the retailers and consumers. This channel is designated as Channel – II. The price spread in garlic marketing is analyzed and the results are presented in Table 4 and Table 5.

The Garlic imported from China is received by the wholesalers cum importers at Rajasthan and other states through Tibet, the same are distributed to Vadugapatti, Tamil Nadu, the same are received by the commission agents cum wholesalers of Garlic Merchants Association at Vadugapatti and then it was subjected to open auction, other traders received the produce and distributed to other pockets.

Sources of Supply of Garlic to the Vadugapatti Wholesale Market

Garlic is procured from different parts of the state (Madhya Pradesh, Uttar Pradesh, Gujarat, Bihar and Himachal Pradesh) and also as imports from China through Nepal. Farmers also bring their produce to this market for sale. Wholesalers get seven per cent commission from the farmers and help them to sell their produce. Trucks are the main sources of transportation and in general trucks with 16 tonnes of capacity are employed for economic handling. Cost of transportation ranges from Rs 10 to 12 per kg depending upon the distance from the buying markets. Trucks take a long trip and reach the market yard a day prior to the auction and male laborers are actively engaged in unloading the garlic received from different states. A bag may weigh around 100 kg and the different types of packing materials like gunny, polythene and transparent nylon net bags are used. The quantity of garlic on an average supplied per annum to Vadugapatti wholesale market is analyzed and the details are presented in Table 3.

Table 3 Sources of Supply of Garlic to the Vadugapatti Wholesale Market

Sl. No	Sources of Supply	Quantity Supplied per Annum in Tonnes	Price per Kg	Percentage to Total
01	Kodaikanal, Tamil Nadu	1912.000	90.00	64.33
02	Ooty, Coonoor, Tamil Nadu	244.000	70.00	08.21
02	Madhya Pradesh	288.000	80.00	09.69
03	Himachal Pradesh	192.000	80.00	06.46
04	Gujarat	096.000	60.00	03.23
05	Rajasthan	080.000	60.00	02.70
06	Other States	160.000	50.00	05.38
		2972.000	70.00	100.00

Table 3 reveals that the garlic received at Vadugapatti wholesale market is from five important sources. The potential sources are Kodaikanal, The Nilgiris, Madhya Pradesh, Himachal Pradesh, Gujarat, Rajasthan and other states. The total quantity of garlic at the Vadugapatti market is about 2972 tonnes out of which 64 per cent of the supplies are made from Kodaikanal alone and the same is followed by the Niligiris contributing only to the level of eight per cent. Put together, the supply of garlic to the Vadugapatti wholesale market arrived at 72 per cent and the remaining 28 per cent was contributed by different states like Himachal Pradesh, Madhya Pradesh, Gujarat and Rajasthan.

Immediately after the arrival, garlic bags are untied and the process of cleaning and grading is started for next day's sale (Plate – 1). In general, female laborers are engaged in cleaning. The cleaned whole and uniform garlic lots are displayed on the floor and kept ready for the bi-weekly auction.

Grading and Pricing of Garlic in Vadugapatti Wholesale Market

Two types of garlic are dealt in the market in terms of its color. They are smoked and white. Among the types, there are different grades of garlic starting from malformed, small, broken cloves to premium garlic. During the visit to Garlic market during the first week and third week of December 2013, the price range for different grades and their price are analyzed and the details are presented in Table 4.

Table 4 Grading and Pricing of Garlic in Vadugapatti Market

Sl. No	Name of the Garlic Grades	Garlic Grade	Price Range of Grade per Kg
01	Kodaikanal – Malai Poondu	I	170 – 220
02	The Nilgris – Malai Poondu	II	140 – 160
03	Himachal Pradesh	IIIA	090 – 120
04	Himachal Pradesh	IIIB	080 – 100
05	Rajasthan and Gujarat	IVA	060 – 080
06	Rajasthan and Gujarat	IVB	050 – 060
07	Broken Cloves and Malformed	V	030 – 040

Table 4 reveals that five grades of garlic are prevailing in the Vadugapatti garlic wholesale market. Kodaikanal Malaippoondu was ranked as number one grade and it is capable of fetching Rs 170 to 220 per kg. Though there were different grades available within the Malaippoondu, it is decided based on the size of the cloves. But the same was not discussed here as the wholesale market has different grades at national level. The Grade – II goes to the Nilgiris Malaippoondu which is less in size when compared to the Malaippoondu of Kodaikanal and it is capable of fetching Rs 140 to Rs 160 in the wholesale market. The garlic of Himachal Pradesh ranked three and allotted the Grade – III which is capable of fetching the price of Rs 80 to Rs 120 for two different grades viz., IIIA and IIIB, Grade IV Garlic is drawn from Rajasthan and Gujarat which is priced at Rs 50 to 80 for two grades viz., IVA and IVB and the broken cloves and the malformed ones were priced at Rs 30 to Rs 40 and being ranked the last grade. Every grade of Garlic has its own consumers and hence such grading is categorized and practiced.

Price information published by the AGMARKNET is based on Theni and Koyambedu (Chennai) reference markets for garlic in Tamil Nadu.

However, the data consistency and periodicity are high for Theni market. Theni market price is influenced by the price prevailing in Vadugapatti wholesale garlic market.

PRICE SPREAD IN GARLIC MARKETING

Price spread is said to be a composite of various costs incurred and margin of intermediaries in the various processes, such as those of assembling, processing, storage, transport and retailing. How these are varying between the identified marketing channels in respect of garlic marketing is analyzed and the results are presented in Table 5.

Table 5 Price Spread in Garlic Marketing Through Channel - I

Sl. No	Particulars of Cost	Amount (Rs/Kg)	Per Cent
Producer			
01	Producers' Price	100.00	83.33
1.1	Marketing Cost		
1.1.1	Cleaning and Drying Charges	02.60	
1.1.2	Packing and Loading Charges	00.80	
1.1.3	Cost of Labor in Selling	16.00	
1.1.4	Miscellaneous Expenses	00.60	
	Marketing Cost at Producers' Level	20.00	16.67
1.2	Marketing Margin	00.00	
02	Consumer		
2.1	Sale Price to the Consumer	120.00	100.00

Table 5 reveals that the Channel – I is capable of offering the produce directly to the consumers. In this channel, the producer is able to get Rs 100 per kg of Garlic which accounted for 83 per cent to the consumers' rupee. The remaining 17 per cent is incurred towards the marketing cost per kg of garlic. The farmer arranges the produce for direct selling in the town areas and he earns little income for immediate cash need. The remaining lots were released slowly to the Vadugapatti wholesale market established for the purpose. In the marketing channel – II, the garlic is sent to the Vadugapatti wholesale market. The produce is received by the commission agent cum wholesalers, then the wholesalers do the cleaning (Plate – 1), grading and packing in gunnies and finally make the produce available for the open auction on Thursdays and Sundays. The wholesaler cum traders from different places assemble at the market on those days and bid the produce and take the title of goods to their locale for further distribution and sale (Table 6).

Table 6 Price Spread in Garlic Marketing Through Channel - II Producers → Commission Agent (Vadugapatti) → Wholesaler cum Retailers → Retailers → Consumers

Sl. No	Particulars of Cost	Amount (Rs/Kg)	Per Cent
01	Producer		
01	Producers' Price	80.00	47.06
1.1	Marketing Cost		
1.1.1	Packing Charges	00.60	
1.1.2	Loading Charges	00.70	
1.1.3	Transportation Cost	08.30	
1.1.4	Miscellaneous Expenses	00.40	
	Marketing Cost at Producers' Level	10.00	05.88
1.2	Marketing Margin	10.00	05.88
02	Commission Agent		
2.1	Procurement Price by the Commission Agent	100.00	58.82
2.2	Marketing Cost		
2.2.1	Assembling Charges	02.50	
2.2.2	Auctioning Charges	02.00	
2.2.3	Miscellaneous Charges	01.50	
	Marketing Cost at Commission Agent cum Wholesaler Level	06.00	

Sl. No	Particulars of Cost	Amount (Rs/Kg)	Per Cent
2.3	Marketing Margin to the Commission Agent	14.00	08.24
03	Wholesaler cum Retailer		
3.1	Purchase Price Incurred by the Wholesaler cum Retailer	120.00	67.65
3.2	Marketing Cost		
3.2.1	Loading and Unloading Charges	03.00	
3.2.2	Transportation Charges	03.50	
3.2.3	Watch and Ward Charges	01.00	
3.2.4	Miscellaneous Cost	02.50	
	Marketing Cost to the Wholesaler cum Retailer	10.00	
3.3	Marketing Margin to the Wholesaler cum Retailer	20.00	11.76
3.4	Sale Price to the Retailer	150.00	85.29
04	Retailer		
4.1	Procurement Price by the Retailer	150.00	85.29
4.2	Marketing Cost to the Retailer		
4.2.1	Transportation Charges	03.00	
4.2.2	Miscellaneous Charges	02.00	
4.2	Marketing Cost to the Retailer	05.00	
	Marketing Margin to the Retailer	15.00	08.82
5	Sale Price to the Consumer and the Consumer's Price	170.00	100.00

Table 6 reveals that the producer is receiving Rs 80 per kg of garlic in the Channel – II while the consumer's price per kg of garlic is at Rs 170 which accounted for 47 per cent. The wholesaler cum retailers' purchase price is arrived at Rs 120 per kg which accounted for 67.65 per cent to the consumer's rupee and the retailer takes the good at Rs 150 per kg and arranges to sell the garlic at Rs 170 per kg. Even at Vadugapatti market itself, the commission agent cum wholesalers used to sell the produce locally at Rs 170 per kg.

With respect to marketing cost, the commission agent cum wholesaler does assembling and auctioning of the produce and incur Rs 6 per kg of garlic as marketing cost and the margin generated by him was arrived at Rs 14 per kg. The wholesaler cum retailer has incurred Rs 10 as marketing cost and taken the margin of Rs 20 per kg. Among different functionaries, the wholesaler cum retailer takes higher margin in distribution of garlic to the consumers.

Marketing Efficiency in Garlic Marketing Channels

Marketing efficiency is defined as the ratio of output (Consumers' Price) to the input (Marketing Cost). In this study, the consumer price varies from channel to channel, hence the same is worked out using the Calkin's Index and the results are presented in Table 7.

Table 7 Marketing Efficiency in Distribution of Garlic

Marketing Channel	Value of Output (Consumers' Price)	Input (Marketing Cost)	Marketing Efficiency Ratio (O/I)
Shepherd Method			
I	120.00	20.00	06.00
II	170.00	31.00	05.48
Calkin's Index			
Marketing Channel	**Marketing Margin**	**Marketing Cost**	**Marketing Efficiency**
I	10.00	10.00	02.00
II	59.00	31.00	02.90

Table 7 advocated that an increase in the marketing efficiency ratio would represent improved efficiency and vice versa. A reduction in the cost for the same level of satisfaction or an increase in satisfaction at a given cost would result in an improvement of efficiency. When comparing the data delineated in the Table 3.7 one could understand that Channel – II is found to be efficient because it fulfilled the requirements of several functionaries in the marketing channel and keeps them satisfied.

GARLIC TRADING PRACTICES IN INDIA

Trading of Garlic included both import and export from India. In this section, the export and import scenario of spices, export from and import of garlic into India and the state wise importers available in India are analyzed and the details are presented in this section.

Export of Spices from India

While discussing on the status of Garlic export, the overall status of spices need to be assessed as a prelude so that the percentage of garlic export on total spices trade could be understood and the special efforts can be taken to bridge the demand-supply gap if any which will in turn facilitate minimizing the productivity gap. In this respect, the export of spices from India is analyzed and the results are presented in Table 8

Table 8 Quantity of Spices Traded from India during 2012-13

Sl. No	Particulars of Spices	Quantity Traded in Lakh Tonnes	Percentage to Total
01	Garlic and Methi	19.61	27.01
02	Green Chilli	10.16	13.99
03	Dried Chillies	02.18	03.00
04	Red Chillies	01.45	02.00
05	Coriander Seed	07.26	10.00
06	Turmeric	02.90	03.99
07	Ginger	02.18	03.00
08	Betel Leaves	01.45	02.00
09	Cumin Seeds	01.45	02.00
10	Others	23.97	33.01
	Total	72.61	100.00

(Source: NHM Update – 2012-13)

Table 8 reveals that the quantity of spices traded from India is arrived at 72.61 lakh tonnes. Among this, Garlic and methi accounted for 27 per cent to the total export of spices followed by green chilli exported from India to the tune of 10.16 lakh tonnes which accounted for 14 per cent to the total spices export from India. The third important commodity is coriander seed exported to the tune of 7.26 lakh tonnes that accounted for 10 per cent to the total spices export from India. Though there is scope for further promotion of export on spices, the Spices Board should conduct the study on enhancing the export potentials of spices from India through the Research and Development efforts with reputed firms involved in horticultural research especially the spices.

The status of Indian export and import of spices is further discussed in the ensuing section. When we compare the quantity of spices exported during 2002-03 and during the year 2011-12, an increase in the spices export is found to be to the level of 175 per cent in terms of quantity of spices exported and the export value during the same period is registered with 706 per cent increase (Table 9). The efforts taken by the Spices Board for enhancing such quantum jump in the export is commendable.

Table 9 Indian Export and Import of Spices during Different Periods

Sl.No	Year	Export Quantity in Thousand Tonnes	Export Value in Crores	Import Quantity in Thousand Tonnes	Import Value in Crores
01	2002-03	277.02	1635.49	121.28	586.44
02	2010-11	762.71	8043.47	113.38	1556.11
03	2011-12	931.26	13175.52	124.33	2102.17
% Change in 2011-2012 over 2002-03		(+) 175.33	(+) 705.60	(+) 02.51	(+) 258.46

Source: Agricultural Statistics at a Glance – 2012

Table 9 reveals that though the export is increased on one side, the import of spices which are essential for the Indian consumers is also found to be increased between the period 2002-03 and 2011-12 which arrived at 2.50 per cent and the import value is found to be increased to the tune of 258 per cent. The import value of spices between these periods exorbitantly rose from the level of 586 crores mainly due to the inflationary problems. It is true while examining the quantity of spices imported and hence the Indian spices board should concentrate more on the need based spices to be produced in the Indian environ to minimize the import of spices. Similarly, the researcher is interested in assessing the quantum of garlic imported into India over a period of time and hence these details are analyzed and the results are presented in Table 10.

Table 10 Quantity of Garlic Imported into India over a Period of Time

Sl.No	Year	Quantity in Tonnes	Value Rs in Lakhs	Percentage to Total Spices Import
01	2008-09	185.000	048.25	00.22
02	2009-10	125.000	073.50	00.18
03	2010-11	115.000	116.10	00.13
04	2011-12	310.000	220.05	00.28
05	2012-13	095.000	105.30	00.07
Percentage Change during 2012-13 over 2008-09		(-) 48.65	(+) 118.75	

(Spices Board, Cochin 2012)

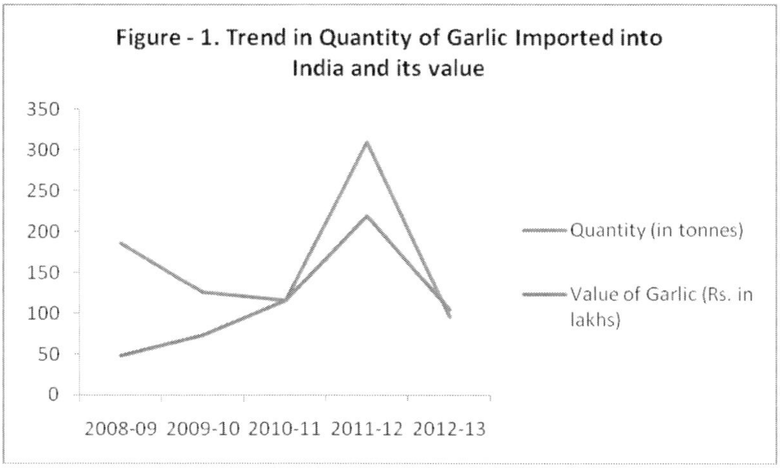

Table 10 reveals that the garlic imported into India is found to be decreased between the period 2008-09 and 2012-13. The quantity of garlic imported into India showed a declining trend to the tune of 49 per cent. It is a good sign that India is progressing towards fulfilling the essential spice like garlic towards self-sufficiency. Thanks to the Spices Board and the Indian farmers who have really dedicated to fulfill the nation's requirement in respect of Garlic. Though the import of Garlic in terms of value is on the increase, it is not burdening the nation due to smaller amount. However, the spices board in future should take active steps to reduce the import of spices and save the foreign exchange and the benefit of import should be made available to the Indian farmers through announcement of better prices for the produce particularly the Garlic so that the farmers will be motivated to produce more.

MAJOR EXPORTING COUNTRIES OF GARLIC IN THE WORLD

It is right time to assess the major countries who are exporting garlic in the globe. This information will reveal the status of which nation is leading in production of garlic so that the surplus quantity be exported to other countries. The details of export quantity country wise are analyzed and the results are presented in Table 11.

Table 11 Major Exporting Countries of Garlic in the World during 2011

Sl. No	Name of the Country	Quantity Exported in Tonnes	Value of Export in Thousand US $	Percentage to Total
01	China	1254382.00	1222300.00	61.75
02	Argentina	89123.00	233406.00	11.79
03	Spain	66983.00	211638.00	10.69
04	Netherlands	17606.00	61706.00	03.12
05	France	12263.00	55374.00	02.80
06	Chile	10041.00	27590.00	01.39
07	Mexico	12221.00	23225.00	01.17
08	Italy	4380.00	20698.00	01.05
09	United States of America	6949.00	17544.00	00.89

Sl. No	Name of the Country	Quantity Exported in Tonnes	Value of Export in Thousand US $	Percentage to Total
10	Germany	3379.00	17309.00	00.87
11	Egypt	5656.00	11916.00	00.60
12	United Kingdom	3154.00	8138.00	00.41
13	India	6916.00	7820.00	00.40
14	Other Countries	49885.00	60715.00	03.07
	World Total	1542938.00	1979379.00	100.00

Source : UN Comtrade

Table 11 reveals that there are several countries exporting garlic. Total garlic exporting countries is found to be more than 100 nations across the globe. However, special effort has been taken to assess the top 10 nations contributing higher quantity of export of garlic to different nations. Among the Garlic exporting nations, China is the leader capable of exporting 12.54 lakh tonnes of garlic accounting for 62 per cent of the total garlic export. Second place goes to Argentina which is capable of exporting 0.89 lakh tonnes accounting for 11.80 per cent of the total export. Third rank goes to Spain in terms of exporting garlic to different nations. Spain has exported 0.67 lakh tonnes accounting for 10.69 per cent to the total export of garlic in the globe.

In terms of exporting Garlic, India is ranked in 13th place capable of exporting to the tune of 0.50 lakh tonnes accounting for 0.40 per cent to the total global export in garlic. Though it is too low, India should take appropriate efforts to enhance the productivity of Garlic and the export share should be increased. India should learn the progressiveness of China in Garlic production and export and the same should be infused in the Indian context to reach the level of number one nation in garlic export. The area expansion, technology adoption, and technology dissemination are the important areas the Indian institutions should concentrate on a large scale.

IMPORTERS OF GARLIC IN INDIA

The exporters and the importers are the pillars in meeting the national needs of any commodity. The exporters are of product specific or food commodity specific, spices specific and other non-edible and engineering product related traders. In our study, the spread of the importers state wise and over the length and breadth of our nation is analyzed and the results are documented in Table 12.

Table 12 State wise Importers of Garlic in India

Sl. No	Name of the State	Number of Importers	Percentage to Total
01	Andhra Pradesh	02	06.06
02	Gujarat	08	24.24
03	Madhya Pradesh	02	06.06
04	Maharashtra	05	15.15
05	Rajasthan	03	09.09
06	Tamil Nadu	07	21.22
07	Uttar Pradesh	01	03.03
08	New Delhi	02	06.06
09	West Bengal	02	06.06
10	Karnataka	01	03.03
	Total	33	100.00

Figure - 2. Price of Garlic in Chennai Market during April - June 2014 (Rs./Quintal)

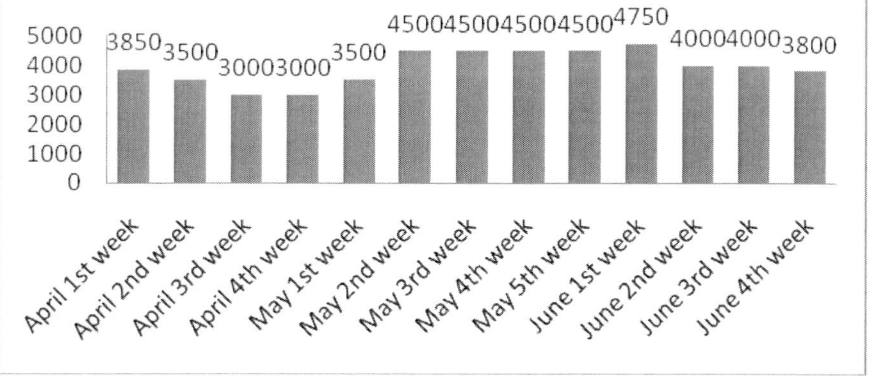

Table 12 reveals that there are 33 importers involved in importing the garlic into India which is spread in over 10 different states. The state of Gujarat had eight importers followed by Tamil Nadu which is blessed with seven importers who are actively involved in importing the garlic into India. During the year 2012-13, only 95 tonnes of garlic have been imported into India due to the policy measures in India. India has restricted the import quantities to a great extent and permits only the most essential one. During the year 2011-12, the garlic import was pegged at 310 tonnes; the Government of India has restricted the same and has imposed conditions to the importers to restrict the garlic import.

SUMMARY AND CONCLUSIONS

With respect to markets, India is having 21 important organized assembling wholesale markets for garlic. Among these organized wholesale markets, large number of private and unorganized markets is functioning in all the major garlic producing states. These markets handle a significant quantum of garlic bulbs and acts as a wholesale market in those regions. One such market functioning over a long period of time in Tamil Nadu is the Vadugapatti of Theni District and the other one is Mettupalayam market wherein the Nilgiris Cooperative Marketing Society, Mettupalayam is one of the institutions that facilitates garlic marketing and supplies to Vadugapatti also. Madhya Pradesh state is enriched with seven wholesale markets for garlic (33 per cent) followed by Uttar Pradesh that had four wholesale markets for garlic. All these markets contribute to the Vadugapatti garlic market.

The garlic marketing has commanded its distribution through six important channels. Among the six channels, the channel – I commands direct marketing in the local town area of Kodaikanal. During the peak tourist arrival, the garlic is distributed directly to the consumers. The producers of Garlic from Kodaikanal send the garlic through the regular service so as to reach the same to the Vadugapatti market. The commission agents at Vadugapatti receive the produce and conduct open auction on Thursdays and Sundays every week. The wholesalers cum retailers used to attend the open auction, take the title of goods to their place and the same was distributed to the retailers and consumers (Channel – II).

The garlic received at Vadugapatti wholesale market is from five important sources. The potential sources are Kodaikanal, The Nilgiris, Madhya Pradesh, Himachal Pradesh, Gujarat, Rajasthan and other states. Five grades of garlic are prevailing in the Vadugapatti Garlic wholesale market. Kodaikanal Malaippoondu is ranked as number one grade and it is capable of fetching Rs 170 to 220 per kg.

When we compare the quantity of spices exported during 2002-03 and during the year 2011-12, an increase in the spices export is found to be to the level of 175 per cent in terms of quantity of spices exported and the export value during the same period is registered with 706 per cent increase. Though the export is increased on one side, the import of spices which are essential for the Indian consumers is also found to be increased between the period 2002-03 and 2011-12 which arrived at 2.50 per cent and the import value was found to be increased to the tune of 258 per cent. The quantity of garlic imported into India showed a declining trend to the tune of 49 per cent. It is a good sign that India is progressing towards fulfilling the essential spice like garlic towards self-sustainability.

RECOMMENDATIONS

- The Vadugapatti wholesale garlic market is suffering from space constraints and the road infrastructure also needs a top priority and hence the Town Panchayat of Vadugapatti can explore the possibilities of creating integrated storage cum market yard for garlic separately with the support of Garlic Merchants Association, Vadugapatti.

- The Farmers' Retail Market in the foothill town may be established for the producer-consumer advantage by establishing Commodity Interest Groups and the CIGs can run the retail outlet on their own for higher percentage share by minimizing the role of middlemen in the name of product specific outlet (Spices Retail Outlet) and the fruits and vegetable store on the lines of Pazhamudhir Nilayams established in different cities which had been accepted by all sections of the consumers and hence the same may be practiced in the other foothill towns.

References

- Government of Tamil Nadu (2008). "***Hort-Stat 2008***" (Directorate of Horticulture and Plantation Crops: Chennai) pp. 1-493
- Government of Tamil Nadu, "Season and Crop Report of Tamil Nadu – 2010" (Directorate of Economics and Statistics: Chennai)
- Government of India, "Agricultural Statistics at a Glance – 2011-12" (Directorate of Economics and Statistics: New Delhi)
- Government of India (2011). ***Horticulture Data Base 2011*** (National Horticulture Board: New Delhi) pp. 01-278
- Government of India (2012). Spice India (Spices Board, Kochi) July 2012
- Sekhar, C., R.C. Panda, R.S.Vinaya Rai and P. Devaraj (1998). "Marketing of Forest Produce from Farm Land – A Case Study in Palani, Tamil Nadu, India", ***Journal of Tropical Forest Products***, 4(1): Pp. 68-73

Chapter 23

TRADING OF VEGETABLES FROM INDIA

INTRODUCTION TO VEGETABLES

Vegetable cultivation has a checkered history in India. Mention of different kinds of vegetables, like bottle gourd (*Suraikkai, alabu*), watermelon (*Dharbusani, kakinga*), onion (*Ulli, palandu*), brinjal (*Katharikkai, vartaku*), Hyacinth bean and radish are available in ancient Sanskrit literature and in the epic like Ramayana. Vegetable cultivation took momentum during the Aryan invasion (1800 BC) and was later developed by the Mughal rulers. It was flourished by the colonist traders from Portugal, France and Britain, who introduced some new vegetables like cabbage, cauliflower, carrot, potato, knol-khol, chilli, tomato, asparagus, etc., in India. At present, India is the second largest producer of vegetables in the world after China.

ETYMOLOGY

The word vegetable was first recorded in English in the early 15th century from Old French which was taken from Medieval Latin *'vegetabilis'* means "growing, flourishing". In 1767, the meaning of the term "vegetable" was specified to mean "plant cultivated for food, edible herb or root".

VEGETABLES IN THE ECONOMY

India is the world's second largest producer of vegetables next only to China. Vegetable crops in India occupy only 2.80 per cent of the total cultivated land producing 87.50 million tonnes of vegetables annually including potato from an area of 5.80 million ha.

Vegetable crops give higher yield per unit area as compared to other crops. Most of the vegetables, if properly grown, can give six to ten times more yield than any cereal crop. There exists lot of scope for increasing yield in most of the vegetables by growing high yielding varieties/hybrids and adopting improved production technologies like precision farming. Cultivation of vegetable crops being more labour intensive than cereal crops will be helpful for seasonally under-employed farm workers in increasing their income. Most vegetable crops are of short duration. With proper planning, three or more crops can be grown in the same plot in one year.

EXPORT

Today, horticultural crops contribute about 25 per cent of the total agricultural exports of the country. Of all the fresh vegetables exported, the share of onion is around 75 per cent. Other traditional vegetables which are presently exported are potato, gherkin, okra, bitter gourd, chilli, cauliflower and some root vegetables.

In addition to these traditional vegetables, non-traditional vegetables which have been identified by the Agricultural and Processed Food Products Export Development Authority (APEDA) to have export potential are asparagus, celery, sweet pepper, green pea, French bean and Cherry tomato. Among the processed vegetables, major demand is for dehydrated onion, garlic and preserved gherkin.

Countries which import traditional vegetables like onion, potato, okra, bitter gourd and chilli are Gulf countries, Singapore, Malaysia, Sri Lanka, Bangladesh and Nepal. Non-traditional vegetables are exported to European countries, Australia, Gulf countries and South-East Asian countries. Cole crops and root vegetables are much in demand in temperate countries during winter. In India, these vegetables are grown in the open during winter and thus the cost of production is less as compared to those grown under protective cover. There is lot of opportunity for exporting these vegetables to the European and North American countries from India.

TRADING PRACTICES OF VEGETABLES PREVALENT IN INDIA

Export

According to Wikipedia (2013), 'Export' can be defined as the act of sending a commodity abroad for trade or sale. Export is nothing but the foreign demand for goods produced by the home country. According to On-Line Dictionary (2013), export is to send or transport a commodity abroad, especially for trade or sale. Export is defined as to cause the spread of an idea in another part of the world.

In the present study, export is defined as 'the goods and services that are produced in India are transmitted to the other countries by fulfilling their prescriptions on that commodity'.

Trading of vegetables included both import and export from India. In this section, the export of vegetables in respect of select vegetables is analyzed and the details are presented in this section.

EXPORT OF HORTICULTURAL PRODUCE FROM INDIA

On the back of record production, India being the second largest producer of fruits and vegetables in the world, horticulture exports has helped the country to earn Rs 14,000 crores in 2011-12, according to the Ministry of Agriculture and Food Processing Industries.

India's horticulture production has crossed an all time high of over 240 million tonnes and Sharad Pawar, Agriculture Minister, had announced 2012 as the "Year of Horticulture" to bring horticulture development in the country to the centre stage.

According to Pawar, India has achieved considerable success on the production of horticultural output, but the supply chain issues still need to be addressed in a big way. The Vegetable Initiative for Urban Clusters is an important step in this direction.

The scheme is being implemented with focus on sustained supply of good quality vegetables to urban centres having a population of one million and above and also lays special emphasis on promoting protected cultivation of vegetables in green houses and shade-net houses for enhancing its productivity.

Formation of Farmer Producer Organizations (FPO) and linkage with financial institutions through aggregators is an innovative feature of the scheme. Over 62,000 farmers have already been mobilized into over 3000 Farmer Interest Groups (FIG) and 50 Farmer Producer Organizations (FPOs) are also under this scheme.

Meanwhile, floriculture is another field helping India in generating quick returns per unit of area for the farmer. With the production level at nearly 7000 million cut flowers, there has been significant improvement in their production. These enjoy good demand in the domestic and international markets.

India is producing almost all types of vegetables and other horticultural produce. They are from leafy vegetables to gourds, flowers and fruits. Many countries are exporting vegetables across the globe. It is important to identify the quantity of vegetables and other horticultural output exported from India to other countries in terms of value. In this context, the status of horticultural output exported from India and the same are presented in Table 1.

Table 1 **Export of Horticulture Produce in India *(Quantity in Tonnes; Value – Rs in Lakhs)***

Sl.No	Name of the Horticulture Produce	2008-09		2010-11	
		Quantity	Value	Quantity	Value
01	Floriculture				
1.1	Floriculture	030798.34	36881.41	27776.14	28645.41
1.2	Fruit And Vegetable Seeds	008535.53 (00.25)	11999.09 (01.77)	11182.50 (00.41)	17519.52 (02.52)
01	Total	**039333.87**	**048880.5**	**38958.64**	**46164.93**
02	Fresh Fruits and Vegetables				

Sl.No	**Name of the Horticulture Produce**	**2008-09**		**2010-11**	
		Quantity	**Value**	**Quantity**	**Value**
2.1	Fresh Onion	1670186.29	182752.21	1163472.58	174155.41
2.2	Other Fresh Vegetables	505285.46 (14.89)	068020.32 (10.05)	490914.05 (17.96)	89293.61 (12.82)
2.3	Walnuts	005696.34	014123.63	005244.58	15650.59
2.4	Fresh Mangoes	083703.18	017071.25	059220.77	16292.13
2.5	Fresh Grapes	124627.97	040861.28	099311.83	41206.32
2.6	Other Fresh Fruits	256768.53	043086.84	253850.99	48964.74
02	Total	2646267.77	365915.53	2072014.8	385562.8
03	Processed Fruits and Vegetables				
3.1	Dried & Preserved Vegetables	147861.22 (04.36)	049641.51 (07.33)	110173.91 (04.33)	051697.09 (07.42)
3.2	Mango Pulp	0173013.6	075298.9	171929.43	081400.66
3.3	Other Processed Fruits and Vegetables	387126.42 (11.41)	137179.00 (20.27)	340067.97 (12.44)	131635.53 (18.90)
03	Total	708001.24	262119.41	622171.31	264733.28
	Grand Total (1+2+3)	3393602.88	676915.44	2733144.75	696461.01
Percentage Change in 2010-11				(-) 19.46	(02.89)

(Figures in Parentheses Indicate Percentage of Vegetables to the Total Produce Export)

Table 1 reveals the details of export of horticultural produce from India. The fruits and vegetable seeds, fresh vegetables, dried and preserved vegetables and other processed fruits and vegetables are among the various produce exported,. Among these items, the percentage quantity exported between 2008-09 and 2010-11 is found to be on the increase in respect of only the fruits and vegetable seeds. But, the export of fresh vegetables, dried and preserved vegetables and other processed fruits and vegetables are showing a decrease in the export in terms of quantity between the year 2008-09 and 2010-11. The overall export of horticulture produce between these period in terms of quantity is found to show a decrease which accounted for around 19 per cent to the total quantity of horticultural produce export. The export value between these periods is found to show only a marginal increase which accounted for only around three per cent to the total value of export. In this circumstance, the export of vegetable seed and other fresh vegetable export should be enhanced by increasing the productivity of vegetables by practicing the precision farming techniques in a larger scale so that the quantity produced will be enhanced and the export possibilities could be increased.

EXPORT OF TOMATOES FROM INDIA

The general trend in the export of horticultural produce showed a decline in the quantity exported between 2008-09 and 2010-11. In respect of tomato export, the details of data are analyzed and the results are presented in Table 2.

Table 2 Export of Tomatoes from India ***(Product Code Number: 07020000; Quantity in Tonnes; Value – Rs in Lakhs)***

Sl.No	Name of the Country	2008-09		2010-11	
		Quantity	Value	Quantity	Value
1	United Arab Emirates	011617.30	01425.20	33902.90	7468.60
2	Bangladesh	009166.40	00848.60	19711.90	2048.20
3	Saudi Arabia	000250.00	00035.90	2261.50	539.40
4	Oman	000205.80	00036.80	2203.60	532.00
5	Pakistan	100764.70	10172.00	3708.90	340.60
6	Nepal	001434.30	00111.40	3266.60	227.30
7	Maldives	000483.30	00053.10	2447.70	169.80
8	Bahrain	000112.50	00015.00	0442.30	102.20
9	Kuwait	000155.80	00023.10	0125.90	029.90

Sl.No	Name of the Country	2008-09		2010-11	
		Quantity	Value	Quantity	Value
10	Qatar	000045.70	00006.60	0019.60	008.00
11	Others	000381.50	00038.50	0092.80	014.60
	Total	124617.20	12766.20	68183.70	11480.60
Percentage Change in the Export during 2010-11				(-) 45.29	(-) 12.68

Table 2 reveals that the export of Tomato to different countries between the period 2008-09 and 2010-11 showed a dip which accounted for 45 per cent to the total quantity exported between the periods. The value of Tomato export also showed a decrease of around 13 per cent to the total quantity of Tomato exported to different countries during the year 2010-11. Pakistan was found to be the biggest importer of Tomato from India during 2008-09 which was reduced drastically during the year 2010-11 accounting for 96.32 per cent to the total Tomato export. Similarly, export of Tomato to the Kuwait and Qatar is also reduced which accounted for 19.23 per cent and 56.52 per cent respectively to the total export of Tomato from India during the year 2010-11.

The countries like Bangladesh, Saudi Arabia, Oman, Nepal, Maldives, Bahrain and others have showed positive signals in importing the Indian Tomato during the year 2010-11. It is evident from the Table 2 that the quantity exported between 2008-09 and 2010-11 showed an increase in respect of all these nations other than Pakistan, Kuwait and Qatar.

EXPORT OF ONION FROM INDIA

Onion is an important crop which played significant role in changing the government too due to the increase in its price level. It is mainly due to the export policies in vogue with the Government. In this respect, the quantity of onion exported to different countries from India needs an analysis. Hence the export data in respect of onion was analyzed and the results are presented in Table 3.

Table 3 **Export of Onion from India *(Quantity in Tonnes; Value – Rs in Lakhs)***

Sl. No	Name of the Country	2008-09		2010-11		Percentage Change in Quantity during 2010-11
		Quantity	Value	Quantity	Value	
1	Bangladesh	685447.90	73571.50	391550.20	57233.40	(-) 42.88
2	Malaysia	283892.80	31928.00	273232.70	46343.50	(-) 03.75
3	SRI Lanka	152197.60	15884.20	114450.40	17599.20	(-) 24.80
4	United Arab Emirates	189678.40	21535.40	126228.80	16627.70	(-) 33.45
5	Indonesia	012726.00	01350.20	43061.30	9000.00	238.37
6	Pakistan	128121.00	15761.30	45689.10	5417.60	(-) 64.34
7	Singapore	024302.50	02642.70	20876.30	2920.60	(-) 14.10
8	Nepal	040668.60	02638.00	29977.60	2530.40	(-) 26.29
9	Oman	011977.40	01179.20	17988.60	2266.20	50.20
10	Vietnam Social Republic	000024.50	00002.50	15281.50	2077.50	61024.00
11	Others	141149.30	16259.30	85135.90	12139.40	(-) 39.68
	Total	**1670186.30**	**182752.20**	**1163472.60**	**174155.40**	(-) 30.34

Table 3 reveals that the Onion is exported to different countries from India. The importing countries of onion from India are Bangladesh, Malaysia, Sri Lanka, United Arab Emirates, Indonesia, Pakistan, Singapore, Nepal, Oman and the Vietnam Social Republic. Among all these countries, the quantity of onion exported to Indonesia, Oman and to the Vietnam Social Republic is found to be on the increasing side and the percentage increase in the export is arrived at 238 per cent, 50 per cent and 61024 per cent respectively during the year 2010-11 over the year 2008-09. Export of Onion to all other countries registered a dip which is ranging from 4 per cent to 43 per cent. Pakistan and Bangladesh are found to import poor quantities of onion during 2010-11 which accounted for a decrease of 64 per cent and 43 per cent respectively. It might be due to the poor output in India or the policy guidelines of importing countries. However, the reasons need to be identified and the trade promotion efforts need to be taken to the neighbouring countries to promote onion export.

EXPORT OF FRUITS AND VEGETABLE SEEDS FROM INDIA

India is exporting not only the vegetables but also the vegetable seeds for further development of area under vegetables. In this respect, the quantity of fruits and vegetables seeds exported to different countries is analyzed and the results are presented in Table 4.

Table 4 Export of Fruits and Vegetable Seeds from India *(Quantity in Tonnes; Value – Rs in Lakhs)*

Sl. No	Name of the Country	2008-09		2010-11		Percentage Change in Quantity During 2010-11
		Quantity	Value	Quantity	Value	
1	Pakistan	3,767.90	2,660.40	5,813.60	6,540.20	54.30
2	Bangladesh	1,238.80	1,109.70	2,031.80	1,453.60	64.00
3	United States	091.10	1,350.20	0129.90	1,291.60	42.86
4	Netherland	081.10	1,098.20	0082.40	1,219.40	01.36
5	Nepal	121.70	0148.30	0361.20	0841.70	195.90
6	Malaysia	081.30	0430.80	0094.60	0736.80	07.41
7	Japan	101.90	0847.10	0191.80	0718.80	88.24
8	Italy	160.60	0296.80	0232.80	0383.30	44.72
9	Kenya	087.00	0189.80	0074.80	0355.00	(-) 13.79
10	Singapore	047.30	0148.00	0064.90	0342.50	37.21
11	Others	2,756.90	3,720.00	2,104.80	3,636.80	(-) 23.65
	Total	8,535.60	11,999.10	11,182.50	17,519.60	31.01

Table 4 reveals that the fruits and vegetable seeds were exported to Pakistan, Bangladesh, United States of America, Netherland, Nepal, Malaysia, Japan, Italy and Singapore. The export of fruits and vegetable seeds in terms of quantity has shown an increase between the period 2008-09 and 2010-11. In terms of absolute quantity, Pakistan is found to be the potential country in importing the Indian vegetable seeds. The percentage quantity of vegetable seeds exported between these periods accounted for 54 per cent followed by Bangladesh which had imported the fruits and vegetable seeds considerably next to Pakistan. The percentage of vegetable seeds exported to Bangladesh accounted for 64 per cent during the year 2010-11 over the year 2008-09. United States of America, the leader of the world in terms of technologies had also imported the fruits and vegetable seeds from India considerably which accounted for 43 per cent increase during the year 2010-11.

Among the countries, Nepal is found to import higher quantity of fruits and vegetable seeds from India during the year 2010-11 which accounted for 196 per cent increase over the year 2008-09. Kenya is the only country which had showed a negative trend in importing the fruits and vegetable seeds from India. It might be due to the attainment of self sufficiency in respect of vegetable seeds. However, the APEDA should examine the reasons for the trade dip with Kenya and appropriate strategies have to be taken to bridge the gap.

EXPORT OF DRIED AND PRESERVED VEGETABLES FROM INDIA

Besides export of fresh vegetables to different countries, the dried and preserved vegetables are also exported to different countries from India. The details of those countries importing the preserved vegetables are analyzed and the results are presented in Table 5.

Table 5 Export of Dried and Preserved Vegetables from India *(Quantity in Tonnes; Value – Rs in Lakhs)*

Sl. No	Name of the Country	2008-09		2010-11		Percentage Change during 2010-11
		Quantity	Value	Quantity	Value	
1	Russia	13,445.90	3,574.30	13,477.00	5,962.20	00.23
2	Germany	3,093.80	2,419.20	05,900.00	5,588.80	90.69
3	Malaysia	00237.20	0081.70	07,465.10	4,063.40	3049.79
4	United States	27,128.50	8,888.90	11,164.40	4,037.40	(-) 58.85
5	France	24,241.70	8,341.20	10,693.00	3,869.70	(-) 55.89
6	Spain	14,581.40	3,967.80	10,175.80	2,884.10	(-) 30.21
7	United Kingdom	04,329.80	3,250.80	03,454.50	2,337.10	(-) 20.23
8	Belgium	11,652.10	3,684.60	05,088.40	01,987.10	(-) 56.33
9	Poland	00705.10	00383.10	02,110.20	01,613.50	199.29
10	Brazil	00820.60	00577.10	01,834.10	01,443.50	123.39
11	Others	47,624.10	14,472.80	38,811.50	17,910.30	(-) 18.51
	Total	147,860.20	49,641.50	110,174.00	51,697.10	(-) 25.49

Table 5 reveals that the dried and preserved vegetables are exported to Russia, Germany, Malaysia, United States of America, France, Spain, United Kingdom, Belgium, Poland, Brazil and other countries. Among these countries, export of dried and preserved vegetables showed a positive sign and the percentage change in the export during 2010-11 was arrived at 199 per cent in respect of Poland, 3050 per cent in the case of Malaysia, 90 per cent in respect of Germany and 123 per cent in respect of Brazil. The export of dried and preserved vegetables to all other countries including United States of America and France, during the year 2010-11 showed a decrease which accounted for 59 and 56 per cent respectively.

DETAILS OF ORGANIZATIONS INVOLVED IN PROMOTION OF TRADE

The details of organizations involved in trade promotion activities and its headquarters are delineated in Table 6. There were 12 organizations involved either directly or indirectly in promotion of trade and relevant to the export and import of farm produce.

Table 6 Details of Institutions Facilitating the Trade in India

Sl.No	Name of the Institutions	Year of Establishment	Headquarters
01	Agricultural and Processed Food Products Export Development Authority	1986	New Delhi
02	Export Inspection Council	1964	New Delhi
03	Indian Institute of Packaging	1966	Mumbai
04	Export Credit Guarantee Corporation	1957	Mumbai
05	India Trade Promotion Organization	1992	New Delhi
06	Export Promotion Councils		
07	Federation of Indian Export Organizations	1965	New Delhi
08	National Center for Trade Information	1995	New Delhi
09	Price Stabilization Fund Trust	2003	New Delhi
10	Indian Council of Arbitration	1965	New Delhi
11	GS1 – India	--	New Delhi

Role of APEDA, New Delhi

The Agricultural and Processed Food Products Export Development Authority (APEDA) was established in 1986 as a Statutory Body under an Act of Parliament. The Authority, with its headquarters at New Delhi, is headed by a Chairman. The Authority has five Regional Offices at Guwahati, Hyderabad, Kolkata, Bangalore & Mumbai and is entrusted with the task of promoting agricultural exports, including the export of processed foods in value added form. APEDA has also been entrusted with monitoring of export of 14 agricultural and processed food product groups listed in the Schedule to the APEDA Act. APEDA has been actively engaged in the development of markets besides upgradation of infrastructure and quality to promote the export of agro products. In its endeavor to promote agro products, APEDA provides financial assistance to the registered exporters under its schemes for market development, infrastructure development, quality development, research and development and transport assistance. During the year 2009-10, APEDA has been empowered by an Act of Parliament to enforce the Intellectual Property Right (IPR) of Geographical Indications (GI) for Basmati Rice.

Role of Export Inspection Council, New Delhi

The Export Inspection Council was set up as a Statutory Body on 1st January, 1964 under Section 3 of the Export (Quality Control and Inspection) Act, 1963 to ensure sound development of export trade of India through quality control and inspection and for matters connected therewith. The Council is an advisory body to the Central Government, with its office located at New Delhi and is headed by a Chairman. The Executive Head of the EIC is the Director of Inspection & Quality Control who is responsible for the enforcement of quality control and compulsory pre-shipment inspection of various commodities meant for export and notified by the Government under the Export (Quality Control and Inspection) Act, 1963. The Council is assisted in its functions by the Export Inspection Agencies (EIAs), which are field organizations located at Chennai, Delhi, Kochi, Kolkata and Mumbai and have state-of-art and accredited laboratories with the required logistic support for quality certification activities. These agencies have a network of thirty six sub-offices and laboratories located at different ports or major industrial centres to back up the pre-shipment inspection and certification activities.

Role of Indian Institute of Packaging, Mumbai

The Indian Institute of Packaging was registered in May, 1966 under the Societies Registration Act, 1860. The Institute, with its office located at Mumbai and branch offices at Delhi, Chennai, Kolkata and Hyderabad, is headed by a Director. The main function of the Institute is to undertake research on raw materials for the packaging industry, organize training programmes on packaging technology, consultancy services on packaging problems and stimulate consciousness of the need for good packaging.

Role of Export Credit Guarantee Corporation of India Limited

The Corporation was established in 1957 as the Export Risk Insurance Corporation of India Ltd. Keeping in view the wider role played by the Corporation, the name was changed to Export Credit Guarantee Corporation of India Ltd. (ECGC). The ECGC is the premier organization in the country, which offers credit risk insurance cover to exporters, banks, etc. The primary objective of the Corporation is to promote the country's exports by covering the risk of export on credit. It provides (a) a range of insurance covers to Indian exporters against the risk of non-realization of export proceeds due to commercial or political causes and (b) different types of guarantees to banks and other financial institutions to enable them to extend credit facilities to exporters on liberal basis.

Role of Indian Trade Promotion Organization

India Trade Promotion Organization has been formed by merging erstwhile Trade Development Authority (TDA) with Trade Fair Authority of India (TFAI) with effect from 1st January, 1992. India Trade Promotion Organization is the premier trade promotion agency of India and provides a broad spectrum of services to trade and industry so as to promote India's exports. These services include organization of trade fairs and exhibitions in India and abroad, buyer-seller meets, contact promotion programmes apart from information dissemination on products and markets.

Role of Export Promotion Council

Presently, there are fourteen Export Promotion Councils under the administrative control of the Department of Commerce. Names and addresses of these Councils are given in Annexure 1.2. These Councils are registered as non-profit organizations under the Companies Act/ Societies Registration Act. The Councils perform both advisory and executive functions. The role and functions of these Councils are guided by the Foreign Trade Policy, 2009-14. These Councils are also the registering authorities for exporters under the Foreign Trade Policy 2009-2014.

Role of Federation of Indian Export Organization

The Federation of Indian Export Organizations set up in 1965, is an apex body registered under the Societies Registrations Act XXI of 1860, of various export promotion organizations and institutions with its major regional offices at Delhi, Mumbai, Chennai and Kolkata. The main objective of FIEO is to render an integrated package of services to various organizations connected with export promotion. It provides the content, direction and thrust to India's global export effort. It also functions as a primary servicing agency to provide integrated assistance to its members comprising professional exporting firms holding recognition status granted by the Government, consultancy firms and service providers. The Federation organizes seminars and arranges participation in various exhibitions in India and abroad. It also brings out 'FIEO News', for creating awareness amongst its member exporters and importers.

Role of Indian Council of Arbitration, New Delhi

The Indian Council of Arbitration, India's premier Arbitral Institution, is a Society registered under the Societies Registration Act, 1860 operating on no profit basis, with its head office in New Delhi and eight branches with a pan-India network. The organization originally established in 1965 promotes and administers the use of Alternative Dispute Resolution mechanisms in commercial disputes, thereby expediting dispute resolution and encouraging greater domestic and international commerce. The main objectives of the Council are to promote the knowledge and use of arbitration and provide arbitration facilities for amicable and quick settlement of commercial disputes with a view to maintaining the smooth flow of trade, particularly export trade on a sustained and enduring basis.

Role of National Centre for Trade Information

National Centre for Trade Information was set up in 1995 with a view to create an institutional mechanism for collection and dissemination of trade data and improving information services to the business community, especially small and medium enterprises. NCTI is a Government of India recognized Trade Point in India under the Trade Efficiency Program of United Nations Conference on Trade & Development (UNCTAD). NCTI is the Operational Trade Point in India and is also the recognized Focal Point of Trade Analysis and Information System (TRAINS) of UNCTAD Trade Point Development Centre (UNTPDC). NCTI is promoted by India Trade Promotion Organization (ITPO) and National Informatics Centre (NIC).

Role of Price Stabilization Fund Trust

The Price Stabilization Fund (PSF) Scheme was launched by Government of India in April 2003 against the backdrop of decline in international and domestic prices of tea, coffee, rubber, and tobacco causing distress to primary growers. The growers of these commodities were particularly affected due to substantial reduction in unit value realization for these crops, at times falling below their cost of production. The objective of the Scheme is to safeguard the interests of the growers of these commodities and provide financial relief when prices fall below a specified level without resorting to the practice of procurement operations by the Government agencies. The Scheme is being operationalized through the Price Stabilization Fund Trust. As on 31 October 2009, the PSF Corpus Fund consists of Rs.435.19 crore, out of which Rs.432.88 crore has been contributed by Government of India and Rs.2.31 crore by growers by way of entry fee.

A Personal Accident Insurance Scheme is also under implementation by PSFT through National Insurance Corporation Ltd., which provides insurance cover to the growers in the sectors of tea, coffee, rubber, tobacco and spices (chillies, cardamom, ginger, turmeric and pepper) having plantations up to 4 hectares. The Scheme also covers all plantation workers working on these plantations regardless of the size of holdings. The insurance cover is up to Rs. 1.00 lakh per person. The premium of Rs.17/- is shared between the beneficiary and the PSF Trust in the ratio 50:50. The target coverage is 57.17 lakh growers and workers.

Role of GS 1-India

GS1 India is a not-for-profit standards body promoted by the Ministry of Commerce (GOI) and Indian industry to spread awareness and provide guidance on adoption of global standards in Supply Chain Management by Indian industry for the benefit of consumers, industry, government, etc.

GS1 India is the only organization in India authorized to issue company prefix numbers for use in barcodes, RFID tags, etc., for unique, unambiguous and universal identification of products, cartons, containers, etc. GS1 standards find wide application in supply chains across sectors for unique-yet-universal product, consignment and entity identification, EDI (Electronic Data Interchange), product data synchronization, etc. GS1 standards are the de-facto global standards in identification of consumer products in retail. GS1 India is an affiliate of GS1 Global Office, twin headquartered at Brussels (Belgium) and Lawrenceville, New Jersey (U.S.A.), which oversees operations of a network of over 100 GS1 organizations across the world.

PAST STUDIES ON MARKETING OF HORTICULTURAL PRODUCE

Aileni and Rao (1991) concluded that the cultivators account for 50.77 to 67.66 per cent of marketable surplus, while village traders account for 18.63 to 30.00 per cent of marketable surplus. The major place of assembling is done in regulated markets. They found that the arrival trends in maize declined due to inadequate market prices and market imperfections.

Renuka *et al.,* (1991) revealed that nearly 62 percent of respondents sold their output immediately after harvest, due to lack of storage facilities and to meet the expenses for present consumption. He found that the major problems in marketing of agricultural produce were lack of infrastructure facilities in market yard, poor grading and standardization of products, and unhelpful altitudes of commission agents. Due to illiteracy among the farmers, they were unable to maintain and follow crop accounts and marketing regulations.

Prabhakar Rao *et al.* (1993) studied the marketing costs and margins of jasmine in Guntur district of Andhra Pradesh and they identified the following marketing channel: Producer →Wholesaler → Retailer → Consumer

They found that the producers' share in the consumer rupee was Rs. 700 (50.44%) per quintal. The total cost incurred by the wholesaler was Rs. 45.10 (3.5%) per quintal and his margin was Rs. 320.45 (23.09%) per quintal. Retailers incurred a cost of Rs. 29.45 (2.12%) per quintal and his margin was Rs. 292.92 (21.10%) per quintal.

Jairath (1994) studied production and disposal pattern of sweet potato in India and found that majority of the producers sold their produce to city traders working in the regulated market followed by village merchants and pre-harvest contractors. The following three types of marketing channels were identified.

Farmers → Traders/Commission Agents → Sub Dealer → Retailers → Consumers

Farmers → Village Merchants →Commission Agents →Sub-Dealer → Retailer → Consumers and Farmers → Pre-Harvest Contractors → Commission Agent → Sub-Dealer → Retailer → Consumers.

Vedini (1995) studied the problems in production and marketing of jasmine in Mysore district and stated that inadequacy of financial resources, lack of technical know-how, non-availability of labour and incidence of pests and disease were some of the problems faced by the farmers in production, while high commission charges, inadequate transport facility and high cost of transportation were the major problems faced by them in marketing.

Shyamasundar *et al.* (1995) in their study identified the four important marketing channels followed in disposal of onion:

Producer → Village Level Trader → Wholesaler → Retailer → Consumer

Producer → Wholesaler → Retailer → Consumer

Producer → Commission Agent → Trader → Trader cum Retailer → consumer and

Producer → Commission Agent → Cart Vendor →Consumer.

The producers got the highest net price per quintal in channel-2 (Rs. 163.57) and lowest in channel-1.

Vedani and Gracy (1995) analysed the marketing cost for Jasmine in the Mysore city for the year 1992-93 which was observed to be more than 15 per cent of the value of the jasmine sold by the farmers under this marketing channel. Further they studied the performance of marketing channel which was identified as Producer → Trader → cum → Commission Agent → Retailers → Consumers

They found that producer's net share was as low as 45 per cent. The trader-cum-commission agent and retailer cost and margin were 6.02 and 45.78 per cent respectively. All the farmers opined that non-availability of adequate financial help for cultivating the crop was a major production problem. About 85 per cent of the farmers felt that commission charges were high.

Sarawgi and Beohar (1995) made an attempt to study the marketing of wheat. He found that quantity marketed to the regulated market increased as size of land holdings and total production increases. The disposal of wheat was 100 per cent during March to June (i.e.) peak season period. The problems faced by the producers regarding the marketing of wheat were lack of transport facilities, mandi facilities, storage, late payments, etc., and these increased the marketing cost. He concluded that if transportation and storage facilities are established at proper place, farmers could retain their produce for longer period to get higher price and to some extent can reduce the marketing cost.

Lal *et al.*, (1996) observed that the most important marketing problem is low marketed surplus of produce; the main reason for this is low production per farm due to small size of farms and low productivity. The marketed surplus of maize, paddy, and wheat were 36.87, 14.19 and 24.23 per cent respectively. The two main channels for marketing of maize were channel I → producers → primary wholesalers → secondary wholesalers → consumers and channel II → producers → Primary wholesaler → consumer. The quantities of marketed surplus through this channel are 57 and 42 per cent respectively. The small farmers encountered more problems in marketing of food grain as compared to large farmers due to low marketed surplus. The majority of small farmers were found to be more concerned regarding low price, lack of market information, while large farmers were more concerned about marketing credit and support prices of food grains.

Rajeevan *et al.* (1997) analyzed the performance of bush jasmine cultivation which offered promising opportunities under small holdings situation and where land considered as the most limiting production factor. The study inferred that the small and marginal farmers could undertake its cultivation and marketing as a supplementary enterprise, in order to utilizing under employed/unemployed family labour, especially women labour. It states that bush jasmine cultivation was more feasible because of the initial capital investment required at the initial stage is low.

Thakur *et al.*, (1997) identified the most important problems in marketing of food grains was high cost of transportation, lack of storage and forced sale, low prices in the market, lack of processing plants, malpractices in the market, etc. These problems are faced by around 53 per cent of the small growers, while large farmers reported less of these problems.

STRENGTHS AND WEAKNESSES OF LARGE FARMS

According to FAO (2005), large farms are suitable for mechanized, large scale production of major horticultural crops like vegetables. They can establish vegetable orchards that require large capital investment. They are best equipped to produce and sell produce in large volumes to major buyers. They have access to capital, information and technology.

The weaknesses of large farms are often they have overhead costs. They can be poor at organizing and controlling large number of workers. They cannot easily service small and niche markets. In some countries they may be criticized for using land that could be used by poor farmers.

More profitable production often involves greater risks. Farmers should aim at achieving balance between profit and risk. Prices increase along the marketing chain to cover marketing costs and trader's profit. New products can be more profitable. Reducing marketing costs can have a major impact on profitability, as can control produce losses.

PRICE SPREAD AND MARKETING EFFICIENCY

Arshad (1983) evaluated the efficiency of coconut marketing system by small holders in Malaysia and observed that the efficiency suffered from various inefficiencies in the form of imperfection that existed in market structure, practices and performances. Farm level constraints and lack of marketing facilities had resulted in low quality produce which in turn induced the middlemen to indulge in unethical trading practices.

Srivastava (1984) in his study on the price spread showed that the retailers share increased with an increase in consumer's price, whereas the producer's share decreased with an increase in consumer's price. He concluded that the benefit derived from an increase in the consumer price did not reach the producers. Instead it was absorbed by the retailers.

Narayanan (1991) identified the two marketing channels for the cut-flowers produced in Bangalore. The price spread in marketing of cut-flowers through one channel revealed that producer received a maximum of 59.26 per cent in tuberose and a minimum of 40.0 per cent in rose as their share out of consumer's rupee. In second channel, the share of producers ranged from 16.76 per cent roses to 31.28 per cent in tuberose.

Senthil Kumar (1991) studied the economics of different varieties of banana in Tiruchirapalli district. The cost of production of Poovan banana per bunch was 13.53 and net income per bunch was 14.47. The price spread in marketing channel- I (Farmer – Commission agent – Primary wholesaler – Secondary wholesaler – Retailer and Consumers) was 23.87 per cent and channel II (Farmer – Pre harvest contractor – commission agent – primary wholesaler – secondary wholesaler – retailers --consumers) was 20.72 per cent. The major problem in production was incidence of pests and diseases and wind destruction.

Sahu (1995) in his study on coconut marketing in Odisha identified two marketing channels. For working out the marketing efficiency of the two channels, the Shepherds formula was used. In the regulated market, the share of producer in consumer's price was 62.22 per cent. The marketing cost for 1000 coconuts from producers to consumers worked out to 1700/- which was 37.78 per cent of price paid by the consumer. The maximum share of price spread went to retailer as profit (15.89 per cent) followed by traders (6.67 per cent) and wholesalers (2.23 per cent). In the unregulated markets, the producers share in consumer's rupee was 68.57 per cent and the price spread was 31.43 per cent of the price paid by the consumer. The maximum share of price spread went to the traders at 24.61 per cent of the consumer's price. A comparison of price spread revealed that in regulated markets the producer enjoyed 62.22 per cent of consumer's rupee as against 68.57 per cent in the unregulated markets. The marketing efficiency as per the Shepherds formula was calculated to be 1.65 and 2.18

for the regulated and unregulated markets respectively. This was due to the higher number of intermediaries, higher marketing costs and margins in the unregulated market. However, on the whole, marketing efficiency appeared to be very low in both markets due to higher market margin.

Upendar and Venkateshwarlu (1995) made an attempt to study price spread and marketing margins of paddy in Andhra Pradesh. They opined that in order to increase the marketing efficiency, to rationalize the marketing margins and reduce the costs, the market fees, commission, handling charges and all kinds of unauthorized deductions in the market must be avoided besides reducing the number of intermediaries. This would enable the producer to get a reasonable price for his produce that induces him to produce more on one hand and the consumer pays a lower price on the other. He concluded that marketing costs and margin should be reduced and rationalized to the possible extent, so that the net price received by the producer will reflect his increase in productivity.

Sarawgi and Beohar (1995) in their study on marketing of wheat observed that the average percentage marketed quantity of wheat to total production was found to be 68 per cent. The selling made to the regulated market was highest in respect of large size farm (65%) followed by medium (41%) and small size farm (23%). Regarding the price spread of wheat in the different marketing agencies, producers received highest price by selling their produce in regulated market followed by co-operative society and city market.

Pawar (1996) analysed the marketing cost and margins and observed that these marketing cost and margins varied depending upon the marketing channels. The results showed that one of the marketing channels was more efficient in groundnut marketing, because the producers share in consumer's rupee was 98.33 per cent. The marketing cost was estimated to be 176.05 which was equivalent to 21.67 per cent of consumer rupee. The share of retailer was comparatively more (3.47 per cent) as against wholesaler (91.98 per cent). The study concluded that distress sale of groundnut was found in small farmers due to immediate need of cash and received less price as compared to medium and large farms.

Nagarajan (2006) in his article on the fragrant Madurai Gundumalli endorsed on the amount of labour invested in the cultivation of this flower crop. The farmers in the study area had to leave the farm with produce early in the morning in order to realize premium price. Considering the problems and dangers associated with the jasmine flower harvesting during early morning time, he then suggested to evolve suitable mechanism to minimize the hardship experienced by the women labourers in particular.

Vidhya (2006) studied potential of cut-flower production in India. The result revealed that India had a long floriculture history and flower growing is an age old enterprise however, what lacked was its commercialization. The study suggested to make concerted efforts on the part of the government as well as the private entrepreneurs to develop floriculture on scientific lines. To achieve this greater emphasis need to be laid on the development of suitable practices to ensure quality input supply and better remunerative management to create entrepreneur friendly environment.

Chapter 24

Role of Farmers' Organization In Marketing of Jasmine

INTRODUCTION TO FLOWERS

Flowers have their own importance since ancient times and are being used for decoration, worshiping as well as in satisfying the aesthetic feelings. A language of flowers were nurtured and developed in England during 18th and 19th centuries. Flower speaks a language, which could be universally understood by the people across the world and are provided as the best ambassadors of good will. In recent years, floriculture is emerging as a market demand driven activity throughout the world and considered as a potential money spinner. Under WTO regime, the floricultural activities are being recognized as a lucrative profession with much higher margin of returns per unit area when compared to the field and other horticultural crops.

Among horticultural crops, cut flowers and ornamentals perhaps do fetch higher value. The words of John Keats, "A thing of beauty is a joy forever", connotes fresh flowers have been one of the best medium of expression of greetings, regards, love and care and respect in enhancing aesthetics of daily décor in human habitations. It is being rightly said "flowers speak a millions of unspoken words..." They fit all occasions, be it at birth, marriage or death ceremonies. Hence, floriculture is fast emerging as a major commercial venture in the world scene. (Vanishree, 2007)

The social and economic aspects of flower growing were, however, recognized much later. The offering and exchange of flowers on all social occasions, in places of worship and their use for adornment of hair by women and for home decoration have become an integral part of human living. With changing life styles and increased urban affluence, floriculture has assumed a definite commercial status in recent times and particularly during the past 2 to 3 decades. Appreciation of the potential of commercial floriculture has resulted in the blossoming of this field into a viable agri-business option. Availability of natural resources like diverse agro-climatic conditions helped in producing wide range of temperate and tropical flowers, almost all through the year in some part of the country or other. Improved communication facilities have increased their availability in every part of the country. The commercial activity of production and marketing of floriculture products is also a source of gainful and quality employment to section of people.

Flowers are inseparable from the social fabric of human life. Flowers being an adorable creation of God, befits all occasions, be it at birth, marriage or death. In the past, flowers were not of much economic importance. One would grow flowers to fulfill his or her aesthetic desire. At times, flowers were offered for sale to meet the special requirements of people. The purposes for which the flowers are used are presented in Table 1.

Table 1 Flowers Used for Different Purposes in India

Sl.No	Name of the Flowers	Purpose for which Used
01	Rose	Love
02	Carnation (White)	Women's Love
03	French Marigold	Sorrow
04	African Marigold	Vulgar Mind
05	Pansy	Thoughts
06	Narcissus	Self Esteem
07	Daffodil	Regards
08	Amaryllis	Pride
09	Iris	Message
10	Lily	Purity
11	Stock	Luxury
12	Sweet Pea	Departure

(Source: S.N. Gupta 2013)

With the passage of time drastic changes have come about in the life style of people leading to commercialized cultivation of flowers. Today, flower plants are no longer meant only for window gardening but play an important role in the decoration of the living houses and office establishments. The science and art of commercial floriculture has been recognized as an economic activity with the potential for generating employment and earning valuable foreign exchange. In several countries of the world, floricultural products are amongst the main export items of agricultural origin. For any country to diversify its agricultural base geared towards export, the ornamental crop industry presents one of the most interesting and viable options. The aesthetic value of flowers and ornamental plants, their use in social events, overall satisfaction in working with them and high income generating power are attracting modern entrepreneurs to invest money in the floriculture industry. The demand for flowers and ornamental plants for different needs like religious purposes, official ceremonies, parties, house decoration, weddings, funerals, etc., is on the rise. This demand for fresh flowers and plants is increasing worldwide over the coming years. The recent liberalization policy of the Government of India has given phillip to commercialized agriculture particularly horticultural crops. Growing of flowers is in vogue in India since a long time. Nevertheless, growing of cut-flowers has emerged as an important industry mainly to cater to the needs of the demand in the overseas market. It is being viewed as a high growth industry in our economy. There is a tremendous transformation in our floriculture sector mainly due to the entry of corporates who are producing cut-flowers to meet the emerging demand in the developed countries for floricultural products. The Government of India has also identified floriculture as a niche area with vast potential for export. There are many incentives given by the government for setting up of floricultural units as Export Oriented Units (EOUs).

India is bestowed with diverse agro-climatic and ecological conditions, which are favorable to grow all types of commercially important flowers generally found in different parts of the world. It also enjoys the best climate in selected pockets for floriculture during winter months. India is in an enviable position to become a leader in the world floricultural trade because of the prevailing congenial location, overall favorable climate of liberalization and globalization and also specific incentives by the government for floricultural development. Specific and authentic quantitative data are not available for existing production and area under floriculture in India.

Tamil Nadu is a leading state in area and production of flowers in the country. A large number of flowers like jasmine, tuberose, rose, chrysanthemum, marigold, crossandra, barleria, lily, limonium, alstroemeria, liatris, freesia, iris, lisianthus, calla, carnation, gerbera and anthurium are commercially cultivated in the state. Many hi-tech units with export tie-ups are there in the state. There are several commercial tissue culture laboratories. The daily average trade of cut flower is over Rs. 2 lakh and loose flower over Rs. 5 lakh in Tamil Nadu itself (Sudhagar, 2013).

Floriculture in India comprises both traditional and modern flower crops. Most traditional flowers are grown in open fields while modern flowers are grown under protected conditions. The traditional flowers grown in the open include chrysanthemum, jasmine, crossandra, rose, tuberose, aster, marigold, champaka, etc. A large number of small and marginal farmers as well as small traders (forming the unorganized sector) are seeking out a living in the sector compared to hi-tech floriculture. The traditional flowers are grown in all districts of the state. The traditional floriculture is still predominant in the state in terms of area and production. However, there are a few exceptions to this general practice in Tamil Nadu. For instance, roses (as cut flowers) are also grown by small and marginal farmers under open conditions, while traditional flowers like chrysanthemum are grown by large growers under polyhouse conditions. The chief flowers grown under greenhouse conditions are rose, gerbera, carnation, bird of paradise and anthurium.

Jasmine is one of the major commercial flower crops of India. It has its special place in South Indian flower crops. It is mainly used for extraction of scented oil. India exports this oil mainly to England, United States of America, Holland, Sweden, Japan, Norway and European Union.

PROBLEM FOCUS

In Sathyamangalam of Erode District, the jasmine is cultivated in around 40000 acres and the marketing of jasmine is taken over by the Farmers Flower Growers' Association which is not prevalent in any part of Tamil Nadu or even in South India. The farmers have themselves established an association to protect their produce from being offered with low price by the traders during 2007 and the practice of marketing and distribution of jasmine is taken up by the association. In this context, to document the practice of production and marketing of jasmine by the farmers flowers growers' association, an effort has been taken to assess the same by adopting a case study approach with the following specific objectives.

OBJECTIVES OF THE STUDY

- To assess the area, production and productivity of flowers with special reference to jasmine in Tamil Nadu and India
- To assess the area, production and productivity of flowers exclusively the jasmine in the sample farms
- To assess the marketing practices prevalent in respect of flowers especially the jasmine
- To document the status of export of flowers to different countries from India and
- To document the status of investment made in horticulture especially for flowers if any and to suggest suitable policy measures

DESIGN OF THE STUDY

The main objective of this chapter is to document the area, production, productivity and export of flowers in India and to document the supporting infrastructure available in the state of Tamil Nadu. To document the same, the data were collected from the secondary sources of both central and state government reports. The reports taken into account for collecting the details on area, production, productivity and export of flowers are furnished as follows.

- Agricultural Statistics at a Glance 2012 of Directorate of Economics and Statistics, Ministry of Agriculture, Government of India, New Delhi
- Food and Agriculture Organization Website
- Indian Horticulture Data Base 2011 of National Horticulture Board, Government of India, New Delhi
- Hort-Stat – 2008 of Government of Tamil Nadu, Chennai

The data gathered for different periods with respect to the area production and productivity of flowers, export of flowers, assembling and trading centers for flowers and the importing institutions and states involved in flower export were gathered from the above reports and the data were analyzed using descriptive statistics like mean and the current change in the data in respect of area, production and productivity of flowers over the base period was also documented to assess the technological and extension impact due to the government intervention through its planned programs in horticulture. The results of this analysis will be better utilized by the policy makers and the planners in horticulture and agriculture for further development of these sectors.

The productivity of flowers in the state was found to be marginally low when compared to the National Average Productivity and hence a comparison made with the intensely managed jasmine gardens of Sathyamangalam Farmers Flower Growers' Head Association of Erode district which was established during the year 2007 and their experiences were also incorporated as a case study approach as an illustration for easy implementation by other jasmine garden owners.

PRIMARY DATA

The farmers who are cultivating flowers in the Sathyamangalam Taluk formed the base for the study. In the first stage, the Sathyamangalam Taluk was selected as it has higher area under flowers cultivated by the farmers. The intention of conducting the survey among the flower growing farmers have been conveyed to the President and Secretary of the Sathyamangalam Farmers Flower Growers' Head Association for the project work on production and marketing of flowers. The President of the society has arranged the contact with the farmers from the selected villages. The number of farmers and the list of villages selected are presented in Table 2.

Table 2 Number of Villages Selected for the Study

Sl. No	Name of the Village	Area under Flowers in Ha	Percentage to Total	Number of Farmers Selected	Percentage to Total
01	Nanjappakavundan Pudhur	023.00	05.87	01	05.00
02	Kenjanoor	028.00	07.14	02	10.00

Sl. No	Name of the Village	Area under Flowers in Ha	Percentage to Total	Number of Farmers Selected	Percentage to Total
03	Kembanaickenpalayam	030.00	07.65	02	10.00
04	Phandampalayam	026.00	06.63	01	05.00
05	Akkarai Negamum	032.00	08.16	02	10.00
06	Chickarasampalayam	102.00	26.02	03	15.00
07	Bahuthampalayam	052.00	13.27	03	15.00
08	Badhramangalam	038.00	09.69	02	10.00
09	Chenbagapudhur	014.00	03.58	01	05.00
10	Kothamangalam	019.00	04.85	01	05.00
11	Raj Nagar	028.00	07.14	02	10.00
	Total Area under Flowers	392.00	100.00	20	100.00

From the farmer, the details of land use, crop management for its production and productivity, marketing of flowers, cost of production and the constraints involved are assessed using the structured and pretested questionnaire exclusively designed for the purpose and the same were analyzed and documented with respect to the objectives enshrined in the study by adopting a descriptive statistics like mean and percentage analysis.

With respect to marketing of flowers, the traders are visiting Sathyamangalam Farmers Flower Market, Sathyamangalam every day for getting their requirement of flowers. Sathyamangalam which is one of the leading wholesale farmers' market functions every day from early morning to the noon. The Sathyamangalam Farmers Flower Growers' Head Association has around 1480 members to its credit who are involved in producing and selling the flowers in the Farmers' Flower Market,

Sathyamangalam. Among the 1480 members, the active members who are involved in regular transactions of flowers numbers around 1100 and the traders from different parts of Tamil Nadu and nearby states visit the market and participate in the open auction of flowers and take the flowers to their zones for consumption. The price that prevailed during the study period of 8th and 9th November 2013 during the open auction is presented in the Chapter V. The traders visiting the Farmers Flower Market regularly are around 50 in number. Among the 50 traders, only five traders have been interviewed to identify the important markets in Tamil Nadu and other states, marketing channels in which the produce is moved and the price spread in marketing of flowers. These are measured using the following formulae.

PRICE SPREAD ANALYSIS

Price spread in general, is referred to as the difference between the price paid by the consumer and the price received by the producers per unit of the commodity. Price spread analysis would estimate the share of different market intermediaries in the consumer rupee and this would often facilitate the understanding of the relative efficiencies otherwise of alternative channels in marketing.

To analyze the price spread in the distribution of flowers, the following formula is used.

$$Ps = (Pp / Cp) \times 100$$

Where,

Ps = Producer's share in the consumer money expressed in per cent

Cp = Consumer's price per unit of flowers in rupees

Pp = Producer's price per unit of flowers in rupees. Similar approach was used by Sekhar *et.al.* (1998).

The price spread analysis involved computation of different marketing cost and marketing margin at each stage and their expression as a percentage to the consumer's rupee. The various costs incurred in the marketing process were considered for each of the identified channels and separate price spread was worked out. The profit margin for each market functionaries in the different marketing channels was computed by subtracting the price paid and the marketing cost incurred by them from the price received by him on the sale of the commodity. Thus the marketing costs and marketing margins were distinguished.

ESTIMATION OF MARKETING EFFICIENCY

Marketing efficiency is the degree of market performance. The movement of goods from producer to the ultimate consumers at the lowest possible cost consistent with the provision of services desired by the consumer is termed as marketing efficiency or efficient marketing. To analyze the marketing efficiency, the formula adopted by Shepherd is used.

$$ME = (V/I) - 1$$

Where,

ME = Marketing Efficiency Index

V = Value added to the output in the marketing system. Here it is referred to as the consumers price per kg of flowers

I = Input used in the marketing process. Here it is referred to as the marketing costs incurred per kg of flowers. Similar methodology was used by Sekhar *et.al.* during 1998.

The lower the value of marketing cost higher would be the efficiency.

Calkin's index

Calkin's index of marketing efficiency was estimated using the following formula:

$$ME = 1 + [\text{Sum of profit or margin/Sum of marketing cost}]$$

The lower the value of the index, higher would be the efficiency.

MARKETING OF JASMINE IN TAMIL NADU

Marketing is the process of transferring the jasmine flower buds from producer-farmer to the ultimate consumers. This section deals with marketing channels through which the produce is transferred, price spread and marketing efficiency among the different channels. Price spread addresses on the producers' share in the consumers' money whereas marketing efficiency addresses on the efficiency aspects that which channel is transferring the commodity at the least marketing cost and fulfills different role players by meeting their margin and gets satisfied.

Marketing Channels Prevalent in Distribution of Jasmine in Tamil Nadu

Here the producers are the cultivators of jasmine and they transfer the commodity to the Sathyamangalam Farmers' Flower Market and the traders are taking the produce through open auction and then it reaches to consumers through different middlemen. These details are analyzed and the results are presented in Table 3.

Table 3 Marketing Channels Prevalent in Distribution of Jasmine Flowers

Sl.No	Channel Number	Marketing Channel Prevalent in Marketing of Jasmine
01	I	Producers → Farmers' Flowers Market → Consumers
02	II	Producers → Farmers' Flower Market → Commission Agent (Mettupalayam) → Industrial Consumers
03	III	Producers → Farmers' Flower Market →Wholesaler cum Commission Agent (Coimbatore) → Retailers → Consumers
04	IV	Producers → Farmers' Flower Market → Wholesaler cum Exporters → Wholesaler cum Importers → Wholesaler cum Retailers →Retailers → Consumers

Sl.No	Channel Number	Marketing Channel Prevalent in Marketing of Jasmine
05	V	Producers → Farmers' Flower Market → Primary Wholesalers (Tamil Nadu) → Secondary Wholesalers (Other States) → Wholesaler cum Retailers → Retailers → Consumers
06	VI	Producers → Farmers' Flower Market → Commission Agents (Coimbatore) →Wholesaler cum Exporters → Wholesaler cum Importers (UAE) → Retailers → Consumers

Table 3 reveals that the jasmine flower is reaching the consumers of regional, national and international through some intermediaries in the six important channels. Among the six important channels, only the channel – II and channel – III alone are traceable from producer level to the consumer level and the price that prevailed among the different functionaries were analyzed and the results are discussed in the section dealing with price spread. The second channel identified is Producers → Farmers' Flower Market → Commission Agent (Mettupalayam) → Industrial Consumers. Here, the commission agent at Mettupalayam used to visit the farmers' flower market and they take the title of goods to the Mettupalayam town and supply the same to M/s. Encee Aromatics Limited, a floral concrete extractor at Mettupalayam.

The third marketing channel identified is Producers → Farmers' Flower Market →Wholesaler cum Commission Agent (Coimbatore) → Retailers → Consumers. Here, the commission agent at Coimbatore city visits the Farmers' Flower Market and takes the title of goods and sends the produce to the retailers at Coimbatore as well as the traders of Kochi. These are discussed and analyzed and the results are presented in Table 4 and 5.

Price Spread in Marketing and Distribution of Jasmine

Price spread is said to be a composite of various costs incurred and margin of intermediaries in the various processes, such as those of assembling, processing, storage, transport and retailing. How these are varying between the identified marketing channels in respect of marketing of jasmine is analyzed and the results are presented in Tables 4 and 5.

Price Spread in Jasmine Marketing Through Channel II The price spread for the marketing through channel – II is worked out and the details are presented in Table 4.

Table 4 Price Spread in Jasmine Marketing Through Channel – II Producers → Farmers' Flower Market → Commission Agent (Mettupalayam) → Industrial Consumers

Sl. No	Particulars of Cost	Amount (Rs/Kg)	Per Cent
Producer			
01	Producers' Price	40.00	80.00
1.1	Marketing Cost		
1.1.1	Packing Charges	00.45	
Producer			
1.1.2	Transportation Charges	01.25	
1.1.3	Cost of Labor for Auctioning and Measuring	02.40	
1.1.4	Farmers' Market Commission Charges	00.60	
1.1.5	Miscellaneous Charges	00.30	
	Marketing Cost at Producers' Level	05.00	10.00
1.2	Marketing Margin	00.00	
02	Commission Agent		
2.1	Commission Charges	05.00	10.00
03	Procurement Price by the Industrial Consumer	50.00	100.00

Table 4 reveals that the Channel – II is capable of offering the produce directly to the industrial consumers through the commission agent. Farmers' flower market receives the produce and sends the same to the commission agent by open auction. In this channel, the producer is able to get Rs 40 per kg of jasmine which accounted for 80 per cent to the consumers' rupee. The cost of production per kg of jasmine buds is arrived at Rs 25 and the producer's price per kg is arrived at Rs 40. A net return of only Rs 15 is able to be achieved by the producer. The remaining 20 per cent is incurred towards the marketing cost per kg of flowers. The commission agent shares Rs 5 per kg as commission which is accounted for 10 per cent to the consumers' price and the same has been transferred to the industry like M/s Encee Aromatics Limited, a floral concrete extracting firm positioned at Vellipalayam of Mettupalayam Taluk.

Price Spread in Jasmine Marketing Channel - III The price spread in jasmine marketing through the channel – III is analyzed and the results are presented and discussed in Table 5.

Table 5 Price Spread in Jasmine Marketing Through Channel – III Producers → Farmers' Flower Market →Wholesaler cum Commission Agent (Coimbatore) → Retailers → Consumers

Sl. No	Particulars of Cost	Amount (Rs/Kg)	Per Cent
01	Producer		
01	Producers' Price	130.00	59.09
1.1	Marketing Cost		
1.1.1	Packing Charges	00.45	
1.1.2	Transportation Charges	01.25	
1.1.3	Cost of Labor in the Process of Auctioning and Measuring	02.40	
1.1.4	Farmers' Market Commission Charges	00.60	
1.1.5	Miscellaneous Charges to the Producer	00.30	
	Marketing Cost at Producers' Level	05.00	02.27
1.2	Marketing Margin	00.00	00.00
02	Wholesaler cum Commission Agent		
2.1	Procurement Price by the Commission Agent	135.00	61.36
2.2	Marketing Cost		
2.2.1	Assembling Charges to the Coimbatore Market	02.80	
2.2.2	Auctioning Charges	01.20	
2.2.3	Miscellaneous Charges	01.50	
	Marketing Cost at Wholesaler cum Commission Agent Level	05.50	
2.3	Marketing Margin to the Commission Agent	19.50	08.86

Sl. No	Particulars of Cost	Amount (Rs/Kg)	Per Cent
03	Wholesaler cum Retailer		
3.1	Purchase Price Incurred by the Wholesaler cum Retailer	160.00	72.73
3.2	Marketing Cost		
3.2.1	Loading and Unloading Charges	02.00	
3.2.2	Transportation Charges	03.50	
3.2.3	Cost of Labor for Distribution to Flower Shops	01.00	
3.2.4	Miscellaneous Cost including Rent	02.50	
3.2	Marketing Cost to the Wholesaler cum Retailer	09.00	
3.3	Marketing Margin to the Wholesaler cum Retailer	21.00	09.55
3.4	Sale Price to the Retailer	190.00	86.36
04	Retailer		
4.1	Procurement Price by the Retailer	190.00	86.36
4.2	Marketing Cost to the Retailer		
4.2.1	Transportation Charges	03.00	
4.2.2	Cost of Labor in Distribution of Flowers to Flower Shops	02.00	
4.2.2	Miscellaneous Charges Including Rent	03.00	
4.2	Marketing Cost to the Retailer	08.00	03.64
	Marketing Margin to the Retailer	22.00	10.00
5	Sale Price to the Consumer and the Consumer's Price	220.00	100.00

Table 5 reveals that the producer is receiving Rs 130 per kg of jasmine in the Channel – III while the consumer's price per kg of jasmine is at Rs 220 which accounted for 59 per cent. Prabhakar Rao *et al.* (1993) studied the marketing costs and margins of jasmine in Guntur District of Andhra Pradesh and they identified the following marketing channel:

Producer →Wholesaler → Retailer → Consumer

They found that the producers share in the consumer rupee was Rs. 700 (50.44 per cent) per quintal. The total cost incurred by the wholesaler was Rs. 45.10 (3.50 per cent) per quintal and his margin was Rs. 320.45 (23.09 per cent) per quintal. Retailers incurred a cost of Rs. 29.45 (2.12 per cent) per quintal and his margin was Rs. 292.92 (21.10 per cent) per quintal.

The wholesaler cum commission agent in our study (Table 5) has received the jasmine flower buds at the rate of Rs 135 per kg, he is incurring 5.50 as marketing cost, earns marketing margin to the tune of 19.50 per kg of jasmine and then he is distributing to the wholesaler cum retailer at the rate of Rs. 160 per kg.

The wholesaler cum retailer is incurring marketing cost towards transportation, packing and forwarding and distribution to the flower shops to the tune of Rs 9 per kg and earns Rs 21 as marketing margin for the services he rendered in the distribution process of jasmine and the same has been sold to the retailer at the rate of Rs 190 per kg of jasmine.

The retailer is also incurring certain marketing costs including transportation and labor and spends an amount of Rs 8 per kg and earns the marketing margin of Rs 22 per kg of jasmine and the consumer price is fixed at Rs 220 per kg.

To sum up, the flower buds of jasmine during the summer gave only Rs 40 per kg and capable of fetching 80 per cent share of consumers' rupee through the marketing channel – II. Whereas, the marketing Channel – III is capable of fetching only 59 per cent as producers' share in the consumers' rupee per kg of jasmine. Therefore the profit gained is mainly due to the number of intermediaries involved in facilitating the trade of jasmine to meet the consumers of different destinations in India and abroad.

Marketing Efficiency in Jasmine Marketing

Marketing efficiency is defined as the ratio of output (Consumers' Price) to the input (Marketing Cost). In this study, the consumer price varied from channel to channel and hence the same was worked out using the Calkin's Index and the results are presented in Table 6.

Table 6 Marketing Efficiency in Distribution of Jasmine in Tamil Nadu

Marketing Channel	Value of Output (Consumers' Price)	Input (Marketing Cost)	Marketing Efficiency Ratio (O/I)
Shepherd Method			
II	040.00	10.00	03.00
III	220.00	22.50	08.78
Calkin's Index			
Marketing Channel	**Marketing Margin**	**Marketing Cost**	**Marketing Efficiency**
II	05.00	05.00	02.00
III	72.50	22.50	04.22

Table 6 advocated that an increase in the marketing efficiency ratio would represent improved efficiency and vice versa. A reduction in the cost for the same level of satisfaction or an increase in satisfaction at a given cost would result in an improvement of efficiency. When comparing the data delineated in the Table 6, one could understand that the Channel – III is found to be the efficient one. Because, the Channel – III fulfilled the requirements of several functionaries in the marketing channel and satisfied them.

MARKETING CONSTRAINTS FACED BY SATHYAMANGALAM FARMERS' FLOWER MARKET

Marketing of jasmine and its process attracted many intermediaries into the market operations. They are producers, commission agents, retailers, wholesalers and the consumers. Among the many intermediaries present in the market, the needs of producers, the consumers and the market operators need to be fulfilled immediately for smooth functioning/transfer of jasmine. In this respect, the constraints faced by different personnel in the marketing are analyzed and the results are presented in Table 7.

Table 7 Marketing Constraints Faced by the Sathyamangalam Farmers Flower Growers' Head Association

Sl. No	Marketing Constraints	Rank
01	Absence of Land for Positioning the Farmers' Flower Market near Bus Stand	I
02	Higher Rent to be paid for the open Land for market operations	II
03	Fluctuation in the Prices of Jasmine Flower	III
04	Transportation Infrastructure needs to be upgraded	IV
05	Transportation Hire Charges are Exhorbitant	V

Table 7 reveals that the farmers' market is having a set of problems to be solved on priority basis. Among the constraints, absence of land for positioning the Farmers' Flower Market near the bus terminus of Sathyamangalam town is very important to them. The vacant land of Public Works Department available at the backside of the bus stand was first used by the personnel of the farmers' market. Later on, the department denied the occupation of their space and even after several efforts by the society, the land was unable to be transferred for flower transactions and hence they suffer a lot for want of space. Currently, the society has occupied a vacant land for farmers at the backside of the municipal bus stand on rental basis and the marketing of jasmine is being practiced by the society from this place .

In this regard, the government through the Public Works Department or revenue department or the town panchayat should come forward to identify a land suitable to the Sathyamangalam Farmers' Flower Growers Head Association for their efficient functioning. Around 3000 farmers depend on the society for their livelihood and they are involved in direct marketing of flowers through the society. This novel effort should be supported by the government at the earliest by providing place and erecting an integrated market yard for the flower procurement and trade including parking and storage facilities.

The third important constraint is fluctuation in price of jasmine flower buds called for uncertainty in the income of the jasmine producers. Since the Government of Tamil Nadu is poised for enhancing the income of the farmers by following a phrase "Doubling the Production and Tripling the Income of the Farmers", to fulfill this goal, the government should come forward to erect an integrated farmers flower market at Sathyamangalam, Nilakottai, Thovalai, Madurai and Thirunelveli following the pattern of Koyambedu market at Chennai by getting some share from the farmers society and the rest borne by the government for enhancing further trade development and if need be a special economic zone of flowers may be announced.

Another important constraint is transportation infrastructure. Here what is needed is providing a direct flight service from Coimbatore to Mumbai will readily enhance the transportation of flower buds on time which will in turn facilitate the producer traders to explore new markets keeping Mumbai as the base. The last but not the least constraint is exorbitant transportation charges and hence the society can come forward to buy some pick up vans to collect the produce from different production sites and the same can be transported to the airport and to the demanding ends at the earliest.

SUMMARY AND CONCLUSION

Among the six important channels, only the Channel – II and Channel – III alone are traceable from producer level to the consumer level and the price that prevailed among the different functionaries were analyzed and the results are discussed in the section dealing with price spread. The second channel identified is Producers → Farmers' Flower Market → Commission Agent (Mettupalayam) → Industrial Consumers. Here, the commission agent at Mettupalayam used to visit the farmers' flower market and they take the title of goods to the Mettupalayam town and supply the same to M/s. Encee Aromatics Limited, a floral concrete extractor at Mettupalayam. The third marketing channel identified is Producers → Farmers' Flower Market →Wholesaler cum Commission Agent (Coimbatore) → Retailers → Consumers. Here, the commission agent at Coimbatore city visits the Farmers' Flower Market and takes the title of goods and sends the produce to the retailers at Coimbatore as well as the traders of Kochi.

The flower buds of jasmine during the summer gave only Rs 40 per kg and capable of fetching 80 per cent share of consumers' rupee through the marketing channel – II. Whereas, the marketing channel – III is capable of fetching only 59 per cent as producers' share in the consumers' rupee per kg of jasmine. The main reason for the difference in the share of the consumers' rupee is due to the number of intermediaries involved in facilitating the trade of jasmine to meet the consumers demand of different destinations in India and abroad.

A reduction in the cost for the same level of satisfaction or an increase in satisfaction at a given cost would result in an improvement of efficiency. When comparing the data delineated, one could understand that the Channel – III is found to be the efficient one.

The farmers' market is having a set of problems to be solved on priority basis. Among the constraints, absence of land for positioning the Farmers' Flower Market nearby the bus terminal of Sathyamangalam town is very important to them. The vacant land of Public Works Department available at the backside of the bus stand was first used by the farmers' market personnel. Later on, the department denied the land and even after several efforts by the society, the land was unable to be transferred to the society for flower transactions and hence they suffer a lot for want of space. Currently, the society is using a vacant land of a farmer to the backside of the municipal bus stand on rental basis and the marketing of jasmine is being practiced by the society through this land.

Chapter 25

Guidelines for Formulation of Bankable Projects

INTRODUCTION

"Projects are the cutting edge of development" (J P Gittinger). Project means a plan or scheme for achieving some objective. The whole complex of activities involved in using up of resources to gain the benefits constitutes the project. A Project is an activity for which money will be spent in expectation of return and which logically seems to lend itself to planning, financing and implementing as a unit. It is the smallest operational element prepared and implemented as a separate entity in a national plan or program. It is a specific activity, with a specific starting point and a specific ending point, intended to accomplish specific objectives of the firm or organization (Subramanian and Senthilnathan 2004).

Project approach could be applied for small investments like purchasing of buffaloes as also huge dams involving several crores of rupees. It is a systematic approach to development. Each project is considered to be an independent unit having its own costs and benefits. In the liberalized economic environment and in the wake of financial sector reforms, the importance of project approach has further increased.

PROJECT APPROACH

Project means a location specific activity with specific objectives, time and cost limitations and of non-repetitive nature. In banking, bankers refer to 'Project' as an activity in which financial resources are expended to create capital assets that produce benefits over an extended period of time and which logically lends itself to planning, financing and implementing as a unit.

Project Includes

- Capital investment
- Creation of Capital Assets
- Generation of Benefits over an extended period of time

Project - An Acronym Stands For

P - Production of goods and services

R - Resources mobilization like men, materials, methods and money

O - Organization which is initiating and implementing the project

J - Justification – Fulfilling the social benefits like income, employment, community needs, meeting the ecological and environmental requirements, etc.

E - Economic viability, development of quality output, profit generation, etc.

C - Continuity - Planning, execution, monitoring and follow-up, research and development, etc.

T - Time bound implementation and operations

PROJECT CYCLE

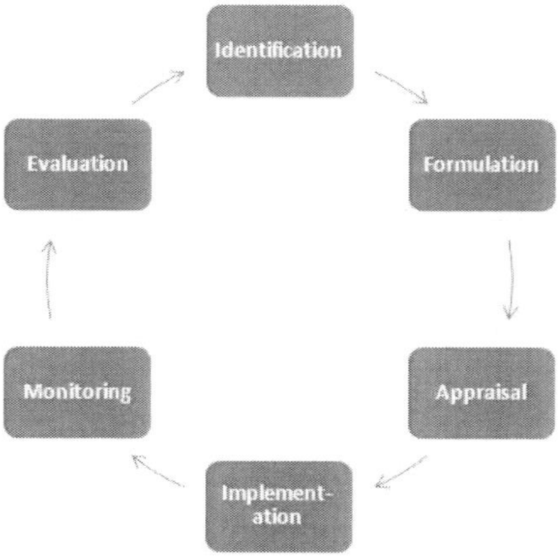

IDENTIFICATION OF THE PROJECT

It is a first step in the project cycle. Several sources are available for identification of projects. They are progressive farmers, successful entrepreneurs/agripreneurs, technical experts, bankers, media personnel, newspapers, research and annual publications of national and regional research institutions. In this proposal, the successful sheep breeding unit was taken into account and a model proposal developed.

FORMULATION OF A PROJECT PREPARATION STAGE OF THE PROJECT

At times, the banker may have to form, state the project himself if the potential entrepreneurs are too scattered. Feasibility study is the first step in formulation of the project. In case of complex projects, even a pre-feasibility study may have to be conducted. Specialized consultancy firms offer services for conducting feasibility studies. Advance techniques of Project Planning like PERT (Program Evaluation and Review Techniques) and CPM (Critical Path Method) are used in, intensive and complex projects which involve long gestation periods.

PROJECT APPRAISAL

Includes analysis and scrutiny of each and every aspect of project preparation, details and assumptions that need to be postulated in this appraisal. Banks undertake appraisal of projects before financing. The aspects of appraisal may involve technical, institutional/managerial/organizational, social, commercial, financial, economic, ecological and environmental analysis. These aspects in short can be pronounced as "TISCOFEE".

Essence Of Project Appraisal

Appraisal is a comprehensive and systematic review of all aspects of a project:

- It is a second look at the project report by a person who is in no way involved in its preparation.
- No cut and dry formula to straight away decide a project to be acceptable or unacceptable.
- Appraisal highlights the weak areas in the project with the ultimate objective of strengthening them adequately to ensure final success of the project.
- Project appraisal is an exercise in future, based on certain assumptions. Hence, it is necessary to understand the environment in which the project has to sustain itself.
- Appraisal is a joint exercise by promoters of agribusiness and the financial institutions. Speed of appraisal therefore, depends very much on the speed with which the information is forthcoming from the entrepreneurs.
- Project should be appraised within a time bound program. Any undue delay distorts profitability projections and success of the project becomes a difficult proposition.
- In case of technocrat and new entrepreneur, a tolerant and understanding attitude is very much necessary.

The assumptions made in respect of establishing an agribusiness venture namely the Sheep Rearing Unit in Periyakulam, Theni District of Tamil Nadu are delineated in Table 1.

Table 1: Assumptions Made in the Agribusiness Proposal

Sl. No	Particulars of Assumptions Made	Description / Remarks
01	Unit Size	500 Female Sheep + 25 Male Sheep
02	Rearing System	Zero Grazing
03	Breeds used for Rearing	Meacheri and Mylambadi
04	Shed Area	10000 Sq. Feet
05	Wandering Zone	15000 Sq. Feet
06	Pillars	Pillars Made out of Cement with Iron rod of 12 Feet Height (10X10 Feet Spacing)
07	Feeding Systems	Stall Feeding by cut and carry Method
08	Maize Crop	3 crops per acre per annum
09	Leucerne on the Fence or in one acre area separately	First cut after 50 Days; Later for every 25 Days
10	Construction of Silos	Below the Ground or Above the Ground Level
11	Labor Required	5 Labor @ one per 100 sheep
12	Cost of Insurance	Four Per cent of the Total Cost of the Animal or the Prevailing Rate with Insurance Agencies
13	Expected Growth per Sheep	Five Kilograms per Month
14	Veterinary Health Care	Rs. 100 Per Annum per Sheep
15	Manure or Pellets Output	1.50 Kg of Pellets per Day per Adult Sheep

Sl. No	Particulars of Assumptions Made	Description / Remarks
16	Duration of Breeding of Sheep	3 Cycles in a Couple of Year (24 Months)
17	Duration Required for Fattening of Kids	Six Months (2 Months on Mothers Milk + 4 Months on Feed)
18	Mortality Rate	10 Animals per Annum
19	Purchase Price of Breeder Sheep	Rs 250 per Kg of Live Weight
20	Selling Price of Animals	Rs 300 per Kg of Live Weight
21	Life of the Sheep	07 Years
22	Productive Life	05 Years
23	Maturity Age for Reproduction	12 Months

Steps And Methodology For Appraisal

- Entrepreneur seeking financial assistance should furnish detailed information about the project in prescribed format.
- An economist and / or an engineer and a financial analyst should do the appraisal of the project.

The First Step In Project Appraisal

- The first step in detailed appraisal is with regard to information submitted by the entrepreneur, i.e., the appraiser has to check for the following.
- Adequate data has been furnished against every item.
- Prima facie information furnished is correct
- Consistency of data and information furnished should be cross-checked.
- Both in the scrutiny of preliminary information and processing of application, personal discussions between the officials of the financial institution and the borrowing concerns are both necessary and desirable. This will reduce avoidable correspondence and facilitate expeditious disposal of applications.

- Clarification on additional information called for wherever necessary.
- A project is likely to receive favorable consideration and detailed appraisal is taken if it has priority according to government guidelines.
- The promoter inspires confidence.
- The technology to be adopted is a proven one.
- The product to be manufactured or produced has higher market potential.
- The project cost is not unreasonably high.
- The promoter's contribution is not unduly low.
- Profitability estimates are conservative and indicate repayment of the loan within reasonable period.
- Preference is given to proposals/projects which utilize any industrial waste or agricultural surplus as raw materials.
- The project is export-oriented or import substitutes;
- The project is capable of generating higher employment
- The project is located in specified backward or less backward areas or drought prone areas.
- The project is able to help in setting up of ancillary and small-scale units.

A proposal without detailed appraisal is rejected if it has some of the following features:

- Banker's report on the promoters is found unsatisfactory.
- Promoters reported to have indulged in illegal and anti-social activities.
- Financial position of the promoter company is not satisfactory.
- Cost of the project is unduly high.
- Promoter's contribution is unusually low and promoter has declined to increase it.
- Debt-Equity Ratio (DER) is abnormally adverse.
- Location of the proposed unit has apparent disadvantages, i.e., far away from sources of raw materials, market, etc.
- Collaborators have inadequate experience.

- Equipment supplied by the proposed machinery suppliers to some other units has not been performing satisfactorily.
- Second hand equipment to be acquired is too old and will not have trouble free residual life.
- Process know-how / technology to be adopted have not been proved successful on a commercial scale.
- Process know-how / technology have become obsolete.
- Availability of raw material is found to be inadequate in quantity and its sustained supply is doubtful.
- There is no certainty that utilities like power, water, etc., will be available by the time the project will need them.
- Products to be manufactured do not have sufficient market potential.

The Second Step In Appraisal

- Involves site inspection by appraisal team.
- Sufficient notice of site inspection to be given to the business promoters.
- Inspection team to pay attention regarding suitability of the site, distance from railway station, national highway, sources of raw materials, market for end-products, timely availability of utilities.
- Scanning the environment protection arrangements
- Ascertaining the sources of skilled and unskilled man-power
- Availability of social infrastructure
- Most importantly, collect market reports, promoters' financial strength, credibility, capability and capacity to repay the loan to be availed.

Third STEP in Appraisal

Before providing financial assistance to an agribusiness unit, it becomes necessary for a financial institution to examine:

- Viability of the project—both financial and economical
- It should ensure that the project will generate enough funds

Aspects of Appraisal of the Projects

In this stage the various aspects of appraisal of the projects should be taken into consideration. In short, it is designated as "**TISCOFEE**". They are

- **T**echnical Appraisal (T)
- **I**nstitutional/Managerial/Organizational Appraisal (I)
- **S**ocial Distributive Appraisal (S)
- **Co**mmercial Appraisal (Co)
- **F**inancial Appraisal (Cost of the Project) (F)
- Financial Appraisal - Means of finance
- Financial Appraisal - Cash Flow Ratio Analysis
- Financial Appraisal - Break Even Analysis.
- Management Appraisal
- **E**conomic Appraisal (E) should furnish the results of Benefit Cost Ratio, Net Present Value and Internal Rate of Return. The results of these parameters should be higher than the recommended one.
- **E**nvironmental/Ecological Appraisal (E)

1. *Technical Appraisal* Technical appraisal forms the basis and other aspects of project preparation and analysis can only proceed in light of this (technical) analysis. Care must be taken in selection of Technical Personnel or Consultant. Whether the technical expert should be chosen locally or from other states or the other country must be charted out first and his technical soundness should be checked based on his previous project preparation experience. Besides this, the following points also need to be considered in technical appraisal of the project.

Technical appraisal will focus in the following areas.

- Manufacturing process/technology proposed
- Project inputs (supplies) and outputs (production) of real goods and services.
- Technical arrangement like parting know-how, etc.
- Size of the plant (minimum economic size) or installed capacity of the plant

- Product mix (various items to be manufactured by the unit)
- Selection of the plant and machinery (whether any unit purchased, reputed - no trouble, service back up)
- Procurement of plant and machinery (seller's name, cost, etc.)
- Plant layout and location of the project
- Land availability and its suitability for installation of the project
- Raw material market and its capacity to fulfill the industrial needs
- Labor availability and the prevailing wage rate
- Utilities such as, soil, water, power, fuel and other inputs like seed, fertilizer mix, soil analysis, ground water analysis, collection of hydrological data and analysis, previous farming practices, current farming methods/practices, proposed cropping pattern and crop mix, field trails and demonstration, likely impact in yield/productivity.
- Effluent/waste disposal mechanism
- Transportation of inputs/outputs.
- Development of other line industries/cropping pattern in liaison with market potential

The project analysis must continue to make sure that the technical work is thorough and appropriate, that the technical specifications, estimates and projections are related to realistic conditions and that farmers/ manufacturers using the proposed technology on their fields can realize the results projected.

Background Details of the Promoters of Agribusiness

The promoter is a farmer belonging to the Ramanathapuram district who migrated to Periyakulam 15 years back for maintaining the sheep flocks available with them. Sheep rearing is an ancestral business to them. They have 3 acres of dry land with one bore well capable of providing partial irrigation to 2 acres of land. The leader of the business is having an educational background of 10th standard. Now they have planned to establish one unit (500 female + 25 male) of sheep in Kamatchipuram village of Periyakulam taluk. The cost of 3 acres of land will be around 75 lakhs. The value of 200 numbers of sheep flocks currently available with the case farmer is arrived at 25 lakhs. Put together, the asset value of the farmer is arrived at 100 lakhs. Meat market is an attractive one, the animals are of good demand and hence agripreneur wanted to increase the number of sheep flocks. For that an amount of Rs 50 lakhs may kindly be sanctioned. The technical aspects of sheep rearing are furnished in the following Table. 2

PERFORMANCE Appraisal OF THE SECTOR

The performance of the agricultural and allied sectors like horticulture, forestry, animal husbandry, poultry and its current position in the country in general and in the respective state in particular to which the concerned project pertains has to be analyzed keeping the potential and the problems of the sector in view and suggesting remedial measures needed, if any for development in future with the special reference to exports.

TECHNO-FINANCIAL STUDY

Apart from other details, the techno-financial study should indicate the dates on which the studies were conducted and the names of the officers who were associated with the study. The report should invariably mention and highlight whether the joint appraisal was done along with bank's representatives.

Table 2 Techno Economic Features in Sheep Rearing

Sl. No.	Description	Units	Number of Units / Animals	Total Cost (Rs)
1	Unit size			
1.1	Female Breeders (Ewes) (Meacheri Breed)	Numbers	500	
1.2	Male (Rams) (Meacheri and Mylambadi)	Numbers	025	
1.3	Rearing System	Zero Grazing	000	

Sl. No.	Description	Units	Number of Units / Animals	Total Cost (Rs)
2	Breeds			
2.1	Breeds purchase plan @ 8 month old	Ewes	500	1800000
		Ram	25	090000
		Total	**525**	**1890000**
2.2	Average Weight of Ewes and Ram on purchase	Kgs	12	
2.3	Average weight of Lamb on sale	Kgs	20	
2.4	Average Weight of Adult (Ewes & Ram) on sale	Kgs	28	
Sl. No.	**Description**	**Units**	**Number of Units / Animals**	**Total Cost (Rs)**
2.5	Sale / purchase Rate of Adult (Ewes & Rams) and lamb (Rs. / kg)	Rs/ Kg Live Weight	300	
2.6	Life of Sheep	Years	07	
2.7	Productive Life	Years	05	
2.8	Maturity Age for Reproduction	Months	12	
2.9	Frequency of Reproduction	3 Times in Two Years	03	
2.10	Production in the first year	Per Cent	12	
2.11	Production in the later years	Per Cent	80	
2.12	Unproductive Sheep every year	Per Cent	17	

Sl. No.	Description	Units	Number of Units / Animals	Total Cost (Rs)
2.13	Sheep Death rate every year	Per Cent	03	
2.14	Shed Area	Square Feet	6735	
2.15	Exercise paddock	Square Feet	5000	
3	Veterinary Expenses			
3.1	Adult	Animals Per Year	10	
3.2	Young	Animals / Year	05	
3.3	Lambing interval	Months	12	
3.4	Lambing percentage	Per Cent	75	
4	Mortality			
4.1	Lambs	Per Cent	10	
4.2	Adults	Per Cent	05	
4.3	Sale of Lamb (Ewes & Ram)	Months	12	
4.4	Culling Rate of Ewes from Third Year Onwards	Per Cent	20	
4.5	Ram Culling (The old ram will be replaced after every two years by fresh purchase)			
4.6	Grazing Charges per Adult per Year	One	04	
4.7	Cost of Concentrates	Rs per Kg	10	
5	Insurance	Per Cent	04	

Sl. No.	Description	Units	Number of Units / Animals	Total Cost (Rs)
6	Wool (Shearing will be done twice a year)			
6.1	Adult	Kg per Year	02	
6.2	Lamb	Kg per Year	0.60	
6.3	Shearing Charges	Rs per Kg	02	
6.4	Price of Wool	Rs per Kg	40	
7	Cost of Maize Silage in - house Production	Rs per Acre	8000	
8	Breeding Percentage	Per Cent	90	
9	Lambing Rate	Lambs per Ewe	01.20	
10	Mortality Rate in New born	Per Cent	10	
11	Sale Rate of Adult (Ewes & Rams)	Rs/Kg Live Weight	300	
12	Sale Rate of Lamb	Rs/Kg Live Weight	300	
13	Cost of Mineral Feed supplements	Paise per Day	00.75	
14	Cost of Shed Construction	Rs per Sq.Ft	100	
15	Cost of Chain Link Mesh Fencing	Rs per Sq.Ft	25	
16	Cost of Feeders with Drinkers	Per Unit	5000	
17	Average Weight of Lamb	Kgs	18	
18	Average Weight of Adult (Ewes & Ram) (kg)	Kgs	28	
19	Sale of Manure			
19.1	Manure from 1 sheep per annum	Kgs	365	
19.2	Sale rate of manure	Rs / Ton	2000	

Sl. No.	Description	Units	Number of Units / Animals	Total Cost (Rs)
20	Total Power Cost	Rs per Year	39210	
21	Wages and Salary Cost	Rs per Year	270000	
22	Repairs and Maintenance	Rs Per Year	24000	
23	Product Mix			
23.1	Sheep	Per Cent	100	
24.1	Depreciation(WDM) on Buildings	Per Cent	10	
24.2	Machinery	Per Cent	25	
24.3	Vehicles	Per Cent	25	
25	Income Tax rate	Per Cent	00	
26	Interest Rate on Term Loan	Per Cent	12	
27	Repayment Period	Years	06	
28	Moratorium	Years	01	
29	Other Factors			
29.1	Maize crop	3 crops/Acre/Annum		
29.2	Labour	One labourer for 100 sheep.		
29.3	Insurance Cost	Prevailing norms in customer area.		
29.4	Silo sizes	Case to case design is tailor made		
29.5	Feeding systems	Stall feeding		
29.6	Manure output (kg/adult sheep/year)	365		
29.7	Breeder cycle duration	8 months		
29.8	Lamb fattening cycle duration	6 months -2 months on mother's milk + 4 months on feed		
29.9	Lambs per Breeding Cycle (on 24 months duration)	3	No	
	R-Ram E-Ewe ML-Male lamb FL-Female Lamb			

Institutional/Managerial/Organizational Appraisal The most important aspect of project appraisal is the management appraisal. Some of the points to be considered while appraising the management are the qualities of an agripreneur and his capability in running the project.

- Honesty and integrity of an agripreneur
- Educational background of an agripreneur
- Extent of involvement and commitment in the project
- Financial resources in case of over-run, whether he/she can arrange his/her own resources
- Details of competence, risk taking, initiative, intelligence, drive and energy, self-confidence, frankness, patience
- Does the project design take into account the farmers who participate in the project?
- Type of communication systems prevalent to infuse new informations and to teach new skills
- Time given to the farmers to accept new technologies/procedures
- Land tenure arrangements
- Details of form of organization viz., sole proprietorship, partnership, co-operative form of organization
- Board of Directors and their role in managing the project
- The role of a Chief Executive of the company/project
- Other Executives—Role and functional clarity of line managers
- Organizational structure and line of or hierarchy of reporting, management problems and strategies adopted to solve the same are needed
- Project authority and managerial levels who influence the farmers
- Linkage of authority and responsibility is proper or not need a mention
- Training needs and provision for conduct of training
- Accounts and accounting procedures
- Type of extension agency involved in diffusion of knowledge to the farmers/stakeholders

Talent Appraisal

In a nutshell, the practice of managing the agribusiness involves five perspectives which are delineated as follows (5Ms).

1. Managing Self: The Reflective Mindset
2. Managing the Organizations: The Analytical Mindset
3. Managing the Context: The Worldly Mindset
4. Managing the Relationships: The Collaborative Mindset
5. Managing the Change: The Action Oriented Mindset

The Mindsets as Threads and the Manager as Weaver: Effective performance means weaving each mindset over and under the others to create fine, shred cloths (Jonathan Gosling, 2003). While analyzing the promoters' capability, the manager of the financial institution should listen to the mindsets of the Owner Manager of the business too for deciding effective financial sanction and advice.

APPRAISAL OF INFRASTRUCTURE AND RELATED ASPECTS

- Adequacy or otherwise of infrastructure right from procurement of inputs, production processing, transport and export up to the point of marketing in the global markets has to be appraised with special reference to maintaining a continuous cold chain for several products exported to the world markets.
- While discussing the availability of the requisite infrastructure such as, power, water, etc., the actual quantity required for running the unit and the arrangements that are being made vis-a-vis requirements will have to be clearly indicated. For example, in the case of water requirement, daily quantity of water required and its availability will have to be commented upon as also quality of water. In case, water is to be treated, this also should find a mention in the paragraph.

Air Cargo Facilities

- Air lifting facility and capacity in the concerned airport for the particular product will have to be mentioned in the appraisal report.

Collaborators

- While discussing the aspects relating to import of technology, full details with regard to foreign collaborator, Indian companies for which they have rendered technical assistance, etc., will have to be recommended. As far as possible, the standing of the foreign collaborators will have to be collected and incorporated.

Manpower Requirements

Manpower requirements will have to be classified into 3 categories, namely, professional management at senior level, skilled and un-skilled labour and specific number of personnel in each of these categories will have to be added vis-a-vis availability.

In this project, the agripreneur is sincere, risk taking, interested in taking initiatives, aiming to promote a proprietorship form of business organization, has family labor plus two attached farm servants with the family and his past experience in sheep rearing for 15 years is an added qualification.

Social Distributive Appraisal The following aspects are considered in this appraisal:

- Details of employment opportunities to be created by the projects,
- Measurement of the distribution of income due to the project, and
- Identification of the impact on the basic needs of the society.
- Social analysis is concerned with two aspects:
- Who will benefit from the project and by how much?
- Who will pay for the project and how much will they pay?
- At the end of the project, adverse effects of technology if any need a special mention, e.g., introduction of genetically transformed seeds and its adverse effects. If there are adverse effects, what alternatives exist to reduce the adverse effect of genetically modified seeds?
- Introduction to mechanization may reduce employment to women and the men. How it can be improved to provide employment to them.

- Determine and spell out the ways to improve the quality of life of the stakeholders of the project. What are the special efforts added in the project to improve the "Quality of work life" need a special mention.
- Financially unviable, but socially and economically viable. In such a situation, govt. may decide in favor of the project, if it believes that the positive externalities (net benefits) outweigh the negative financial cash flow.

In the model proposal taken into account, the project is capable of providing 5 man-days of direct employment per day for 365 days and the indirect employment is for another 10 man-days especially in maintaining livestock health care, transportation of animals to the market centers, culling of animals, meat extraction and distribution, sale of hides, processing of hides and leather making, etc., for 185 man-days. Put together the employment generated is arrived at 550 man-days.

Besides the well being of the promoters of the project, the stakeholders are also benefited because of implementation of the project. The consumers are day-by-day increasing and their preference to meat and meat products is also increasing and hence scope exists in implementation of such useful project to the society. It also helps the agriculture sector directly by providing them higher quantity of manure which is being supplied from the project. This adds to the productivity of agricultural and horticultural crops, hence the same may be worthy to take up and the same is not contributing to any environmental damage or ecological damage but contributes to environment building.

Commercial Appraisal In the commercial appraisal the following aspects will be looked into:

- Demand and supply aspects of the output and the input by using the techniques of forecasting of demand.
- Type of marketing channels that exist for input supplier.
- Import substitution.
- Special arrangements made to provide marketing channels for new inputs.
- Procurement procedures and specification determination for procurement.
- Past trends in supplies.

- Market segments - whom you are going to cater?
- Export market and market trends.
- Pricing Policy- whether price will be lower when compared to the competitors price.
- Life cycle of the product.
- Brand name of the product.
- Packaging and forwarding details.
- Sales force backup.
- Advertising and sales promotion methods aimed at servicing and follow-up details.
- Sources of market information and publications useful to study various aspects of marketing.

APPRAISAL OF MARKETS AND MARKETING

In respect of sophisticated high-value projects, the products are usually marketed in foreign countries. At times, a part of the production is marketed in the domestic market also. Therefore, the demand and supply of the products in the global market as also in the domestic market has to be assessed. Local demand and supply position will have to be highlighted. While discussing aspects relating to marketing, mere mention about buy-back arrangements with foreign collaborator would be inadequate. Analysis has to be made about the global demand for exportable products and supply by different countries in this regard. The percentage of growth of the particular product over a period in the international market should find a mention. In case where selling arrangements have been made with the foreign parties the details in this regard along with the selling commission to be paid should find a mention. Since the demand for meat and meat products are increasing both in international and domestic markets, the demand for animals for multifarious end uses is recognized and hence an agribusiness unit of establishing sheep rearing becomes vital. The cost of production of meat is found to be only Rs. 150 per kg of mutton but the sale price of animal by live weight alone is arrived at Rs. 300 per kg revealing the better income generation possibilities had motivated the author to launch similar projects.

Financial Appraisal In the financial appraisal, the following aspects will be kept in view:

- Participants of the project need to be clearly spelt out. In agricultural projects, participants include farmers, private sector firms, public corporations, project implementing agency, etc.
- For each of these, separate budgets needs to be furnished
- In forestry and horticultural projects with long generation periods, the details of adequate income to live on during those periods need a mention.
- When the funds are needed for investment?
- How often the funds are needed? Spell out the project phases in which the fund is required
- Operation and maintenance cost year by year
- Return to be realized over years, price and sale arrangements need a mention.
- In case of agribusiness, the following points need a special mention. Capital cost of the project: Land and site development, building, plant and machinery. Technical know-how and engineering fees, miscellaneous fixed assets, preliminary and pre-operative expenses, provision for contingencies, margin money for working capital (Table 3)
- Source of Finance, Financial Projections, Profitability Estimates (Table 4), Cash-Flow Statements (Table 5), Projected Balance Sheet (Table 6)
- Ratio analysis, Debt equity ratio, Current ratio, Debt Service Coverage Ratio (DSCR)
- Margin on security (Table 5)
- Break-Even Analysis (Table 7)
- The results of Discounted Cash Flow analysis which are proving that NPV, BCR, IRR are on positive side (Table 8) to show that the projects are financially viable.
- Treat taxes as a cost.
- Treat subsidies as a return.
- Interest payable to external suppliers of money may be deducted to derive the benefit stream available to the owners of capital. The financial aspects are discussed in the following Tables 3 and 4 for the project on sheep rearing as an example.

Table 3 Fixed Cost Involved in Sheep Rearing

Investment Cost					
Sr.	**Items**	**Specifi-cations**	**Phys-ical Units No.**	**Unit Cost (Rs./ Unit)**	**Total (Rs.)**
	Land	Own land			
	Shed	Ewes	500	85	572,475
		Ram	25		
		Lamb	415		
	Total shed area		**6735**		
	Chain link mesh fencing		5000	25	125,000
	Bore well for Drinking Water Facility		1	75000	75,000
	Electrical motor and other fittings		1	50000	50,000
	Chaff Cutter for Feed Sizing		1	100,000	100,000
	Feeders, Drinkers, equipments, etc.		4	5000	20,000
	Cost of Animals	Ewes	500	3600	1,800,000
		Ram	25	3600	90,000
	Construction of Quarters for Laborers		1	74120	74,120
	Store Room cum Office with a Retail Unit		1	58860	58,860
	Contingencies	3% on animal			56,700
		2% on others			21,509
				Sub total	**3,043,664**

Preliminary & Pre-operative expenses

Investment Cost

Sr.	Items	Specifi-cations	Phys-ical Units No.	Unit Cost (Rs./ Unit)	Total (Rs.)
	Interest during first year				360,000
	Others				615,881
	Total Cost				**4,019,545**
	Means of Finance				
	Total Cost				**4,019,545**
	Subsidy			0%	0
	Promoters Contribution			25%	1,019,545
	Bank loan		Say	75%	**3,000,000**
Investment Cost					
Sr.	**Items**	**Specifi-cations**	**Phys-ical Units No.**	**Unit Cost (Rs./ Unit)**	Total (Rs.)
	Note:				
*	including replacement of died animals				
	Fodder cultivation area		2.00	acre	
	Ewes	500	10	5000 (sq.ft)	
	Ram	5	15	75	
	Lamb	415	4	1660	
	Misc			0	
				6735	

Table 3 reveals that the total cost of the project is arrived at Rs 40 lakhs out of which the promoters' contribution is 10 lakhs accounting for 25 per cent to the total project cost. The bank loan eligible is arrived at Rs 30 lakhs accounting for 75 per cent of the total project cost.

Project Outlay The project outlay and the source from which this is being met will have to be incorporated. The percentage of individual items to the total outlay has to be worked out.

ECONOMIC APPRAISAL

- Objective is to examine the project from the entire economy's point of view to determine whether the project will contribute to the economic welfare of the country.
- Generally conducted by the World Bank and development agencies of the government for the projects with huge investment and profound implication for the whole economy.
- Projects where economic analysis is required are big dams, horticultural and forestry projects, big industrial projects, etc.
- Document the expected returns to be realized by the society.
- Provides the details of taxes and subsidies that arose due to new income generation.
- Treat government subsidy as a cost of the society.
- Remuneration details to labor and other service inputs either at market prices or shadow prices need to be included.
- Expected outcome of the project and indicators of development like increase in the employment, improvement in the literacy rate, improved community assets, etc.
- Impact created by the project over a period of project life and beyond.

Table 4 reveals that the net sale of animals is arrived at Rs 25 to 26 lakhs every year. The average cost of production is arrived at Rs 8.50 lakhs. The average retained profit after deducting the taxes is arrived at around Rs 13 lakhs. Table 4 also reveals that the producer is able to repay the loan amount of Rs 30 lakhs at the rate of Rs 5 lakhs per annum following the repayment period of six years. It reveals that the enterprise is found to be a sound one in terms of cash incentives to the promoter of the agribusiness unit.

Table 4 Projected Profit and Loss Account in an Agribusiness Enterprise (Rs. in lakhs) Continued

	2014	2015	2016	2017	2018	2019	2020
SALES:							
Deduct: Closing Stock In Process	0.00	0.00	0.00	0.00	0.00	0.00	0.00
Cost Of Production:	0.00	8.56	8.36	8.31	8.37	8.52	8.75
Add: Op Stock Of Finished Goods	0.00	0.00	0.00	0.00	0.00	0.00	0.00
Deduct: Cl Stock Of Finished Goods	0.00	0.00	0.00	0.00	0.00	0.00	0.00
Cost Of Sales:	0.00	8.56	8.36	8.31	8.37	8.52	8.75
Selling, General & Admn. Expenses	0.00	0.00	0.00	0.00	0.00	0.00	0.00
	2014	2015	2016	2017	2018	2019	2020
Profit Before Interest & Tax (Pbit)	2.21	16.50	16.69	17.94	17.88	17.73	17.50
Interest & Other Financial Charges	0.00	3.60	3.00	2.40	1.80	1.20	0.60
Profit Before Tax / Loss	2.21	12.90	13.69	15.54	16.08	16.53	16.90
Tax Paid	0.00	0.00	0.00	0.00	0.00	0.00	0.00
Provision For Taxes	0.00	0.00	0.00	0.00	0.00	0.00	0.00
Retained Profit	2.21	12.90	13.69	15.54	16.08	16.53	16.90
Cash Profit	5.00	14.99	15.26	16.72	16.96	17.19	17.40
Cumulative Cash Profit	5.00	19.99	35.25	51.97	68.93	86.12	103.52
Retained Profit	2.21	12.90	13.69	15.54	16.08	16.53	16.90
Interest							

	2014	2015	2016	2017	2018	2019	2020
SALES:							
Term Loan	0.00	3.60	3.00	2.40	1.80	1.20	0.60
Total interest	0.00	3.60	3.00	2.40	1.80	1.20	0.60
Repayment Obligations							
Towards Term Loan	0.00	5.00	5.00	5.00	5.00	5.00	5.00
Total Repayment Obligations	0.00	5.00	5.00	5.00	5.00	5.00	5.00
Adult	0.50	0.50	0.50	0.50	0.50	0.50	0.50
Lamb	0.00	0.19	0.19	0.17	0.17	0.17	0.17

Table 5 reveals that the cash inflow to the project arrived on an average at Rs 16 lakhs. The average cash outflow arrived at Rs 5 lakhs every year barring the initial year of investment because it attracts higher investment cost and hence an amount of Rs 30 lakhs is to be borrowed from the bank for running the agribusiness on sustainable basis. The loan amount to be borrowed will be repaid in a period of six years. The details of payment of loan along with interest are delineated in Table 7.

Table 5 Projected Cash Flow Statement (Rs in Lakhs)

PROJECTED CASH FLOW STATEMENT							
(Rs. in lakhs)							
Particulars	2014	2015	2016	2017	2018	2019	2020
Inflow							
Cash Profit	5.00	14.99	15.26	16.72	16.96	17.19	17.40
Term Loan	30.00						
Promoters contribution	10.20						
Total	45.20	14.99	15.26	16.72	16.96	17.19	17.40
Outflow							
Capital Expenditure	30.44						
Preliminary Expenses	9.76						
Repayment-Instalments	0.00	5.00	5.00	5.00	5.00	5.00	5.00
Total	40.20	5.00	5.00	5.00	5.00	5.00	5.00
Opening Balance	0	5.00	9.99	10.26	11.72	11.96	12.19
Surplus/(Deficit)	5.00	9.99	10.26	11.72	11.96	12.19	12.40
Closing Balance	5.00	14.99	20.25	21.98	23.68	24.15	24.58

Table 6 Projected Balance Sheet (Rs in Lakhs)

Particulars	2014	2015	2016	2017	2018	2019	2020
Liabilities							
Reserves and Suplus	2.21	12.90	13.69	15.54	16.08	16.53	16.90
Promoters contribution	10.20	10.20	10.20	10.20	10.20	10.20	10.20
Secured Term Loans	30.00	25.00	20.00	15.00	10.00	5.00	0.00
Total	42.41	48.09	43.89	40.73	36.28	31.72	27.09

Particulars	2014	2015	2016	2017	2018	2019	2020
Assets							
Gross Block	11.16	11.16	11.16	11.16	11.16	11.16	11.16
Less: Accumulated Depreciation	2.79	4.88	6.45	7.63	8.51	9.17	9.67
Net Block	8.37	6.28	4.71	3.53	2.65	1.99	1.49
Other Current Assets	18.90	18.90	18.90	18.90	18.90	18.90	18.90
Cash & Bank Balance	15.14	22.91	20.28	18.30	14.73	10.83	6.70
Total	42.41	48.09	43.89	40.73	36.28	31.72	27.09
TOL/TNW	2.42	1.08	0.84	0.58	0.38	0.19	0.00
DER	2.94	2.45	1.96	1.47	0.98	0.49	0.00
Current Ratio	1.13	1.67	1.96	2.48	3.36	5.95	0.00

Table 7 Repayment Schedule to be followed by an Agripreneur

Repayment Schedule		
Amount of Loan	3,000,000	
Annual Interest Rate	12.00%	
Number of Payments	6	Yearly
Yearly Payment	517,645	

Year	Opening Balance	Interest	Principal	Closing Balance
0	3,000,000	360,000	0	3,000,000
1	3,000,000	360,000	500,000	2,500,000
2	2,500,000	300,000	500,000	2,000,000
3	2,000,000	240,000	500,000	1,500,000
4	1,500,000	180,000	500,000	1,000,000
5	1,000,000	120,000	500,000	500,000
6	500,000	60,000	500,000	0
		1,620,000	**3,000,000**	

Table 6 reveals the details of assets and liabilities of the agribusiness enterprise to be promoted by the agripreneur. Here, the assets are comparatively higher than the liabilities. It is supported by higher current ratio which arrived at 1.70 to 5.95 revealing that the firm is strong enough to meet the vicissitudes that arose in the agribusiness.

Table 8 Working for Break-Even Point Analysis

	2014	2015	2016	2017	2018	2019	2020
SALES							
NET SALES	2.21	25.05	25.05	23.26	26.25	26.25	26.25
VARIABLE COSTS:							
Total Cost of Production	0.00	8.56	8.36	8.31	8.37	8.52	8.75
Add Opg Stock-in-process and F G	0.00	0.00	0.00	0.00	0.00	0.00	0.00
	0.00	8.56	8.36	8.31	8.37	8.52	8.75
Less: Clg Stock-in-process and F G	0.00	0.00	0.00	0.00	0.00	0.00	0.00
Total Variable Costs	0.00	8.56	8.36	8.31	8.37	8.52	8.75
CONTRIBUTION	2.21	16.50	16.69	17.94	17.88	17.73	17.50

	2014	2015	2016	2017	2018	2019	2020
Contribution / Sales Ratio [PV Ratio]	100.0%	65.8%	66.6%	68.3%	68.1%	67.5%	66.7%
FIXED COSTS:							
Interest on							
Term Loan	0.00	0.11	0.11	0.12	0.12	0.13	0.13
Working Capital Loan	0.00	0.00	0.00	0.00	0.00	0.00	0.00
Direct Labour and Wages (25% Fixed)							
Power Cost	0.00	0.11	0.11	0.12	0.12	0.113	0.13
General and	0.00	0.00	0.00	0.00	0.00	0.00	0.00
Administrative Expenses	0.00	3.71	3.11	2.52	1.92	1.33	1.73
Break Even Sales [Fixed Cost/PV Ratio]	0.00	5.63	4.67	3.68	2.82	1.97	2.60

Table 8 reveals that the break even sales arrived at Rs. 5.63 lakhs during the year 2015 which gradually reduced to Rs 4.67 lakhs during 2016; 3.68 lakhs during 2017; 2.82 lakhs during 2018; 1.97 lakhs during 2019 and 2.60 lakhs during 2020. The contribution from the enterprise is around 16 lakhs and profit volume ratio is ranging between 66 per cent and 67 per cent revealing the soundness of the project.

Table 9 Discounted Cash Flow Analysis (Rs. in Lakhs)

Particulars of Cost/Revenue	2014	2015	2016	2017	2018	2019	2020
Project Cost	40.20	0.00	0.00	0.00	0.00	0.00	0.00
Total Cost	40.20	08.56	08.36	08.31	08.37	08.52	08.75
Discounted Cost at 15%	40.20	07.44	06.32	05.46	04.78	04.24	03.78
Total Revenue	02.21	25.05	25.05	26.25	26.25	26.25	26.25
Discounted Revenue at 15%	02.21	21.79	18.94	17.26	15.01	13.05	11.35

Particulars of Cost/Revenue	2014	2015	2016	2017	2018	2019	2020
Benefit Cost Ratio at 15%	01.38						
Net Present Value at 15% Discount Factor	99.61						
Internal Rate of Return (IRR) at Assumed Rates	39.00						

Table 9 reveals the details of discounted income and discounted cost using the discount factor of 15 per cent. At this rate, the benefit cost ratio (BCR) is arrived at 1.38 revealing that for a rupee of investment, the return to be generated in present value terms is Rs. 1.38 which indicates the status of soundness of the project. The absolute revenue to be earned is expressed in terms of net present value (NPV) which arrived at Rs. 99.61 lakhs and the internal rate of return (IRR) arrived at 39 percent which is 160 percent higher than the cut off rate chosen for appraisal of the project. These parameters revealed the status of worthiness of the project and hence the loan amount of Rs. 30 lakhs may be released at the earliest.

Environmental / Ecological Appraisal

- One of the concerns of the policy makers is to have sustainable development.
- Importance is now attached to Environmental Impact Assessment (EIA).
- Essence of EIA is prediction of the consequences to the natural environment from the project.
- The impact of the project on the ecology of the area and environmental problems, if any, likely to be generated due to it have to be analyzed and presented in detail. Certain agribusiness firms established on the bank of the river Bhavani has taken away the ecology of the river due to its heavy inflow of pollutants. Such damages if any, the details of processes to treat such pollutants and the mechanism of mitigating them need to be given.
- Project design should clearly spell out the preservation of environment and other resources, e.g., reclamation of land, fish production

by maintaining minimum water level in irrigation projects to tap both crop and fish productivity.

- Elaborate analysis called Environmental Risk Assessment (ERA) is used to assess the probabilistic element of the impact, which is not considered in EIA.
- The detailed project report (DPR) must ensure that the provisions are adequate for fulfilling all the legal requirements obtained in the locality where the project is to come up. This calls for a detailed knowledge of all the state level and national level provisions with respect to environment protection. Many of the provisions call for obtaining specific clearances from controlling bodies. The DPR should also include a time schedule for obtaining all the required clearances. (IGNOU – 2009)

In the sheep rearing agribusiness venture proposed, there is no ecological or environmental damage. Instead, allowing the sheep to graze in the forest floor enriches the soil with manure and the urine is also found to be rich in nutrients; these act as a boon to the plants, shrubs and trees in the forest floor. Moreover, it takes the weeds as feed and permits easy growth of dominant and emerging plants which are assets to the environment and they are asset promoters and builders. Hence the project is environmentally feasible and ecologically acceptable.

APPRAISAL OF ENTREPRENEURSHIP AND MANAGERIAL SKILLS

- When the projects are promoted by first generation entrepreneurs, the success or failure depends not only on the projects and products proposed to be produced but also the entrepreneurial capabilities and managerial skills of the promoters. Therefore a comprehensive appraisal of their strengths and weaknesses has to be done and the ways by which the weakness can be made into strengths needs a special mention in the project.
- While discussing the credentials of the promoters, a paragraph should be incorporated giving details with regard to the various companies/ undertakings in which the promoters are interested and the turnover and the working results of these companies.
- With respect to the companies, which are quoted in the stock exchange, the latest share price also will have to be incorporated in the paragraph.

Table 4 Projected Profit and Loss Account in an Agribusiness Enterprise (Rs. in lakhs)

	2014	2015	2016	2017	2018	2019	2020
SALES:							
Sale of Ewes	0.00	0.00	0.00	3.49	3.49	3.49	3.49
Sale of Lamb (Ewes & Ram)	0.00	21.25	21.25	19.01	19.01	19.01	19.01
Insurance claim Proceeds	1.32	2.44	2.44	2.44	2.44	2.44	2.44
Sale of Manure	0.38	0.67	0.67	0.64	0.64	0.64	0.64
Sale of wool	0.50	0.69	0.69	0.67	0.67	0.67	0.67
NET SALES:	2.21	25.05	25.05	26.25	26.25	26.25	26.25
COST OF SALES:							
Fodder cultivation	0.00	0.50	0.53	0.56	0.58	0.61	0.64
Insurance	0.00	0.79	0.83	0.88	0.92	0.96	1.01
Power & Fuel	0.00	0.41	0.43	0.45	0.48	0.50	0.53
Labour wages	0.00	2.84	2.98	3.13	3.28	3.45	3.62
Feed	0.00	1.49	1.56	1.64	1.72	1.81	1.90
Repairs & Maintenance	0.00	0.25	0.26	0.28	0.29	0.31	0.32
Veterinary care	0.00	0.08	0.08	0.08	0.09	0.09	0.10
Consultants & other Fee	0.00	0.11	0.11	0.12	0.12	0.13	0.13
Depreciation	2.79	2.09	1.57	1.18	0.88	0.66	0.50
SUB TOTAL:	2.79	8.56	8.36	8.31	8.37	8.52	8.75
Add: Opening Stock In Process	0.00	0.00	0.00	0.00	0.00	0.00	0.00

FIRST GENERATION ENTREPRENEURS

While preparing appraisal reports with respect to bigger projects for commercial banks, the promoter's experience in the particular line should be kept in view. Besides, if projects are to be implemented by first generation entrepreneurs, even if they have the required background in the proposed line of activity, bigger outlays should not be considered for such entrepreneurs. To start with, the projects could be implemented in phases by the entrepreneur, in view of the inherent risk involved in implementation of large out-lay projects.

Building A Data Bank

For realistic appraisal of the projects on hi-tech, export oriented, agro-processing, etc., there is a need to collect, update and maintain adequate data on the size of the domestic/foreign markets, product prices and fluctuations, the sources and credibility of technology suppliers, demand and supply of products in domestic as well as foreign markets, availability of alternative technologies and relative costs, raw materials, cost of various components of the investments, freight rates, import duties, market fees, etc. One has to collect reliable data on product prices for the past 2 to 3 years in the international market by writing to various agencies like APEDA, MPEDA, Ministry of Commerce, export/import related agencies to build up a reliable data base for day-to-day use for project appraisal.

IMPLEMENTATION

Project implementation involves three phases viz.,

- Pre-Development Phase – During this phase of implementation, the project team has to be identified. That is the line departments who are participating in the project need to be highlighted and the convergence activities if any need to be delineated in the open forum, discussed clearly and get approved for furtherance.
- **Development Phase** – During this phase, the interim adjustments and need based changes need to be delineated. In the case of construction project, all details regarding the project approval are necessary to occupy the building.

- **Operation Phase** – During this phase, the operation manual need to be prepared in liaison with the project goals, the impact factors need to be highlighted and the mission approaches to achieve the impact are to be clearly spelt out.

Monitoring

- On-going process of keeping track of the project - Banks may undertake Desk Monitoring and /or Field Monitoring.
- Monitoring team should be established with their vision and mission directives to fulfill the goal of the project

Evaluation

- Banks will normally conduct ex-post evaluation for the purpose of learning lessons of success and failure from the project. Several organizations like National Bank for Agriculture and Rural Development (NABARD) and its Training Establishments may conduct such evaluation studies and publish them for wider circulation.
- The need for mid term evaluation need to be checked and if need be, clear directives and the mission based actions need to be given to the evaluation agency or the team of evaluators of the project

DOCUMENTS TO BE PRODUCED BY THE AGRIPRENEUR

i. Copies of land records regarding land owned /to be purchased, certified by the concerned Revenue authorities.

ii. No due certificate from Co-operative Credit Society / Commodity Boards / Financial Institutions. (This can be waived provided sanctioning authority is satisfied that the applicant does not have liability with society / financial institutions for loans up to Rs.1.00 lakh)

iii. Documents of title and other relevant documents like Revenue records, Encumbrance Certificates etc. to establish the right over presently held landed property is to be produced.

iv. Valuation report of the asset to be purchased from Bank's Panel Valuer.

v. Detailed Project Report with details on the developments proposed / implementation schedule, management, repayment of loan etc need to be submitted.

vi. Agreement and Detailed Guidelines for Tie-up in case of projects under tie-up scheme with industries.

CHALLENGES FACED BY FINANCIAL INSTITUTIONS IN FINANCING FARM PROJECTS

1. Inadequate authentic technical data on the production standards, economic viability, technical feasibility of Agricultural and allied sector Projects.
2. In adequate co-ordination between research, user industries, farmers, Financial Institution and Govt. Departments in development and promotion of agribusiness projects.
3. Problems faced by Financial Institutions in recovery of the loans already extended and failure of Tie-up mechanism with the user industries.
4. Failure of certain agricultural and allied sector projects due to lack of infrastructure, interest, market price etc, resulting in poor recovery of loans.
5. Unexpected market failure and absence of adequate value addition efforts lack financial repayment

Hence, one should be careful in designing the project proposal for funding with the financial institutions like banks. On prompt payment and successful run of the project automatically becomes eligible for loans for the purposes of expansion and tie up

SUMMARY

Recently many private investments are forthcoming from companies rising Plantations with or without involving public equity. Though it was a welcome effort to meet the objectives laid down in agriculture and forest policies, the tall claims / promises made by such companies are apprehended to be unrealistic. Hence, it is essential to look into the claims made by these companies to prevent exploitation of investors / farmers as well as to provide adequate safeguards that the investment raised are properly utilized and the confidence of investors in agricultural and allied sector activities does not get a jolt. A valid proposal meeting the requirements delineated above will automatically accepted for finance with valid security.

The Project Proposal on Sheep Rearing under Arid and Semi Arid Conditions of Tamil Nadu has attracted the capital expenditure of around 70 lakhs for establishing one unit of Sheep Farm. Assuming that the commercial bank is capable of providing Rs. 30 lakhs as loan, the same will be repaid in Seven years by generating 258 lakhs as Gross Revenue and the surplus gained over a period of Seven years of the project was arrived at 75.74 lakhs after repaying the loan with interest. The output – input ratio revealed that for a rupee of investment, the project is capable of generating Rs. 1.65 as return and the cost of production per kg is found to be Rs 236 per kg of output. If the farmer / Agripreneur initiate marketing with a retail arrangement on his own, he could reap higher benefits out of this project and hence the proposal can be accepted and the loan amount may be released to the farmer.

References

- Jonathan Gosling and H. Mintzberg (2003). " The Five Minds of a Manager" ***Harvard Business Review.*** pp. 01-12
- Gittinger, P.J. (1984). "Economic Analysis of Agricultural Projects" (Economic Development Institute: The World Bank)
- Subramanian, P. and S. Senthilnathan (2004). " Forest Project Analysis and Management" (International Book Distributors: Dehra Dun)
- IGNOU (2009) "Formulation of Detailed Project Reports" in: Project Management – Ms 52 Course Material (IGNOU: New Delhi) PP. 57-76.

Index

D

E

F

G

H

I

J

K

P

Q

R

S

T

U

V

W

Y

Z

Made in United States
North Haven, CT
04 November 2026

21760245R00298